VOLUME SIXTY ONE

ADVANCES IN
INSECT PHYSIOLOGY
Sound Communication in Insects

VOLUME SIXTY ONE

ADVANCES IN
INSECT PHYSIOLOGY

Sound Communication in Insects

Edited by

RUSSELL JURENKA
Department of Entomology
Iowa State University
Ames, Iowa, USA

ELSEVIER

ACADEMIC PRESS
An imprint of Elsevier

Academic Press is an imprint of Elsevier
125 London Wall, London, EC2Y 5AS, United Kingdom
The Boulevard, Langford Lane, Kidlington, Oxford OX5 1GB, United Kingdom
525 B Street, Suite 1650, San Diego, CA 92101, United States
50 Hampshire Street, 5th Floor, Cambridge, MA 02139, United States

First edition 2021

Notices
Knowledge and best practice in this field are constantly changing. As new research and experience broaden our understanding, changes in research methods, professional practices, or medical treatment may become necessary.

Practitioners and researchers must always rely on their own experience and knowledge in evaluating and using any information, methods, compounds, or experiments described herein. In using such information or methods they should be mindful of their own safety and the safety of others, including parties for whom they have a professional responsibility

To the fullest extent of the law, neither the Publisher nor the authors, contributors, or editors, assume any liability for any injury and/or damage to persons or property as a matter of products liability, negligence or otherwise, or from any use or operation of any methods, products, instructions, or ideas contained in the material herein.

ISBN: 978-0-323-90283-0
ISSN: 0065-2806

For information on all Academic Press publications
visit our website at https://www.elsevier.com/books-and-journals

Publisher: Zoe Kruze
Acquisitions Editor: Sam Mahfoudh
Developmental Editor: Naiza Ermin Mendoza
Production Project Manager: Abdulla Sait
Cover Designer: Christian Bilbow

Typeset by STRIAVE, India

Contents

Contributors

Marion Hall
The Open University, School of Environment, Earth & Ecosystem Sciences, Faculty of Science, Technology, Engineering and Mathematics, Milton Keynes, United Kingdom

Marc W. Holderied
School of Biological Sciences, University of Bristol, Bristol, United Kingdom

Valerio Mazzoni
Research and Innovation Centre, Fondazione Edmund Mach, Italy

Thomas R. Neil
School of Biological Sciences, University of Bristol, Bristol, United Kingdom

Rachele Nieri
Laboratory of Bio-inspired, Bionic, Nano, Meta Materials & Mechanics, Department of Civil, Environmental and Mechanical Engineering, University of Trento, Trento, Italy

David Robinson
The Open University, School of Environment, Earth & Ecosystem Sciences, Faculty of Science, Technology, Engineering and Mathematics, Milton Keynes, United Kingdom

Johannes Strauß
AG Integrative Sensory Physiology, Institute for Animal Physiology, Justus-Liebig-Universität Gießen; Center for Mind, Brain and Behavior (CMBB), University of Marburg and Justus Liebig University Gießen, Gießen, Germany

Nataša Stritih-Peljhan
Department of Organisms and Ecosystems Research, National Institute of Biology, Ljubljana, Slovenia

Bijayalaxmi Swain
Danish Research Institute of Translational Neuroscience (DANDRITE), Aarhus University, Aarhus C, Denmark

Meta Virant-Doberlet
Department of Organisms and Ecosystems Research, National Institute of Biology, Ljubljana, Slovenia

Anne C. von Philipsborn
Danish Research Institute of Translational Neuroscience (DANDRITE), Aarhus University, Aarhus C, Denmark

CHAPTER ONE

Acoustic signalling in Orthoptera

Marion Hall and David Robinson
The Open University, School of Environment, Earth & Ecosystem Sciences, Faculty of Science, Technology, Engineering and Mathematics, Milton Keynes, United Kingdom

Contents

Advances in Insect Physiology, Volume 61
ISSN 0065-2806
https://doi.org/10.1016/bs.aiip.2021.09.001

Abstract

Acoustic communication is one of the most well-known behavioural traits of the Orthoptera. Orthopteran insects and the sounds they produce are both extremely diverse, and species-specific sounds are an extremely important tool in orthopteran taxonomy and systematics. For most species, acoustic signalling is the most important means of communication. It plays a vital role in mating, mate choice, intrasexual competition, interspecific interactions with predators and parasitoids, and the divergence of populations and species. The enormous diversity of the orthopterans has provided researchers with a wealth of model systems for studying anatomy, physiology, neurobiology, bioacoustics, communication, life-history traits, behaviour, evolutionary ecology, and speciation, all areas in which acoustic communication is important. We first reviewed orthopteran sound signalling nearly 20 years ago (Robinson and Hall, 2002), since when there has been an enormous amount of further work. This second review will look mainly at research published since that first review.

Abbreviation

AN ascending neuron
CPG central pattern generator
CF characteristic frequency
CHC cuticular hydrocarbon
CNS central nervous system
EP elevated process
EPSP excitatory postsynaptic potential
FM frequency modulation
IID interaural intensity difference
ITD interaural time difference
LN local interneuron
MR metabolic rate
ON omega neuron
RU repeated unit
SPL sound pressure level
SB styliform body
TN T-fibre neuron

1. Introduction

The order Orthoptera is extremely diverse biologically, morphologically, behaviourally, and ecologically, with more than 26,000 species occupying every possible terrestrial habitat, and some semi-aquatic ones, outside the polar regions (Song, 2018). It is divided into two main branches: the Caelifera and the Ensifera. The Caelifera mostly consist of the 12,000 or so species that make up the grasshopper family (the acridids) and the groundhopper family (the tetrigids). The Ensifera consist of about 15,000 species, of which around 7000 are bush-crickets, or katydids as they are called outside Europe (the tettigoniids) and about 5000 are 'true' crickets (the gryllids); there are also several other smaller families such as the mole crickets in the gryllotalpid family. The classification of the orthopteran insects is much disputed and often revised but here we will follow the taxonomy used in the online Orthoptera Species File database (Cigliano et al., 2021).

Acoustic communication is one of the most well-known behavioural traits of the Orthoptera. The sounds produced by orthopterans are very diverse and, for most species, the most important means of communication. They play a vital role in mating, mate choice, intrasexual competition, interspecific interactions with predators and parasitoids, and the divergence of populations and species. They are an extremely important tool in orthopteran taxonomy and systematics and differences in song are often key to separating otherwise morphologically identical, cryptic species (e.g. Liu et al., 2020; Tan et al., 2020; Tian et al., 2019; Weissman et al., 2021).

Orthopteran species are of huge importance to humans. They include several crop pests such as grasshoppers and locusts (Fig. 1) but on the other hand they provide a major source of nutrition, with well over 250 species known to be consumed by humans (Song, 2018). The enormous diversity

Fig. 1 The Egyptian grasshopper *Anacridium aegyptium*, which can be a pest of fruit trees.

of the orthopterans has also provided researchers with a wealth of model systems for studying anatomy, physiology, neurobiology, bioacoustics, communication, life-history traits, behaviour, evolutionary ecology, and speciation. Acoustic communication is important in all these areas.

Since we published our review of sound signalling in Orthoptera nearly 20 years ago (Robinson and Hall, 2002), there has been an enormous amount of research in this area, which has not only greatly expanded knowledge in previously existing areas of study but has also developed several exciting new topics. We therefore felt the time was ripe for this second review, which will look mainly at research published since 2002.

2. Mating systems

2.1 Variation in mating systems

In the typical orthopteran mating system, males are speculative singers. Using airborne sound they produce a long-distance *calling song*, in the absence of the female, to advertise their presence to potential mates. The calling song is usually loud and often continuous over long periods. In most species, females are silent. They are attracted to the male's song and locate him by moving towards the source of the sound (*phonotaxis*). Some species use vibrational advertisement signals, as well as or largely instead of calling song (e.g. Römer et al., 2010; Sarria-S et al., 2016). Most ensiferan males provide a nuptial gift during mating that the female eats (see review by Vahed, 1998). This is usually the spermatophore (Fig. 2) but can be in the form of other body parts (e.g. tegmina), haemolymph, or secretions from external glands.

However, there is a lot of variation in this basic pattern (summarized in Table 1). The female often sings as well as the male. Apart from the calling

Fig. 2 The male (underneath) transfers a spermatophore to the female during mating in the bush-cricket *Leptophyes punctatissima* (left) which the female then eats (right).

Table 1 Summary of mating systems found among Orthoptera[a] (reviewed in Robinson and Hall (2002) unless otherwise referenced).

Orthopteran group	Mating system (all involve male calling song to which silent female performs phonotaxis unless otherwise stated)	Songs or vibrational signals other than calling song found in at least some species
Suborder Ensifera		
Gryllidae (true crickets)	Diverse but tends to involve aggression between males Common for there to be several copulations within a single mating event (e.g. deCarvalho and Shaw, 2010; Narvaez and Robillard, 2012)	Courtship song Courtship vibrational signal Inter-copulation song Male aggressive song Post-copulation song See e.g. Brown (2016); Narvaez and Robillard (2012); Singh and Jain (2020)
Mogoplistidae (scaly crickets)	Female usually approaches but male may also approach female Usually several copulations within a single mating event See e.g. Glen-Worrall (2003); Saidy and Gray (2004)	Courtship song Courtship vibrational signals Inter-copulation song See e.g. Glen-Worrall (2003); Saidy and Gray (2004)
Phalangopsidae (spider crickets and relatives)		Courtship song Courtship vibrational signals by either male or female Male aggressive song See e.g. Acosta et al. (2020); Fianco et al. (2018); Zefa et al. (2008)
Trigonidiidae (sword tailed crickets)	Some species have several copulations within a single mating event (Centeno et al., 2020; Funk, 2016; Tan, 2020)	Courtship song Courtship vibrational signal Courtship interruption song Post-copulation song Post-copulation vibrational signal Male aggressive song See e.g. Centeno et al. (2020); Funk (2016); Mays (1971)

Continued

Table 1 Summary of mating systems found among Orthoptera[a] (reviewed in Robinson and Hall (2002) unless otherwise referenced).—cont'd

Orthopteran group	Mating system (all involve male calling song to which silent female performs phonotaxis unless otherwise stated)	Songs or vibrational signals other than calling song found in at least some species
Gryllotalpidae (mole crickets)	At least one species (*Gryllotalpa major*) has lek mating system (Howard et al., 2011; Keane et al., 2018)	Courtship song Male aggressive song Territorial song (Zhantiev et al., 2003) Female song; function unclear but possibly aggressive (Ulagaraj, 1976)
Prophalangopsidae (haglids, humped-wing crickets or grigs): *Cyphoderris* species	In *C. monstrosa*, female approaches singing male but hearing sensitivity does not match frequency of song and unclear if females perform phonotaxis as such (Morris et al., 2002) In *C. strepitans* females do perform phonotaxis to male calling song (Ower et al., 2013)	Male song in *C. monstrosa* may be territorial song and/or it may signal fighting ability (Morris et al., 2002) In *C. strepitans* it may function as an aggressive song as well as a calling song (Ower et al., 2013)
Anostostomatidae: New Zealand tree wetas (*Hemideina* species; Fig. 3)	Male has calling song but unclear whether functions to attract females or repel males	Courtship interruption song Aggressive song by both males and females Territorial song See Ewers and Cowley (2005)
Anostostomatidae: New Zealand ground wetas (*Hemiandrus* species) and giant wetas (*Deinacrida* species)	*Hemiandrus* unlikely to be able to use sound in mating but is sensitive to vibrations (Strauß et al., 2017a) *Deinacrida* do not appear to use sound or vibrational signals in intersexual encounters. Usually several copulations within a single mating event. See e.g. Kelly and Gwynne (2016); Watts and Thornburrow (2011) In both, males are attracted to females and follow them, possibly via pheromones	Male *Hemiandrus* use vibrational signals ('drumming') to attract mates and females may also drum (Gwynne, 2004) In *Deinacrida rugosa* (Fig. 4) aggressive vibrational signals by both male and female (Howard et al., 2018)

Table 1 Summary of mating systems found among Orthoptera[a] (reviewed in Robinson and Hall (2002) unless otherwise referenced).—cont'd

Orthopteran group	Mating system (all involve male calling song to which silent female performs phonotaxis unless otherwise stated)	Songs or vibrational signals other than calling song found in at least some species
Anostostomatidae: Indian weta (*Gryllacropsis* species)	Both male and female sing, often at same time. Unclear if both sexes have independent calling songs or whether they are duetting (Diwakar and Balakrishnan, 2006; Tomar and Diwakar, 2020)	
Anostostomatidae: King crickets of Australia and South Africa	Most species do not have ears and males rely on vibrational or chemical signals to attract females	Courtship song (possibly) Male aggressive song
Gryllacrididae (leaf-rolling or raspy crickets)	Can produce sounds but poor sensitivity to sound so unlikely to play important role in mating (Scholtz et al., 2018) Male produces vibrational signal to attract mates; in some species the female responds in a duet. Unclear which sex approaches (Field and Bailey, 1997)	Courtship vibrational signals by both male and female (Ingrisch and Rentz, 2009)
Stenopelmatidae (Jerusalem crickets)	Able to stridulate but unclear if sound used in mating. Instead both male and female use vibrational signal to attract mates (Weissman et al., 2021)	Courtship vibrational signal (Weissman et al., 2021)
Tettigoniidae (bush-crickets/ katydids): Bradyporinae: Bradyporini and Zichyini	Either male or female can produce a calling song and the other sex then performs phonotaxis (Elaeva and Korsunovskaya, 2012; Korsunovskaya, 2008)	

Continued

Table 1 Summary of mating systems found among Orthoptera[a] (reviewed in Robinson and Hall (2002) unless otherwise referenced).—cont'd

Orthopteran group	Mating system (all involve male calling song to which silent female performs phonotaxis unless otherwise stated)	Songs or vibrational signals other than calling song found in at least some species
Tettigoniidae: Bradyporinae: Ephippigerini	Some species perform duets similar to those in Phaneropterinae, see below (Pfau and Pfau, 2015)	Aggressive song by both males and females (Hartley, 1993)
Tettigoniidae: Phaneropterinae	Most species perform a duet in which the male sings and the female responds (Bailey, 2003). Males may change the content or pattern of their song once they get a reply from a female (Heller et al., 2021) Sex that approaches varies between and within species: can be male, female, either or both. Which sex approaches may depend on the size of the nuptial gift the female receives (McCartney et al., 2012) or on the balance of risks incurred by moving (e.g. predation) for each sex (von Helversen et al., 2012) Stationary female system is common; in these species the male usually only performs phonotaxis if female response falls into narrow time-window after his song See review by Heller et al. (2015)	Male aggressive song (Spooner, 1964) 'Ticking' song that may function as territorial song (Spooner, 1968)
Tettigoniidae: Phyllophorinae	Males of *Siliquofera grandis* do not have a calling song; the same may apply to other species (Korsunovskaya et al., 2020)	Courtship song Courtship vibrational signals by both males and females Post-copulation vibrational signal Territorial vibrational signal

Table 1 Summary of mating systems found among Orthoptera[a] (reviewed in Robinson and Hall (2002) unless otherwise referenced).—cont'd

Orthopteran group	Mating system (all involve male calling song to which silent female performs phonotaxis unless otherwise stated)	Songs or vibrational signals other than calling song found in at least some species
Tettigoniidae: Tettigoniinae		Territorial song Courtship song See Zhantiev and Korsunovskaya (2019)
Tettigoniidae: Zaprochilinae	The male's calling song may also serve to repel other males (Simmons and Bailey, 1993)	
Suborder Caelifera		
Acrididae (true grasshoppers): acridid grasshoppers other than locusts	Male calling song may function both in sexual communication and territorial defence In some species, especially gomphocerine grasshoppers, female replies to male song and male then approaches (e.g. Wirmer et al., 2010) Male also approaches in some other species (Garcia et al., 2015)	Courtship song (Vedenina et al., 2020) Courtship vibrational signals by both males and females (Benediktov, 2015) Male aggressive song used in territorial defence Possible female aggressive song used in interactions with other females and when rejecting a male (Larrosa et al., 2008) Vibrational advertisement signal instead of calling song (Ingrisch and Rentz, 2009)
Acrididae: locusts (*Schistocerca gregaria* and *Locusta migratoria*)	Mate location relies on visual/chemical cues rather than acoustic signals (Ely et al., 2006) Courtship minimal but signals exchanged while male mounted before actual copulation, especially in solitary phase compared with gregarious phase (Golov et al., 2018a,b; Wang et al., 2014)	Courtship song and courtship vibrational signals by both male and female (Golov et al., 2018b)

Continued

Table 1 Summary of mating systems found among Orthoptera[a] (reviewed in Robinson and Hall (2002) unless otherwise referenced).—cont'd

Orthopteran group	Mating system (all involve male calling song to which silent female performs phonotaxis unless otherwise stated)	Songs or vibrational signals other than calling song found in at least some species
Pamphagidae (stone grasshoppers)	Males and/or females can produce sounds but unknown how or if sound used in mating system (Massa, 2012, 2020) In at least some species, males do not sing at all; females sing to advertise their reproductive receptivity to males (Lopez et al., 2008)	
Pneumoridae (bladder grasshoppers)	Females respond to calling song of male in a duet. Male performs phonotaxis and female always remains stationary. No courtship before copulation (Couldridge and van Staaden, 2006)	
Tridactylidae (pygmy mole crickets)	At least some species can produce sound and vibrational signals but unknown how these are used in mating system (Benediktov, 2012; Woo, 2020)	

[a]For those taxonomic groups not included in this table, we have been unable to find any evidence that sound signals are used in their mating system apart from, in some groups, the existence of a calling song.

song, several other songs relating to the mating system have been identified, with crickets having a particularly rich repertoire (reviewed by Centeno et al., 2020). Once the pair meet, there may be a courtship phase before copulation with a specific *courtship song*. Where species copulate several times within a single mating event, there may be an *inter-copulation song*. In some species the male guards the female for a time after copulation and this sometimes involves a specific *post-copulation song*. Some males produce a *courtship interruption song* after a failed mating attempt. In some species, males may also produce a *territorial song* that functions in male–male spacing and either sex

Fig. 3 Tree weta *Hemideina thoracica* (female left, male right).

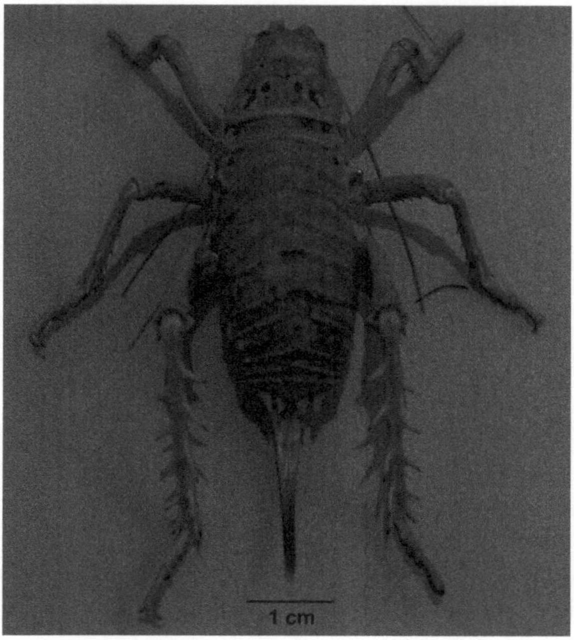

Fig. 4 Giant weta *Deinacrida rugosa*.

may sing an *aggressive song* directed at rivals during intrasexual encounters. We use the term 'song' here only for signals that use airborne sound; vibrational signals are referred to as such.

This terminology is based on that found most commonly in the recent literature. However, usage varies greatly (though attempts have been made to standardize it: see e.g. Baker and Chesmore, 2020). For example, call is frequently used instead of song, advertisement song or mating song instead of calling song, mating song instead of courtship interruption song, and rivalry or agonistic song instead of aggressive song. The distinction between aggressive song and territorial song is not always clear.

2.2 Alternative mating tactics

Not only do orthopteran species vary in their mating systems, many also show differences in mating behaviour within species, with individual behaviour depending on, for example, their stage of life, their condition or, in polymorphic species, their morphology. Individuals do not necessarily show consistent behaviour, instead switching tactics depending on their circumstances. As summarized in Rotenberry et al. (2015) and Torsekar and Balakrishnan (2020), such alternative mating tactics are widespread across the animal kingdom and may result from factors such as predator pressure (see Section 8) or intrasexual competition (see Section 7).

Orthopteran males commonly adopt alternatives to singing for at least some of the time, including switching to vibratory signals (see Section 2.3). Males may search silently for females and/or show satellite behaviour, whereby they move towards and stay near a singing male to try and intercept any females approaching him. Examples include the tree cricket *Oecanthus henryi* (Torsekar and Balakrishnan, 2020) and the field crickets *Teleogryllus oceanicus* (Bailey et al., 2010; Balenger and Zuk, 2015) and *Gryllus texensis* (Bertram et al., 2006); for other examples see Robinson and Hall (2002). A study of 16 Panamanian species of katydid showed a trade-off across species between the number of calling songs produced per day and the number of males silently searching (Symes et al., 2021). In duetting species, males may eavesdrop on duetting pairs and perform phonotaxis towards the responding female without singing themselves, as in the phaneropterine bush-crickets *Caedicia* (Hammond and Bailey, 2003) and *Scudderia pistillata* (Villarreal and Gilbert, 2014). Females may also adopt related alternative tactics in which, instead of performing phonotaxis, they remain stationary and rely on silently searching males to find them (Torsekar and Balakrishnan, 2020). In *Poecilimon affinis*, the female replies to a distant male with a soft response that he is unable to hear: such responses may be aimed instead at closer eavesdropping males, by which strategy the female avoids having to approach to within hearing distance of the calling male (von Helversen et al., 2015).

Males of several foliage-dwelling cricket species have been observed using baffles made of leaves to amplify the sound of their calling song (Forrest, 1982). Studies in *Oecanthus* tree crickets (reviewed in Deb et al., 2020) have shown that males modify leaves to make baffles more efficient and that the use of baffles is an alternative mating strategy used by smaller males and/or those with lower-intensity songs. Males in poor condition and/or with low dominance status can increase their mating success by using baffles; for more attractive males, using baffles makes no difference to their mating success.

Similarly, in the cricket *Anurogryllus muticus*, some males use anthropogenic sites such as small depressions in a wall instead of singing from their natural grassland habitat, allowing them to produce a louder calling song without using more energy (Erregger and Schmidt, 2018).

Wing dimorphism is common among Orthoptera and different wing morphs represent alternative mating tactics. In the cricket *Macroanaxipha macilenta*, a rare narrow-forewing male morph is almost silent, allowing energy saved from singing to be put into producing a second spermatophore for each female; despite its lack of song, its mating success is equal to the singing wide-forewing morph that only produces one spermatophore per female (Elias-Quevedo and Cueva del Castillo, 2020). In several other species such as the crickets *Velarifictorus ornatus* (Zhao et al., 2017) and *Gryllus firmus* (Mitra et al., 2011), flying long-winged and flightless short-winged morphs are found in both males and females. Long-winged males can disperse from unfavourable environments but must allocate more resources to maintaining flight muscles, leaving less energy to produce attractive song and spermatophores. Short-winged males have greater mating success in favourable environments because they can put more effort into singing, but cannot escape unfavourable environments.

Bladder grasshoppers (Pneumoridae) also have two distinct male morphs, each showing alternative mating tactics (Donelson et al., 2008). Most males are 'inflated': large, fully-winged and with a bladder—a conspicuous inflated abdomen that hugely amplifies their calling song. The rest are 'uninflated':

small, flightless and without the bladder. Inflated males engage in long-distance duets with females and can perform phonotactic flight over a large area; uninflated males adopt an eavesdropping strategy in which they intercept responding females before the inflated male arrives. Inflated males mate more frequently but use a lot of energy in singing, inflating their bladder and flying; they are probably also more prone to predation. Bladderless males have lower energy costs and predation risks and live twice as long so have more opportunities to mate over their lifetime. The two morphs arise during development and the proportion of uninflated males increases at high population density: uninflated males should do better at high densities and inflated males at low densities.

2.3 Relationships between sound and other signal modalities

Until relatively recently, work on orthopteran mating has often focused solely on the sound signals involved in mate attraction, location and courtship. The relationships between sound and other types of signal, such as those visual, chemical or vibrational cues that may also be necessary if mating is to be successful, have been less frequently studied. Sound and vibrational signals in particular have often been studied in isolation despite the close functional and evolutionary links between them (Stritih-Peljhan and Virant-Doberlet, 2021) and the fact that it is almost impossible, as Benediktov (2018) has pointed out, to draw a clear boundary between them, since sound signals themselves cause vibrations in solid media.

Before 2000, work in this area was relatively limited, with the possible exception of grasshoppers. But, for example, in crickets, it was shown by Bell (1980) for *Oecanthus nigricornis* and by Balakrishnan and Pollack (1997) for *Teleogryllus oceanicus* that, while the female will not mount the male at all if he does not produce courtship song, the likelihood of her mounting also depends on the presence of appropriate chemical cues from the male. Studies on bush-crickets showed that vibrational signals can be important: for example *Copiphora rhinoceros* males sing only until they make contact with the female, then switch to vibrational signals (Morris, 1980). In many grasshoppers, courtship can be prolonged and complex, consisting of elaborate visual displays in combination either with courtship song, as in *Gomphocerus rufus*, or vibrational signals, as in *Galidacris variabilis* (Riede, 1987). This complex courtship is thought to play an important role in stimulating the female to change from a semi-receptive to a fully receptive state (e.g. *Myrmeleotettix maculatus*: Bull, 1979), probably mediated by the release of juvenile hormone (see review by Heinrich et al., 2012).

More recently, however, the amount of research looking at the relationships between sound signals and other types of signal has expanded greatly. Courtship behaviour has been described in much more detail, especially in grasshoppers (e.g. Berger and Gottsberger, 2010; Ostrowski et al., 2009; Vedenina et al., 2020). Additional signalling modalities have been identified in courtship, for example vibration in the crickets *Adelosgryllus rubricephalus* (Zefa et al., 2008), *Cranistus colliurides* (Centeno and Zefa, 2019) and *Ornebius aperta* (Andrade and Mason, 2000), chemical cues in the grasshoppers *Chorthippus biguttulus* and *C. mollis* (Finck and Ronacher, 2017), and probably chemical and tactile cues in the bush-cricket *Mecopoda elongata* (Dutta et al., 2018). In bush-crickets visual cues can be important during phonotaxis. In *Poecilimon affinis*, if the male's only cue is the female song, his course towards her is meandering but if visual landmarks are available that he can integrate with the acoustic cues, his course is much more direct (von Helversen and Wendler, 2000). In *Requena verticalis*, the female performs phonotaxis towards the male. If another male sings while she is approaching she may switch to him, but if she has a visual cue she is much more likely to stay faithful to the original male, even if he stops singing (Bailey et al., 2003). This benefits the female because it reduces the time she spends in getting to the male and the predation risk associated with movement.

Signals in different modalities may simply send different messages. For example, in *T. oceanicus* (Simmons et al., 2013), the male's cuticular hydrocarbon (CHC) profile and his courtship song have equal weight in predicting his mating success but they are not correlated with each other. Song may signal the male's quality while the chemical cue from CHCs may signal genetic compatibility. However, things are not quite so simple. Thomas et al. (2011) showed that the expression of a male's CHCs is affected by his acoustic experience—if he does not hear other males singing, he increases his concentration of those CHCs that are attractive to females. Similarly, a *Gryllus integer* male signals his quality (e.g. ability to provide a spermatophore) via his calling song but his fighting ability via his CHC profile (Leonard and Hedrick, 2010). Female responses to the two cues are not correlated, but if she is previously exposed to the calling song of an attractive male, she accepts a courting male more quickly, regardless of whether he has an attractive calling song or not. So signals encountered earlier in the whole mating sequence may bias females' assessment of later close-range cues.

Sound and vibration are often used as alternative signals, especially in bush-crickets. In *Docidocercus gigliotosi* (Römer et al., 2010), males use either sound or vibration for long distance signalling. They switch between the two modalities frequently, moving from sound to vibration when for example the risk is high from predators who eavesdrop on sound signals, or there are high levels of background noise that mask calling songs. In *Ephippiger bitterensis* and *E. ephippiger* (Busnel et al., 2019), the female performs phonotaxis towards the male's calling song. But once the female gets close and he perceives vibrations caused by her movements, he starts to alternate sound signals with very brief vibration signals. The female then responds with similar vibration signals. She only actually moves towards the male when he signals, whether by sound or vibration.

In the bush-cricket *Onomarchus uninotatus* (Rajaraman et al., 2015) and in lebinthine crickets such as *Ponca hebardi* (Benavides-Lopez et al., 2020), male and female perform a multimodal duet, where the female responds to the male's calling song by tremulating (making body movements that create vibrations in the substrate). These are true duets, with a clear phase relationship of alternation between the chirps of the male's song and the female's vibrational response. In lebinthine crickets, the female does not perform phonotaxis at all and instead the male performs vibrotaxis towards her. In *O. uninotatus*, if the female is some distance from the male, she will begin phonotaxis after tremulating but if she is within about 4 m of the male, and on the same tree, she never performs phonotaxis and then the male performs vibrotaxis towards her. He can only locate the source of her vibrations, however, if he can hear his own sound element of the duet (Rajaraman et al., 2018).

Singing and vibrational signals are not always used as alternatives but may instead send different messages. While a *P. hebardi* male is singing in his duet with the female, he also simultaneously tremulates in response to the songs of other males, sending a vibrational signal similar to that of the female (Benavides-Lopez et al., 2020). The male's vibrations probably serve to mask the female response so that rival males cannot respond to it. Similarly the bush-cricket *Nesonotus reticulatus* sings and tremulates simultaneously (Stumpner et al., 2013). Again the vibrational signal may be directed mainly at other males, since other males tremulate in reply.

The interactions between signals sent via different modalities can also play an important part in maintaining reproductive isolation between species (e.g. Finck and Ronacher, 2017; Moran et al., 2019; Mullen et al., 2007).

Given the increasing evidence, not only in Orthoptera but across many taxa, that multimodal signals are common, that females often base their choice of mate on several different cues, and that individual components of such signals may interact in a functional way, a number of authors have addressed the general issue of how and why such signals have evolved. Greenfield (2002) argued that multimodal mate recognition signals could have originated in arthropods by default because of an inherent preference shown by some females for complex songs per se (though see Reichert et al., 2017). Candolin (2003) and Hebets and Papaj (2005) both reviewed the various hypotheses that had so far been proposed and proposed a framework for future research. Candolin concluded that, while there was some evidence for other hypotheses, there was good evidence that multiple cues provide additional information and send different messages that either indicate general male quality or allow females to choose the most suitable mate for them. The extra costs of sending multiple signals may be counterbalanced by the saving in time and energy spent assessing potential mates.

3. Songs and signals

The Orthoptera Species File Online contains 28,899 valid species (as of May 2021) a substantial number of which produce sound (Cigliano et al., 2021). Studies of the sounds reveals a rich diversity covering a wider range of frequencies than those audible to humans. Songs produced as part of sexual and intrasexual behaviour contain a sub-structure that requires a descriptive terminology. Such a terminology has yet to be formalized and although individual attempts to set out a standard exist, there is no universally agreed set of terms.

In their comprehensive description of the sounds of 170 European species of Tettigoniidae and Acrididae, Ragge and Reynolds (1998) suggest a terminology based on movements of the sound-producing apparatus. Thus, the sound produced by one complete to-and-fro movement of the stridulatory apparatus is a *syllable*. They adopt the proposal of Broughton (1976) that a discrete group of syllables should be described by a new term, *echeme*. A few authors have embraced the new term, for example Garcia et al. (2014), Heller et al. (2011), Hershberger (2021) and Pfau and Pfau (2015). The term *pulse*, which Broughton (1963, 1976) criticized, continues to be used in song descriptions, but it does not have a consistent definition. For recent use of the term see Cole (2010) and ter Hofstede et al. (2020).

Recently a standardized terminology, together with controlled vocabularies, has been proposed for insect bioacoustics, driven in part by the development of automated identification systems (Baker and Chesmore, 2020). Their terminology structure is based on three major components: echeme sequence, echeme and syllable. This matches the structure used by Ragge and Reynolds (1998). Using this terminology, a song consists of one or more echeme sequences, each echeme being composed of syllables. The sound produced by each unidirectional movement of the stridulatory apparatus (opening, closing, upward or downward), is a *hemisyllable*. If sound is produced by both opening and closing movements, the syllable is described as a *diplosyllable*.

In this review we generally follow the terminology used by the original authors since it is often not possible to indicate the relationship to the terms used by Baker and Chesmore (2020) and Ragge and Reynolds (1998).

3.1 Components of signals

The five main parameters of a sound are: intensity; frequency (the reciprocal of the wavelength); frequency modulation (rise or fall in frequency, FM); intensity modulation (increase or decrease in intensity); and temporal patterning (length of syllable, length of interval between successive syllables, number of syllables, etc.). A sound signal may contain a combination of sounds at different frequencies (the frequency spectrum).

Sound intensity may be measured in units of energy or work, or in units of power. Although loudness is related to sound intensity, it is subjective and depends on the perception of the receiver. Sound intensity on the other hand is objective, being independent of a receiver's hearing, and can be measured. However some authors refer to the loudness of a sound rather than to its intensity so we also use loudness as a general term in this review.

Sound pressure level (SPL) is the average variation in atmospheric pressure caused by the sound. It is measured in decibels (dB), the ratio of the absolute sound pressure against a reference level of sound in the air of 0.00002 pascals. So, sound intensity is expressed as dB SPL.

The various parameters of signals reveal different things about the sender, some of which are not under the sender's control (Endler, 2000). For example, dominant wavelength and wavelength spectrum are determined by the biomechanics of the sender, and biomechanical properties are related to size.

In general, larger male orthopterans produce louder songs and the songs contain lower frequencies. In the tettigoniids, very small species generally have very high frequency songs, for example those of small phaneropterines where the frequencies are well into the ultrasonic range (Heller, 1984; Heller et al., 2015; Robinson, 1990), defined as anything higher than 20 kHz. The signal received by the recipient may differ from the sender's original because of changes that occur during transmission through the environment.

3.2 Intensity and distance

The songs of orthopteran insects are very diverse, covering a wide frequency range from below 1 kHz to over 100 kHz. The duration of the songs ranges from under 1 ms to several minutes and their intensity can be very high. Some of the loudest songs recorded are produced by members of a genus of neotropical Tettigoniidae, *Supersonus*. The male calling songs are ultrasonic and are the highest frequency so far recorded for an arthropod at 150 kHz (Sarria-S et al., 2014). These ultrasonic songs are also very high intensity and can exceed 110 dB SPL at 15 cm. For comparison, typical sound intensities in tettigoniids are 70–100 dB SPL at 10–15 cm (Montealegre-Z et al., 2011b; Montealegre-Z and Morris, 2004).

Mole crickets are also able to produce loud sounds that travel over significant distances. For example, *Gryllotalpa major* has an advertisement song which attracts flying females for mating. In a study of 750 field recordings, Hill (2000) found that the absolute maximum intensity measured at 20 cm from the burrow opening is in the range of 90–104 dB. The song is audible up to 400 m from the male's burrow (Walker and Figg, 1990) and produces low frequency seismic vibrations that are detectable at distances of up to 3 m from the focal male (Hill and Shadley, 1997). The song of *G. vineae* is audible on a still night from at least 500 m (Ragge and Reynolds, 1998).

The longest distance over which acoustic communication has been recorded is 1.9 km. The calling song of the male bladder grasshopper *Bullacris membracioides* travels 100–150 m during the day but at night the change in thermal conditions enables the song to travel a greater distance (van Staaden and Römer, 1997). There is a disparity between the effective communication distances of the male and female, 2 km and 50 m respectively. The males and females duet (Donelson and van Staaden, 2005) but the female only responds if the intensity of the male song is 65 dB SPL or above, which corresponds to a communication distance of 50 m (van Staaden and Römer, 1998).

3.3 Sound production

The production of sound through frictional methods is characteristic of most tettigoniids, gryllids and acridids. Gryllids stridulate by the elytro-elytral method, in which the closing movement of the tegmina (wings) produces sound when a file on the ventral side of a modified vein (Cu2) of the right (upper) tegmen moves over the plectrum sited on the inner edge of the left (lower) tegmen. The file is made up of a number of hooked teeth and, as each tooth is struck by the plectrum, a single cycle of the song is produced (Bennet-Clark, 1999). The catching and then release of each tooth by the plectrum appears similar to the escapement in a clock mechanism and this observation led to Elliot and Koch proposing a 'clockwork cricket' model for sound production in gryllids (Elliott and Koch, 1985; Koch et al., 1988). The harp region of the forewing has been shown to resonate at frequencies similar to those that form the song. This resonance of the harp is analogous to the regulator of a clock. If the clockwork cricket model is valid, it should be possible to analyse the 'ticking' sounds made by the 'escapement' and examine the relationship between them and the dominant sounds produced by the harp 'regulator' when driven by the escapement. Bennet-Clark and Bailey (2002) made a detailed comparison between a clock mechanism and the sound production mechanism in the Gryllidae, validating and extending Koch's studies, thus demonstrating the anatomical features that make the cricket clock tick. However, in a subsequent study, Bennet-Clark (2003) investigated resonance in the wings of *Teleogryllus oceanicus*. He concluded that the harp is not the resonator; the major elastic component of the resonant system is the file plus the first anal vein and the mass component is the combined mass of the file, anal area and harp. A detailed account of the mechanics of cricket stridulation has since been given by Montealegre-Z et al. (2009).

Stridulation in grylllids usually produces pure tones at low frequencies of 2–8 kHz but some species have morphological and mechanical specializations that allow them to produce higher frequencies with rich harmonic content (e.g. some members of the Eneopterinae: Robillard et al., 2013). It has also been shown in *Gryllus bimaculatus* that calling song and aggressive song are produced by similar wing movements, indicating that both song types are probably generated by the same neuronal network, but courtship song is produced using a different wing movement and is probably under different motor control (Lin and Hedwig, 2021b).

Male tettigoniids also stridulate using the tegmina but, in contrast to gryllids, a file of teeth on the lower surface of the left tegmen is struck by a plectrum (the scraper) on the upper surface of the right tegmen. Generally, the sound is produced on the closing stroke of the wings (Montealegre-Z and Mason, 2005), though at least one species, unusually, produces sound on the opening stroke (*Ischnomela gracilis*: Montealegre-Z, 2012). The muscle contractions that move the wings are too slow to generate sounds in the kilohertz range but the rate of impact between the scraper and the teeth on the file generates a vibration in the wings substantially above the frequency of the wing movements. The vibrations produced by the tooth strikes are further amplified by an area of cuticle attached to the scraper which frames the thin cuticle on the right forewing, called the mirror (Baker et al., 2017).

There are three common methods of stridulation used by tettigoniids (Baker et al., 2019). In the first of these the wing and scraper move at an almost identical velocity, maintaining resonance of the mirror and frame by each subsequent tooth strike (Montealegre-Z and Postles, 2010). This method has an upper frequency limit of 40 kHz, so for higher frequencies a more complex process is used (Montealegre-Z et al., 2006). Elastic tissue can store energy when deformed and then release it at a much faster rate than a working muscle (for more detail on storage of energy in elastic tissue see Burrows, 2003).

In the second mechanism of stridulation, during wing closure, when the scraper is caught behind a tooth, it stops and bends, gathering elastic energy. On release from the tooth the elastic energy is suddenly released, and the scraper moves at a greater velocity relative to the wing. Therefore, the scraper can impact on teeth at a much higher frequency, compared with the previous method which has a coupled scraper with the same velocity as the wing (Montealegre-Z et al., 2006).

The third method of stridulation has been termed 'roll-trigger' (Baker et al., 2019; Montealegre-Z and Mason, 2005). The velocity of wing closure gradually increases as the scraper sweeps the file, but the inter-tooth distance of the file is more or less constant. There is no time adjustment of velocity (by tooth density difference) to maintain a constant rate of tooth contact. A solution to this may be to produce scraper jumps that help to compensate for constant distances and increasing speed. A massive vein backs the scraper and serves to trigger its release every time the scraper has reached its maximum bend. This mechanism has only been observed in a few species of Pseudophyllinae and Phaneropterinae (Baker et al., 2019).

Bush-cricket species produce a range of songs from resonant, tonal songs to non-resonant broadband ones, at a huge range of frequencies, from 600 Hz to 150 kHz. Phylogenetic studies of 94 katydid species by Montealegre-Z et al. (2017) showed that, rather than body size per se, frequency is most strongly related to components of the sound generation system: the file and the left-wing and right-wing mirrors. A further phylogenetic study of 40 bush-cricket species by Chivers et al. (2017) showed that parameters of file tooth density and file length are the best-fitting predictors of tonality.

Amongst animals that produce sound, acridids are distinctive in that they utilize two organs simultaneously to generate sound. Each hindleg has a file with a series of pegs (visible in Fig. 1) which, when rubbed against the forewing generates a sound and the sounds produced by the two legs are almost always out of phase with each other. Furthermore, in some species the phase shift is not fixed and can vary such that the leading leg in one song can become the trailing leg in the next (Elsner, 1974; Elsner and Wasser, 1994). Stridulation in many gomphocerine grasshoppers is characterized by specific phase shifts between the hind legs as well as different movement patterns produced by the left and right leg. If the metathoracic ganglion is transected, these differences are abolished and the hindlegs then produce the same pattern (Heinrich and Elsner, 1997).

3.4 Symmetry and asymmetry

The sound producing apparatus in Orthoptera shows asymmetry both in anatomy and in motion. Male crickets produce their calling song by rapid closure of the wings, which induces resonances in the harp and associated wing parts, as described in Section 3.3. The size of the harp influences the mass, and hence the resonant frequency of the wing components that produce sound. The possibility that directional asymmetry may convey information about male quality, encoded in the song, was shown by Simmons and Ritchie (1996).

In a study of four species of cricket, Moradian and Walker (2008) investigated the relationship between the body size of males and the size of the harp. In comparing species, they found consistent directional asymmetry in harp area, but not in total wing area. Within species, larger males typically have larger harps.

Analysis of the song of *Gryllus campestris* showed that each syllable has two parts, starting with a period of sustained frequency generated primarily by the left harp, followed by a period when the right harp takes over, during which the frequency falls (frequency modulation, FM). The greater the difference in area between the left and right harp, the greater the degree of frequency modulation. There is a possibility that females may detect asymmetry in males by the degree of downward FM at the second half of each syllable (Simmons and Ritchie, 1996), though this has been disputed by Montealegre-Z et al. (2011a) in their detailed analysis of sound production and the resonances of the left and right wings. Their measurements of wing resonance and the FM pattern recorded from coupled wings showed that the two wings are simultaneously involved in controlling the frequency of acoustic radiation during stridulation, but that the left wing dominates, particularly in the magnitude of the deflection. So, these findings do not support Simmons and Ritchie's conclusion that the contributions of the two wings take place at different times and at different frequencies.

The role of asymmetry in cricket song has been investigated in *Gryllus bimaculatus* (Hirtenlehner et al., 2013). Using a trackball, female crickets were offered two choices of song model. They only distinguished between the songs of symmetrical and asymmetrical males at a carrier frequency (i.e. the principle frequency in the frequency spectrum) of 4.4 kHz and a large modulation depth of 600–800 Hz and exhibited a preference for the pure tone calling song. No preference was observed at carrier frequencies of 4.9 and 5.2 kHz. A pair of ascending neurons, AN1 (see Section 5.2) are tuned to an average frequency of 4.9 kHz (Kostarakos et al., 2008), which correlates with the preferences in the choice experiments.

3.5 The energetic costs of singing

Studies prior to 2000 of the energetic costs of singing have been reviewed by Robinson and Hall (2002). In species that sing rapidly and loudly, the energetic cost can be at a similar level to that of locomotion. Studies of gryllids and tettigoniids have shown that producing a calling song is energetically costly. For example, in the katydid *Plangia graminea*, metabolic rate during singing is 60% higher than resting metabolic rate (Doubell et al., 2017). In *Mecopoda* sp. and *Anurogryllus muticus*, singing leads to increases in the CO_2 production rate over resting levels of 12 and 16 times respectively. Moreover, singing increases thorax temperature, on average by 7.6 and 5.8 °C respectively (Erregger et al., 2017). Sound production is relatively inefficient at around 1%, with most of the energy input being converted to heat (Prestwich, 1994). The increase in temperature may convert the acoustic signal to a bimodal one in which the energy invested in sound production is not 'wasted' but, instead, is reflected in the elevated thoracic temperature of a male. This thermal cue could provide important, reliable information for females choosing a mate, as it could allow them to evaluate how much cumulative energy has been invested by the male (Erregger et al., 2017).

Male mole crickets excavate a burrow, from which they sing. The shape of the burrow is horn-like and the resonant frequency matches the resonant frequency of the harp, producing amplification of the calling song, which can be heard at a distance of several hundred metres (Bennet-Clark, 1970). The mole cricket burrow has provided an opportunity to study the metabolic cost of singing for an undisturbed insect because the burrow can be used as the respiratory chamber needed for measuring oxygen consumption (Prestwich and O'Sullivan, 2005; White et al., 2008). In their experiments with *Gryllus monanka*, White et al. (2008) showed that the singing metabolic rate (MR) is on average 13.5 times higher than standard MR and 2.2 times higher than MR measured during burrowing in the laboratory. The singing MR of *G. monanka* is similar to that measured for other singing insects but is only 10% of that predicted on the basis of allometry for a similarly sized insect (0.89 g) during flight. A male mole cricket is estimated to consume 0.4 mL of oxygen during construction of a singing burrow and 5.5 mL during a 1-h singing bout; by comparison, a flying female searching for a male would consume a similar volume in less than 6 min.

The energetic cost of producing courtship song in *Gryllus bimaculatus* males has been investigated by Mowles (2014). If a singing male exceeds his aerobic capacity and switches to anaerobic respiration, lactic acid accumulates in the haemolymph and can be assayed. Males that had performed 5 min of courtship were found to have higher amounts of haemolymph lactate relative to the controls. Males who produce more rapid song with shorter intervals between each signal component have higher amounts of lactate in their haemolymph. So sound production in courtship is also an energetically costly behaviour. Males that are compromised physiologically are not likely to be able to maintain as vigorous a courtship as a healthy male (Tregenza et al., 2006).

3.6 Relationship between singing and the immune system

Immunocompetence is used to describe the ability of an individual to resist pathogens and parasites through the response of their immune system.

In insects, immunocompetence can be assessed by measuring the size of the encapsulation response to a standardized antigen such as a nylon filament (Rantala and Kortet, 2003).

The encapsulation process involves the coating of pathogens and other non-self objects with haemocytes. Ryder and Siva-Jothy (2000) looked at immunocompetence in *Acheta domesticus* and, using the relatively simple process of measuring haemocyte load in the haemocoel, showed that haemocyte load is correlated with the number of syllables per chirp, a sexually selected component of the calling song. By selecting males that produce more syllables per chirp females may also be selecting males for higher resistance to pathogens (see Section 6). In *Gryllus campestris*, inducing an immune response leads to reduced daily rate of singing (Jacot et al., 2004).

Singing in turn can affect the immune response: *Teleogryllus oceanicus* males exposed to song during rearing show stronger immune responses as adults than males reared in silence (Bailey et al., 2011). This suggests that such signals play a role in conveying information about population density and the associated risk of pathogen and parasite transmission.

3.7 Effects of age and diet

Mechanical wear in the sound producing organs can cause changes in sound signals related to the age of the individual, though in some species (e.g. *Teleogryllus oceanicus*: Walker and Cade, 2003) there is no evidence of age-related changes in calling song. For example, males of the field cricket *Gryllus pennsylvanicus* show significant changes in their song with age (Judge et al., 2008). Males at 21 days of age (days since the adult moult) have shorter pulse periods, shorter pulse durations, lower pulse peak frequency, more pulses per chirp and more variable pulse periods than when they are younger (Judge, 2011). There is no evidence that males increase their singing effort with age. For a detailed summary of other studies that have looked for age-related changes in field cricket calling song see Judge (2011).

Given the high energetic costs of singing, it is not surprising that diet can affect singing effort (i.e. the proportion of time spent producing calling song). Males on a high-quality diet generally show greater singing effort, as in *G. pennsylvanicus* (Judge et al., 2008) and *Gryllodes sigillatus* (Houslay et al., 2017). Hunt et al. (2004) found that female *T. commodus* on a high-protein diet live longer than those on a low-protein diet whereas the reverse is true for males. This is because males on a high-protein diet invest more energy in singing. Hunt et al. (2004) argue that the relationship between longevity and singing may be either positive or negative depending on local conditions, such as food availability. In some species, females may choose their mates on the basis of age or condition (see Section 6).

3.8 Patterns of singing

The songs of orthopterans are usually species specific, each having a particular temporal pattern of syllables which may be overlaid with modulations of frequency and intensity (e.g. Korsunovskaya, 2009; Kowalski and Lakes-Harlan, 2013; Zefa et al., 2013; Zhantiev et al., 2017). However, this pattern can still vary to some extent within species. The temporal pattern of syllable intervals may differ, for example, depending on whether the song is produced during the day or at night (Hershberger, 2021; Ivkovic et al., 2017). The cricket *Gryllus veletis* often produces so called 'gap' pulses where intensity drops near the middle of the pulse. Gap pulses are increasingly produced through the night and in the morning; they are produced less in the afternoon and evening (Thomson and Bertram, 2014).

A study of male calling songs in 50 species of neotropical katydids (ter Hofstede et al., 2020) found that songs vary greatly in duration, temporal pattern, peak frequency, and bandwidth, both across and within subfamilies. On the whole, katydid species within the subfamilies Pseudophyllinae and Phaneropterinae produce short songs of less than 250 ms at long intervals whereas species in the Conocephalinae produce highly repetitive songs over longer periods of time. However, there are some exceptions to this pattern: a few species in the Conocephalinae produce very short songs at long intervals, and some species in the Phaneropterinae produce relatively long songs of 1–6 s in duration, or sing frequently. A study of 16 Panamanian species of katydid showed a trade-off across species between the number of signals per day and the duration of sound in each signal (Symes et al., 2021).

Symes et al. (2016) looked at the singing patterns of eight neotropical katydids. They found that their calling songs are irregular and extremely short, with an average of around 80 songs per night totalling on average only 6.3 s. Singing may be affected by the acoustic environment, however. In some species males sing more often in response to singing by conspecifics; males in some other species respond similarly to heterospecific singing.

Species usually show diel and seasonal patterns of singing (e.g. Singh and Jain, 2020). Diel patterns can vary greatly between species (e.g. Tan and Robillard, 2021b). In *Phaneroptera falcata*, singing activity peaks about 3 h after sunset and decreases from midnight onwards (Samietz et al., 2014). The start of singing activity may be triggered by decreasing light intensity while the post-midnight decline may be related to a decrease in temperature. Fergus and Shaw (2013) showed that diel singing patterns in the Hawaiian swordtail crickets *Laupala cerasina* and *L. paranigra* result from a circadian rhythm generated by a daily cycling of the *period* gene transcript; they also identified amino acid differences in the *period* gene between the two species which, they suggest, may be responsible for interspecific differences in singing pattern.

Diel singing patterns can vary to some extent depending for example on weather or temperature (e.g. Franklin et al., 2009). The amount of singing activity may also be correlated with the availability of mating opportunities (e.g. Bertram et al., 2013) and/or the risk of attack from predators or parasitoid flies (e.g. Bertram et al., 2004).

Individuals within a species may also differ in their diel singing patterns: in *G. rubens*, for example, total signalling time, trilling bout duration, and hourly bout number are highly consistent within individuals but there is a lot of variation between individuals (Bertram and Bowen, 2006).

In bivoltine species, patterns of singing may differ between spring and autumn generations. For example, the song of the cricket *G. rubens* (Beckers, 2020) differs in frequency, pulse rate, pulse duration and interval duration between seasons. Female preferences shift in parallel. These differences appear to be the result of seasonal differences in temperature during the juvenile phase.

3.9 Environmental influences on signals

Singing and hearing can be affected by environmental factors such as vegetation, temperature, weather and noise. Studies up to 2000 on the effects of the environment on signals were reviewed by Endler (2000) and more recently there has been a wide ranging review of acoustic communication in noise (Brumm and Slabbekoorn, 2005).

The types of vegetation in the habitat affect the degree to which songs become attenuated and distorted with distance, and there is evidence in some species that species adapt their signals to local conditions (e.g. Couldridge and van Staaden, 2004).

In *Teleogryllus oceanicus*, almost every feature of the male calling song is affected by temperature (Walker and Cade, 2003). In the cricket *Acanthogryllus asiaticus* the frequency and temporal pattern of the male song is affected both by ambient temperature and the temperature experienced during development (Singh et al., 2020). In the mole cricket *Gryllotalpa africana*, which signals from a burrow, the syllable period, inter-syllable interval and duty cycle all vary with soil temperature (De Graaf et al., 2005). In tree crickets the frequency of the male's calling song changes with the ambient temperature; females, however, use active mechanical and neural processes to ensure that auditory tuning stays matched to the male's frequency, despite change environmental conditions (Mhatre et al., 2016). Because of the effects of temperature on physical and chemical processes, neurophysiological mechanisms that allow the receiver to process a song are also affected by temperature. It has been shown however, that the auditory system is able to compensate for changes in temperature (e.g. Korsunovskaya and Zhantiev, 2007; Roemschied et al., 2014).

The effects of both weather and anthropogenic noise were studied in the bladder grasshopper *Bullacris unicolor* at two sites close together but differing in noise levels. Passive acoustic recorders were used to monitor calling song from sunset to sunrise over a 3-week period and to record weather conditions. The results show that, as levels of anthropogenic noise increase, *B. unicolor* increases song interval (period between successive songs), decreases song rate, and sings at a lower peak frequency. On the noisier site the total number of songs recorded decreased, and the timing of the songs shifted to later in the night when noise levels are lower (Sathyan and Couldridge, 2021).

4. Hearing and ears

The orthopteran insects have a long history in the fossil record, with the earliest evidence suggesting stridulation coming from the structure of wings found in Triassic strata (Senter, 2008). In a large-scale macroevolutionary study using phylogenomic data, to understand how both hearing and sound production evolved, Song et al. (2020) showed that the Ensifera and Caelifera followed different pathways in the development of sexual communication through sound signals. In the Ensifera, hearing and sound-producing organs co-evolved. By contrast, in the Caelifera, sound-producing structures and hearing organs evolved independently. The ears of Ensifera are sited in the tibia of the forelegs, but in the Caelifera the ears are often in the first abdominal segment (Yack, 2004).

4.1 Structure of hearing organs

4.1.1 The acridid ear

The ears of acridids are located in the first abdominal segment. The anatomy of the ear has been described for a number of species (Jacobs et al., 1999; Yack, 2004). On each side of the abdomen is a tympanic membrane, which is coupled acoustically by air sacs to the contralateral ear. So, each tympanum receives an acoustic input on both the external and internal sides. The tympanal chordotonal organ, named Müller's organ, contains 60 to 100 scolopidia. The organ is attached to a number of sclerotized parts of the inner surface of the tympanum. The scolopidia can be divided into four groups, based on their points of attachment to the tympanum. However, in a study of tympanal receptor cells in the locust *Schistocerca gregaria* using intracellular recordings, Jacobs et al. (1999) showed that the receptor cells can be divided into three functional groups with different characteristic frequencies (CFs). Group 1 contains 20 receptor cells with their dendrites attached to the folded body. Their CFs lie at 400–700 Hz or at 1.5–2.0 kHz. Group II consists of 12–14 high frequency receptor cells (CFs 12–25 kHz) whose dendrites are attached to the pyriform vesicle. Group III receptor cell dendrites are attached to either the elevated process (EP) or to the styliform body (SB); their CFs lie at 3–4 kHz. There are no differences in physiology and central arborizations between those receptor cells of Group III whose dendrites are attached to the SB and those whose dendrites are attached to the EP.

4.1.2 The tettigoniid ear

The ears of a tettigoniid are situated in the tibia of the forelegs (Fig. 5). Each ear has two tympanic membranes which often differ in size and shape. The tympanum may be partially covered by a flap of cuticle, giving an external appearance of being sunk into a slit. Sound enters the large prothoracic spiracle and a pressure wave passes down the trachea to the femoro-tibial joint, where the trachea splits into an anterior and posterior branch that reach the inner surfaces of the anterior and posterior tympanum respectively.

In many species this large trachea is the primary route by which sound reaches the tympanic membranes. The trachea may also act as an exponential horn, amplifying the sound signal. In some other species the prothoracic spiracle is reduced in size and the slits may play a greater role in establishing the pressure difference across the tympanic membranes.

The sense cells of the ear are attached to the dorsal surface of the anterior branch of the trachea and are arranged in a line along the long axis of the

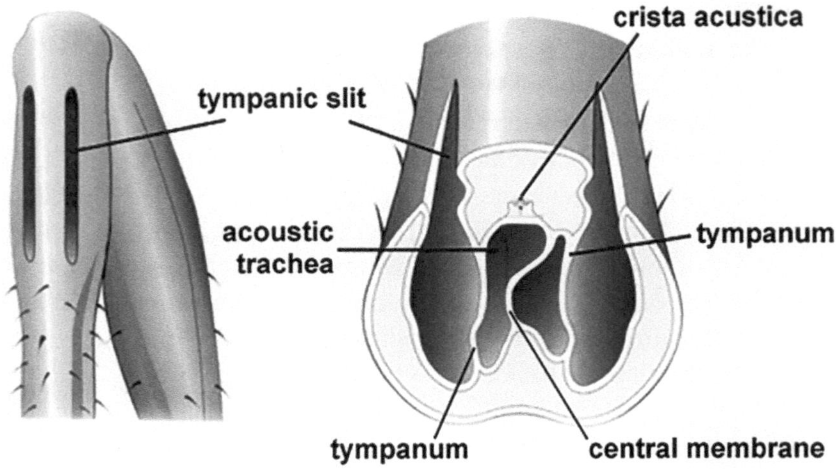

Fig. 5 Structure and location of the tibial hearing organ in a bushcricket.

trachea branch, forming the crista acustica. Unlike the auditory organ in acridids, that of most gryllids and tettigoniids is not attached to the tympanum directly. The auditory sensilla, of which there are between 12 and 116 in tettigoniids (Strauß, 2019), are attached proximally to the anterior trachea, and distally to the tegument of the tibia by one or more attachment cells linked to the tectorial membrane (Yack, 2004). The auditory organ of tettigoniids is part of a complex tibial organ that has three parts, the crista acustica, intermediate organ and the subgenual organ. Each of these contains scolopidia. There have been a number of studies of the complex tibial organ. For details of the neuroanatomy see Rössler et al. (2006) and Strauß (2019).

The scolopidia of the crista acustica are tuned to specific frequencies and are arranged tonotopically, with those tuned to high frequencies at the distal end of the crista. The dendrites of the receptors project into the prothoracic ganglion. The tonotopic structure in the hearing organ is preserved as central neural tonotopy (Stölting and Stumpner, 1998). Strauß et al. (2012) used cobalt chloride backfills of the tympanic nerve to compare the arrangement of auditory receptors in six species in the Tettigoniidae. They found that although the auditory receptors in the crista acustica are generally arranged in a line, some of the auditory receptors deviate from a strictly linear arrangement and are organized as double rows of receptors in two axes: anterior–posterior and dorso-ventral. The proportion of parallel somata is as high as 52% in *Poecilimon ampliatus*. The functional significance of the double row of receptors for frequency tuning is not yet fully understood

(Strauß et al., 2012). It is possible that the double row of parallel somata could enable shorter distances between dendrites, which would give a higher resolution of frequency over a smaller frequency spectrum.

The number of auditory sensilla is species-specific and in general closely related species have similar numbers. The variation in number between individuals of the same species is usually very low. However, the number of auditory sensilla is not directly related to the length of the crista (Strauß et al., 2017b). In most species there is no sexual dimorphism in the number of auditory sensilla (Strauß, 2019), but in the phaneropterine *Ancylecha fenestrata* males have a longer crista acustica than females and a larger number of auditory sensilla:116 in males and 86 in females (Scherberich et al., 2016, 2017).

In a study of the neotropical bush-cricket *Copiphora gorgonensis*, Jonsson et al. (2016) investigated the role of the acoustic trachea, using micro-scanning laser Doppler vibrometry. The male calling songs of *C. gorgonensis* are short pure tone syllables at a frequency of 23 kHz (Montealegre-Z and Postles, 2010). They measured the velocity of sound propagation in the acoustic trachea and found that over the frequency range 10–40 kHz the velocity remained constant. Sound velocity inside the trachea was 255 m s^{-1}, reduced by a factor of 1.35 compared with free field measurements. Thus, the internal sound wave is delayed with respect to the sound wave on the external pathway. The acoustic trachea provides a gain of approximately 15 dB SPL, suggesting that the trachea is acting as an exponential horn. In a study of the outer ear of *C. gorgonensis* using numerical modelling and real 3D geometry of the acoustic trachea, Celiker et al. (2020) showed that the most important factor in generating the gain of 15 dB is the geometry of the acoustic trachea, confirming the function of the trachea as an exponential horn. Similar measurements carried out on gryllids are described in the next section.

4.1.3 The gryllid ear

Crickets have one of the most complicated hearing organs known (Michelsen and Larsen, 2008). The ears are located in the forelegs, just below the joint between the femur and tibia. The structure of each of the acoustic systems is similar to that of the tettigoniids. Sound acts on the external surface of the tympanum but also reaches the internal surface of the tympanum via the prothoracic spiracle and acoustic trachea. Unlike the tettigoniids, where the two ears are separate, the two ears in gryllids are linked (Fig. 6). Each acoustic trachea has a transverse extension, and these extensions meet in the midline and are connected by a thin septum, formed

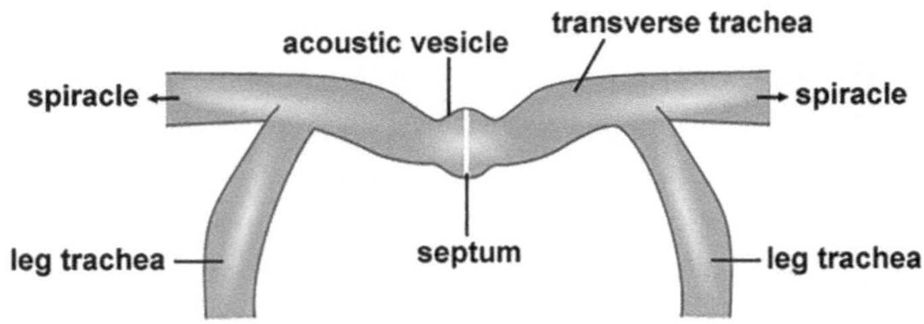

Fig. 6 Schematic diagram of the arrangement of the acoustic trachea in gryllids.

from a double membrane. The septum is contained within an acoustic vesicle. There is considerable diversity in the morphology of the acoustic tracheal system. A study of 40 ensiferan species (Schmidt and Römer, 2013) revealed that those species using intraspecific acoustic communication and possessing tibial tympana have an acoustic vesicle and septum, whereas species that have lost acoustic communication, or that never use acoustic calling songs, lack an acoustic vesicle. The size of the acoustic vesicle varies between species, with field crickets having the smallest vesicle relative to their size (measured as pronotum width), compared with rainforest crickets.

So the two acoustic tracheae are linked both anatomically and, via the thin membrane, acoustically. This connection adds two more sound inputs to the ear from the contralateral ear and from the contralateral spiracle. The cricket ear is thus an acoustic four-input device (Michelsen et al., 1994; Michelsen and Larsen, 2008) and, while the sound inputs from the spiracles and tympana are similar in some respects to those found in tettigoniids, the interactions between the four inputs in crickets add complexity for researchers too.

The properties of the acoustic trachea have been studied as they have in tettigoniids. The propagation velocity inside the acoustic tracheal tubes has been measured in *Teleogryllus commodus* by Larsen (1981) as $263\,\mathrm{m\,s^{-1}}$ and in *Gryllus bimaculatus* by Michelsen et al. (1994) as $264\,\mathrm{m\,s^{-1}}$. These figures are in very close agreement and are well below $344\,\mathrm{m\,s^{-1}}$, which is the velocity of sound in adiabatic propagation (Michelsen and Larsen, 2008).

4.2 Directional hearing

The functional significance of the arrangement of the trachea described above lies in the ability to provide directional hearing. The environment in which Orthopteran insects live will always have background sounds and the function of the auditory system is to enable the insect to pick out the biologically significant sounds from the soundscape and react appropriately.

For an individual avoiding predators or seeking mates, locating the source of those sounds is vital. In many species of insect, the body is smaller than one wavelength of the carrier frequency of the calling song. The carrier frequency of the calling song of *Gryllus campestris* is usually between 4.0 and 5.5 kHz (Ragge and Reynolds, 1998) which corresponds to wavelengths of 8.58–6.24 cm. The adults are 1.8–2.3 cm in length (Benton, 2012). The distance between the tympana is small, around 1 cm or less in crickets. A consequence of this small separation is that both the difference in arrival time of a sound at each ear (the interaural time difference, ITD) and the difference in intensity of the sound arriving (the interaural intensity difference, IID) are very small. Comparing the smallest and largest crickets measured, the ITDs have been calculated as 5–23 μs (Schmidt and Römer, 2013).

In insects that are large in comparison with the wavelength of sounds, the body can cause diffraction. A consequence is that the sound intensity at the contralateral ear is lower than that at the ipsilateral ear, producing an IID. If that difference is large enough to be detectable, sound localization is possible. To measure the minimum IID necessary for determining the location of a sound source, Rheinlaender et al. (2006) examined phonotaxis in the large Asian tettigoniid *Gampsocleis gratiosa*, which has a body length of 50–55 mm and a pronotum width of 0.8–0.9 mm. They used a walking belt to maintain a female in a constant position relative to a loudspeaker playing chirps of natural song. Females that performed phonotaxis were then fitted with a back pack containing two miniature speakers connected to the auditory spiracles by plastic tubes. The intensity of the sound reaching each ear could be independently controlled, allowing the researchers to determine the smallest IID that would enable a phonotactic approach. The results revealed a sensitivity to small IIDs of 1–2 dB.

A study of phonotaxis in *Gryllus bimaculatus* (Schöneich and Hedwig, 2010) showed that ITDs could deliver information about sound direction. The difference in sound intensity on either side of the insect's body is 1.3 dB, when stimulated from 30° (Michelsen et al., 1994). Schöneich and Hedwig (2010) used an open-loop track-ball system to measure the accuracy of female orientation to male song. The sound source was positioned at a range of angles between +30° and −30° to the insect's long axis. At angles of incidence of 1°–6° the females walked precisely towards the sound source. In separate experiments, movements of the posterior tympanic membrane were measured using a laser vibrometer and activity in the tympanal nerve was recorded. An IID corresponding to 12.8 dB at 30° sound incidence was demonstrated. These experiments show that IIDs are reflected in the response latencies of the auditory afferent activity. The overall bilateral latency difference at ±30° was 1.28 ms whereas the actual interaural difference in sound arrival time was less than 15 μs. The results reveal the acoustic orientation of the cricket *G. bimaculatus* as one of the most precise among invertebrates (Schöneich and Hedwig, 2010).

5. Neural control and processing

The cricket, and particularly members of the genus *Gryllus*, has been used as a model organism in neuroethological studies for over 60 years. The cricket continues to serve that function, as exemplified by studies of the neurons and networks underlying singing behaviour (Schöneich and Hedwig, 2017) and of the auditory pathway for processing acoustic signals (Pollack and Hedwig, 2017). Studies up to 2000 are reviewed by Robinson and Hall (2002).

5.1 Control of singing activity

Sound production in crickets and grasshoppers is controlled by descending brain neurons (Hedwig, 2001). The dendritic arborizations of these neurons occupy different brain regions in the two groups. In crickets, the dendrites project within the anterior protocerebrum, between the alpha-lobe and the pedunculus, whereas in grasshoppers the dendrites form a profuse arborization within the posterior dorsal neuropil (Hedwig, 2001).

The motor programmes that drive the wing movements for stridulation are generated in the thoracic and abdominal ganglia of the ventral nerve cord. Recordings of activity in hundreds of brain neurons of *Gryllus bimaculatus* revealed a pair of descending interneurons that are command neurons, in that they are both sufficient and necessary for production of the calling song. There is no activity in these command neurons when the cricket is in a resting state. In active stridulation, there is a linear relationship between the spike frequency of the interneurons and the chirp repetition rate. The soma of the command neuron is on the dorsal surface of the protocerebrum and it has a dendritic aborization located in the ventral protocerebrum. The descending axon runs through the contralateral connective and into the prothoracic ganglion (Hedwig, 2000).

The rhythmic movements involved in stridulation are, like many other repetitive actions in animals, driven by neural networks in the central nervous system (CNS) that form discrete functional units, termed central pattern generators (CPGs). For a description of research on CPGs in animals in a historical context see the review by Mulloney and Smarandache (2010). The idea of innate acoustic behaviours in Orthoptera has been pursued in the search for CPGs that control the stridulation patterns of acridids, gryllids and tettigoniids. Innate releasing mechanisms and fixed action patterns in Orthoptera have been reviewed by Ronacher (2019).

Early work on the production of sound in grasshoppers used lesions of the connectives between ganglions in the ventral nerve cord to determine the location of the CPG (Hedwig, 1986; Huber, 1960). It was suggested that the CPG for calling song in the cricket and grasshopper was sited in the second or third thoracic ganglion. Longitudional lesions of the third thoracic ganglion in the grasshopper *Chorthippus bigutulus* suggested a hemiganglion organization for the CPG (Ronacher, 1989). In the cricket it was proposed that the mesothoracic ganglion is the site of the CPG, co-located with the motor neurons that innervate the wing muscles (Huber, 1960). Subsequently, a number of studies have implicated the abdominal nervous system in generating motor activity that produces the calling song. In *Gryllus bimaculatus*, lesions to the abdominal nerve cord showed that the abdominal ganglia each have a different role in defining the song structure (Jacob and Hedwig, 2016). Data derived from intracellular recordings of activity in singing interneurons provide support for the results obtained by lesioning (Jacob and Hedwig, 2019, 2020; Schöneich and Hedwig, 2011, 2012).

The organization of the singing CPG in crickets has been studied in four species (Lin and Hedwig, 2021a) with different species-specific calling songs (Table 2). Recordings of the normal calling song pattern were made for each species.

The effects on the calling song of lesions to the connectives between abdominal ganglia were compared in the four species. Lesion of the

Table 2 Normal calling song pattern of four species of cricket as described by Lin and Hedwig (2021a).

Species	Description of calling song	Reference
G. rubens	Long trills	Gray and Cade (2000)
G. assimilis	Chirps repeated at rather long intervals	Pollack and Kim (2013)
Teleogryllus oceanicus	Phrases composed of one chirp and several trills	Bailey et al. (2017); Zuk et al. (2008)
T. commodus	Phrases with one chirp, followed by a few trills	Bailey et al. (2017); Simmons et al. (2005)

connectives between the metathoracic ganglion complex T3 and the third abdominal ganglion A3 abolished singing activity in all four species. Males were still able to raise their wings to the singing position. These results are consistent with those of previous lesion experiments (Jacob and Hedwig, 2016; Schöneich and Hedwig, 2011).

Lesion of the connectives between A3 and A4 affected all four species in similar ways. The males showed large decreases in singing activity and loss of song structure. Lesion of the A4 to A5 connectives in *G. rubens* did not change the calling song pattern and neither did lesion between A5 and A6. In *G. assimilis* the effect of lesion between A4 and A5 was that singing activity continued but with extended chirps and a tendency to reduce the chirp interval. The number of pulses (syllables) per chirp increased. No change was observed after lesion between A5 and A6.

The effect of the A4–A5 lesion in the two *Teleogryllus* species was dramatic as it abolished the trill and chirp structure of the song. Instead, males produced repeated short sequences of a variable number of sound pulses. There was no change in singing pattern in either species when lesioned between A5 and A6.

The results of these lesion experiments show a conserved organization of the calling song CPG network, with the generation of pulses controlled by the A3 ganglion and timing information coming from A4 to A5 (Lin and Hedwig, 2021a).

5.2 First order auditory interneurons

In many orthopterans the temporal pattern of the male's song, and especially the relative durations of individual elements, are important in conveying meaning. It has been shown however, that a small set of auditory neurons is sufficient to reliably decode the acoustic signal and predict its behavioural relevance (e.g. Creutzig et al., 2009; Meckenhauser et al., 2014).

The auditory receptors project into the prothoracic ganglion of the crickets and bush crickets and into the abdominal ganglion in grasshoppers. The cricket ear has about 65 receptor neurons. In *Teleogryllus oceanicus* these auditory receptor neurons can be classified into three groups based on their frequency response (Imaizumi and Pollack, 1999). Two of the groups can be linked to behavioural traits. One group contains neurons with a centre frequency of 3.0–5.5 kHz and appears to be involved in acoustic communication. The second group contains neurons with a centre frequency of ≥18 kHz and is probably involved in predator detection. Low frequency receptor neurons comprise about 75% of all the auditory receptors, with the rest divided about equally between these two groups (Imaizumi and Pollack, 2001, 2005; Marsat and Pollack, 2012). The role of the third group, containing neurons with a centre frequency 10–12 kHz, is not clear.

In the prothoracic ganglion the axonal terminals of the auditory afferents synapse with a small set of auditory interneurons (Imaizumi and Pollack, 2005). Auditory information is passed to the cricket brain by two ascending neurons (AN). AN1 is sharply tuned to the centre frequencies of the male song and connects with a network of brain neurons linked to recognition of the species-specific pattern of the male song (Schöneich et al., 2015). It facilitates phonotaxis. Discharge differences in the pair of AN1 neurons drive phonotactic steering behaviour (Trobe et al., 2011). AN1 may also be involved in the recognition of the male aggressive song. There is no evidence for sexual dimorphism in the hearing system of field crickets: the auditory responses of ascending neurons cover the same overall range of sound frequencies with similar sensitivity (Schöneich, 2020).

AN2 is most sensitive to frequencies above 10 kHz, triggering avoidance behaviour in response to bat calls (Marsat and Pollack, 2012). AN2 may also contribute to positive phonotaxis and the processing of the high frequency ticks of the male courtship song (Samuel et al., 2013; Stout et al., 2011). The pattern of the neural response in AN2 is high frequency bursts of spikes. The bursts are triggered by nearly synchronous bursts in the ultrasound receptors. Recent research on the neural basis of ultrasound avoidance in crickets has been reviewed by Marsat and Pollack (2012).

Two types of local interneurons form part of the acoustic processing network. These neurons are called omega neurons and get their name from their shape when stained. The neurons are paired in tettigoniids but there are two pairs in gryllids (Stiedl et al., 1997). ON1 in gryllids responds to sounds over a wide range of frequencies. In *Teleogryllus oceanicus* it is most sensitive to the frequency 4.5 kHz of conspecific songs (Faulkes and Pollack, 2000). The response latency is longest for this frequency by 5–10 ms. The onset of excitatory postsynaptic potentials (EPSPs) is delayed for 4.5 kHz when compared with ultrasound stimuli (Faulkes and Pollack, 2001). ON1 is strongly stimulated by input from the ipsilateral ear but inhibits the contralateral ON1, AN1 and AN2. The inhibition enhances the difference between the sound reaching the two ears, facilitating sound localization (Faulkes and Pollack, 2000; Marsat and Pollack, 2005). In contrast to ON1, ON2 receives excitatory inputs from both ears and there is no inhibitory coupling between the two ON2 neurons. The function of the ON2 neurons is not yet clear (Zhang and Hedwig, 2019).

There are only three auditory interneurons that have ascending axons. The T-fibre TN1 and AN2 are tuned to high frequencies and do not reliably follow the pattern of the calling song. Only AN1 carries information about the calling song towards the brain (Hedwig, 2016).

5.3 Pattern recognition in the brain

In the study of pattern recognition in the brain, the cricket continues to be a model organism, providing not only specific information about crickets but also observations that are more generally applicable in the study of the principles of auditory processing at the level of identified neurons (Zorovic and Hedwig, 2011). Local brain neurons are difficult to identify individually, but systematic intracellular recordings from the cricket brain when stimulated with sound patterns revealed four neuron types that, together with AN1, form an auditory detector network for the pattern of the male calling song (Hedwig, 2016; Kostarakos and Hedwig, 2012; Schöneich, 2020).

The axonal projections of AN1 form a ring-like arborization in the frontal protocerebrum. Four local interneurons (LN2 to LN5) closely match this arborization pattern and form a similar ring-shaped auditory neuropil in the brain (Kostarakos and Hedwig, 2012). These neurons form a local auditory network that has the fundamental characteristics of a coincidence detector circuit (Schöneich et al., 2015). For a detailed account of the filtering processes in the auditory pathway shaping feature detection in crickets see Hedwig (2016). The neuroethology of acoustic communication in field crickets has been reviewed by Schöneich (2020) and a detailed section on song pattern recognition in the brain is included.

6. Choosing a mate

6.1 Recognition of sex and species

Reproductive success depends on choosing a mate of the appropriate sex and the same species. The male calling song is usually species specific (see Section 3.8), so in singing species the first stage of mate recognition usually takes place via the processing of the male's calling song by the female's nervous system (Section 5). The female's evaluation of the male's song may be based on both its temporal parameters and its frequency components (e.g. Schmidt et al., 2008).

In duetting species, the timing of the female response is often species specific too (Heller et al., 2018). Generally females show a strong preference for the calling song of their own species, as in the bush-cricket *Isophya camptoxypha* (Orci, 2007), the grasshoppers *Chorthippus biguttulus* and *C. mollis* (Finck and Ronacher, 2017) and the tree cricket *Oecanthus forbesi* (Symes, 2018), though in closely related species females are not always able to discriminate conspecific from heterospecific songs (e.g. Couldridge and van Staaden, 2006; Kowalski and Lakes-Harlan, 2011). Safi et al. (2006) have shown however that female *C. biguttulus* prefer those conspecific songs that can be more reliably discriminated from heterospecific songs, thus maximizing the efficiency of species recognition.

However, in non-duetting species the male still needs to identify the sex and species of any potential mate that responds to his calling song. In duetting species there may be considerable room for error in species recognition by males (Heller et al., 2018) and females cannot be certain that the male that approaches them is the one that was singing rather than an eavesdropper (Section 2.2). So most species have further means of mate recognition once a pair actually meet. Initial mutual antennation has been widely observed in orthopteran mating (e.g. Hall and Robinson, 2020) and the chemical cues detected in this way have been shown to be important in mate recognition. Ryan and Sakaluk (2009) for example found that antennation is required for sex recognition in the cricket *Gryllodes sigillatus*, while Finck et al. (2016) showed that males of the grasshoppers *C. biguttulus* and *C. mollis* only produce courtship song if they detect appropriate chemical cues from a conspecific female.

Courtship song can have a role in species recognition but is probably more concerned with stimulating the female to mate (see Section 2.3) and/or providing information about the male's quality; it may be particularly important in mate choice if satellite males are common (Rebar et al., 2009; Shestakov and Vedenina, 2015; Simmons et al., 2013; Zuk et al., 2008).

6.2 Sexual selection theory

Sexual selection can be defined as 'differential reproductive success due to the ability to secure matings and/or fertilizations' (Rosenthal, 2017, p. 503). Traditionally it is divided into two categories, though the two do not necessarily operate independently: intrasexual competition, where members of one sex (usually but not always males) compete with each other for access to mates, and mate choice, where one sex (usually but not always females) preferentially mates with individuals that have particular traits. Mate choice is considered in this section and intrasexual competition in Section 7.

Wagner (2011), the recent book by Rosenthal (2017), and Rosenthal's reviewers Makowicz et al. (2020) provide useful reviews of mate choice theory. Mate choice may be the result of active choice. Alternatively, females may simply be passively attracted to the male whose signal is the most stimulating, easiest to locate, or most closely tuned to the female nervous system. Or they may show preferences just because they carry genes that are beneficial in another context but which also happen to influence their responses to male traits in the mating context (*receiver bias*, also called *perceptual* or *sensory bias*); see for example Gray et al. (2016) and Reichert et al. (2017).

In theory, mate choice can have significant effects on female fitness, either through their own reproductive success or through that of their offspring. Preferred males can provide direct benefits (material resources such as nutrients contained in the nuptial gift) and/or indirect benefits (via 'good' genes that increase offspring viability or 'sexy son' genes that increase the attractiveness to females of male offspring). The parasite-mediated sexual selection theory (Balenger and Zuk, 2014) proposes that females choose a male whose signals indicate his ability to resist parasitic infections, thereby providing them with benefits both direct (e.g. reduced risk of sexual transmission of disease) and indirect (e.g. offspring who inherit parasite resistance).

While it is relatively clear how selection can favour preferences for mates that provide direct benefits (though see Wagner, 2011), little evidence has actually been found in support of the good genes/sexy sons/parasite-mediated theories and there has recently been a growing body of opinion that the importance of indirect benefits may be limited (Makowicz et al., 2020; Rosenthal, 2017).

6.3 Choosing a mate based on information in male song

Studies in Orthoptera have often looked at the relationships between song features, male characteristics and female mate preferences. Some of these have shown that song features correlate with certain male characteristics, e.g. with size in the cricket *Acanthogryllus asiaticus* (Singh and Jain, 2020) and the mole cricket *Gryllotalpa major* (Howard and Hill, 2006) or age in the cricket *Gryllus veletis* (Fitzsimmons and Bertram, 2011), without demonstrating that females can perceive the differences between songs or have a preference for males with these characteristics. Others have shown that females prefer certain song features without demonstrating a correlation between song and male characteristics. For example females in the cricket *Gryllus bimaculatus*

Fig. 7 A female two-spotted cricket, *Gryllus bimaculatus* (© Axel Hochkirch).

(Fig. 7) prefer courtship song with louder 'ticks' (very short chirps) and quieter pulses (Shestakov and Vedenina, 2015), but it is unclear what, if any, information about the male these features convey.

Females in many species have, however, been shown to prefer songs with features that provide information about some characteristic of the male that could, at least potentially, affect the female's fitness. Several examples of such preferences were given in Robinson and Hall (2002); more recent examples are shown in Table 3.

There is the potential for direct benefits in several of the examples in Table 3, especially where females choose a mate that is likely to provide a larger or more nutritious nuptial gift, or more sperm, as in *G. sigillatus* (larger spermatophore with higher protein content), *Ephippiger ephippiger* (larger, more nutritious spermatophore, with more sperm), *Poecilimon zimmeri* (larger spermatophore), *Allonemobius socius* (larger nuptial gift and more sperm), and *Velarifictorus ornatus* (larger spermatophore). But of these five species, actual effects of the choice of mate on female fitness have only been demonstrated in the last two. In *V. ornatus* fertilization is more successful. In *A. socius*, immune-compromised males produce less attractive song and smaller nuptial gifts, so that females choosing mates on the basis of their song receive more resources from the male and have higher fecundity (which also provides support for the parasite-mediated sexual selection theory). Direct benefits may also be provided via seminal fluid. For example *G. lineaticeps* females choosing males with longer chirp durations benefit from increased lifespan, even if they only receive sperm and seminal fluid and do not eat the spermatophore. This seems to be due to delayed

Table 3 Species in which females show a preference for parameters of song that provide information about male characteristics.

Species	Parameter of song (refers to calling song unless otherwise stated)	Information contained in song (correlation between parameter and male characteristic)	Reference
Crickets			
Acanthogryllus asiaticus	Lower peak frequency	Larger size	Singh and Jain (2020)
Acheta domesticus	Higher proportion of 3-pulse chirps	Larger size; better immunocompetence	Head et al. (2005); Stoffer and Walker (2012)
Allonemobius socius	Longer pulse duration	Status of immune system	Fedorka and Mousseau (2007)
Gryllotalpa major	More pulses per chirp; lower dominant frequency	Larger size	Howard and Hill (2006)
Gryllus assimilis	Louder	Larger size; better condition; more aggressive; older male	Bertram et al. (2016); Bertram and Rook (2011, 2012); Loranger and Bertram et al. (2016)
	Lower frequency	Larger size; better condition; more aggressive	
	Chirps with longer pulses; longer chirps; more pulses per chirp (i.e. higher duty cycle with more sound per unit of time)	Older male; more aggressive	
	Higher pulse and chirp rates	Older male; better condition	
G. bimaculatus	Louder song	Larger size	Verburgt and Ferguson (2010); Verburgt et al. (2011); Zhemchuzhnikov et al. (2017)
	For songs of equal intensity, higher duty cycle with longer syllables and/or shorter intervals between syllables/chirps	Younger male	

Table 3 Species in which females show a preference for parameters of song that provide information about male characteristics.—cont'd

Species	Parameter of song (refers to calling song unless otherwise stated)	Information contained in song (correlation between parameter and male characteristic)	Reference
	Courtship song with higher duty cycle for high-frequency ticks (very short chirps) with longer ticks at faster rate	Better immunocompetence	Rantala and Kortet (2003)
G. campestris	Lower frequency (given priority over chirp rate)	Older male; better condition as nymph	Jacot et al. (2007); Rodriguez-Munoz et al. (2019); Scheuber et al. (2004);
	Higher chirp rate	Better current condition	
G. lineaticeps	Higher chirp rate	Better nutritional status	Wagner and Harper (2003); Wegehaupt and Wagner (2017)
	Longer chirp duration	Possibly seminal fluid contains fewer oviposition stimulants	
G. pennsylvanicus	Lower frequency, longer pulses	Larger size	Judge (2011); Ting et al. (2017)
	Lower peak frequency, faster pulse rate (shorter pulse period), shorter pulse duration, more pulses per chirp	Older male	
G. sigillatus	Shorter pulses	Larger size	Champagnon and del Castillo (2008)
G. texensis	Lower frequency; higher pulse duty cycle (longer pulses and shorter intervals)	Larger size	McAuley and Bertram et al. (2016)

Continued

Table 3 Species in which females show a preference for parameters of song that provide information about male characteristics.—cont'd

Species	Parameter of song (refers to calling song unless otherwise stated)	Information contained in song (correlation between parameter and male characteristic)	Reference
Oecanthus henryi	Higher frequency	Possibly larger nuptial gift	Mhatre et al. (2012)
Plebeiogryllus guttiventris	Faster chirp rate (shorter chirp period)	Possibly condition	Nandi and Balakrishnan (2013)
	Louder song (overrides other preferences)	Larger size	
Teleogryllus oceanicus	Courtship song with greater trill duty cycle	Better condition: can produce energetically expensive song despite inherent costs (reduced immunocompetence and lower sperm viability)	Simmons et al. (2013)
Velarifictorus ornatus	Specific parameter unknown (song of short-winged males preferred to that of long-winged males)	Larger spermatophore	Zhao et al. (2017)
Bush-crickets			
Ephippiger ephippiger	Longer chirp duration	Younger male	Wedell and Ritchie (2004)
Isophya camptoxypha	Longer syllable duration	Possibly larger size and/or better condition (longer syllables more costly to produce)	Orci (2007)
Orchelimum pulchellum	Louder song; for songs of equal loudness, greater attenuation in the higher frequency components	Larger size or one otherwise capable of more energetically costly song	Harness and Campbell (2021)

Table 3 Species in which females show a preference for parameters of song that provide information about male characteristics.—cont'd

Species	Parameter of song (refers to calling song unless otherwise stated)	Information contained in song (correlation between parameter and male characteristic)	Reference
Poecilimon zimmeri	Specific parameter unknown	Larger size	Lehmann and Lehmann (2008); Reinhold (1999)
Grasshoppers			
Bullacris membracioides	Lower frequency; shorter total song duration; smaller ratio between total duration and final syllable duration	Larger size	Couldridge and van Staaden (2006)
	Longer final syllable duration	Smaller size	
Chorthippus biguttulus	Louder song	Larger size; better condition	Franzke and Reinhold (2012); Klappert and Reinhold (2007); Ronacher and Stange (2013)
	Shorter pauses between syllables (higher syllable to pause ratio); shorter gaps within syllables		
	Greater onset accentuation (syllable louder at beginning)	Larger size: better immunocompetence	
	Larger offset depth (greater decrease in loudness in pause between syllables)	Better immunocompetence	

oviposition, possibly as a result of fewer or less potent oviposition stimulants in the male's seminal fluid. The offspring of *Neoconocephalus ensiger* females who choose a leading male (i.e. one whose song precedes those of other singing males: see Section 7.5).are in better condition (heavier relative to body size), possibly as a result of the proteins provided in seminal fluid (Murphy et al., 2017).

However, even if females prefer larger males, and male size is correlated with spermatophore size, it cannot be assumed that choosy females gain direct benefits. Jarrige et al. (2013) found no evidence that song features in *Ephippiger diurnus* could predict the nutritional content of the spermatophore actually transferred during copulation. Instead males adjusted the size and quality of their spermatophores depending on the age and size of the female. Male choice in this species effectively makes it impossible for females to assess what direct benefits they will get by mating with a particular male.

Evidence for indirect benefits associated with the female's choice is also limited. In *Acheta domesticus* the female's choice results in more attractive sons, more fecund daughters, and a shorter generation time (faster maturation/earlier reproduction) for her offspring. She suffers from direct costs because her lifespan and fecundity are reduced, but these are outweighed by the indirect benefits. In *G. assimilis* females mating with more aggressive males have larger offspring: daughters lay more eggs and sons are more attractive, though there is no increase in offspring viability. In *G. bimaculatus*, successful males (i.e. those chosen by females) have male offspring that are in turn successful (Wedell and Tregenza, 1999). On the other hand, two species provide evidence against indirect benefits. In *G. campestris*, older males attract more females but once the female has approached an older male she is less likely to mate with him than she is with a younger male. There is no relationship between the male's age and the number of his offspring or grand-offspring. In *Chorthippus biguttulus*, there are no differences between females mating with attractive or unattractive males in the hatching success, viability or rate of weight gain of their offspring, or in the attractiveness of their sons. Safi et al. (2006) argue that female choice in this species is probably driven mainly by selection for species recognition.

The female choices listed in Table 3, and the benefits that may be gained from that choice, are also not always as straightforward as might seem. In the cricket *Gryllus lineaticeps*, females prefer higher chirp rates, and benefit from increased fecundity as a result, but only if the male is on a low-quality diet. For males on a high-quality diet, there is actually a negative correlation between chirp rate and fecundity. Females, however, are able to choose males on a poor diet, whatever their chirp rate, so they must be using additional cues to enable them to do so. The benefits gained by females choosing males with high chirp rates are also balanced by costs: these males are more likely to be attacked by lethal parasitoids (see Section 8) and females associating with them are in increased danger too. Females also prefer higher chirp rates in courtship song but there is no correlation between chirp rate in calling song and that in courtship song and females may prioritize courtship over calling song, being more likely to mate with a male that has a higher courtship chirp rate even if the calling song of that male has a lower chirp rate. Courtship chirp rate has no relationship to nutritional status and any benefits of selecting a mate based on courtship song are unknown (Martin and Wagner, 2010; Wagner and Reiser, 2000; Wegehaupt and Wagner, 2017). In *G. pennsylvanicus*, females who mate with a high quality male tend to lay more eggs, but they still do this even if the male they mate with is a low-quality male accompanied by recorded song from a high-quality one (Ting et al., 2017). In *A. domesticus*, females choose initially on the basis of the male's calling song but then reassess the male using proximate signals, which may be non-acoustic (Stoffer and Walker, 2012). Choice may sometimes be more to do with rejecting males that are unsuitable—e.g. heterospecifics, injured males or those with developmental problems—rather than selecting a high-quality mate (Clemens et al., 2017, 2021). The costs and benefits for the female may also be affected by male manipulation of, for example, egg laying rates or longevity, via factors transferred in the spermatophore, rather than her choice of mate (e.g. Wilson and Walker, 2019).

Many studies of mate choice based on song are designed to investigate a single song feature at a time. But different song parameters are not always correlated with each other and an individual male may have a song with a mixture of attractive and unattractive characteristics. Females can, however, make a comparative evaluation of males, integrating different aspects of their song and weighing up negative versus positive features (Clemens et al., 2017; Gabel and Hennig, 2016). Wittmann et al. (2011) developed a neural network model that allows them to analyse courtship songs in *Chorthippus biguttulus* in terms of the relative contribution of different song characteristics to overall song attractiveness and female responsiveness. Bentsen et al. (2006) and Ower et al. (2013) both caution against laboratory studies that draw inferences about sexual selection based on the manipulation of single song parameters, as these underestimate the strength and complexity of sexual selection acting on male signalling characteristics. Morphological constraints are also an important consideration. During mating in haglids from the genus *Cyphoderris*, the female consumes the fleshy hind wings of the male as well as the haemolymph that flows from the damaged wing. Wing morphology is a major determinant of song structure. Sexual selection acting on song structure is much stronger than that acting on wing morphology, suggesting that the latter may constrain the evolution of the former (Ower et al., 2017).

In most of the examples in Table 3, the preferred song feature is either correlated with large male size or with the male's ability to allocate the greater amount of energy required for the production of higher-effort song containing more sound per unit of time (a higher duty cycle). Larger males may provide better direct benefits such as a bigger spermatophore containing more sperm (Whitman, 2008). Females may be passively attracted to song containing more sound or be more likely to hear one that is louder. On the other hand, males singing with a higher duty cycle may be signalling that they are in good condition, though this does not necessarily mean they are able, for example, to provide a larger nuptial gift: high effort song and nuptial gifts are both energetically expensive and the ability to produce one affects the ability to produce the other (see e.g. Copeland and Fedorka, 2012; Lehmann, 2012; Wagner, 2005). Harrison et al. (2013) have also shown that songs with a high duty cycle are not always an honest indicator of quality.

Similarly, males who show greater singing effort (i.e. who spend a higher proportion of their time producing calling song) tend to have greater mating success (e.g. Bentsen et al., 2006; Holzer et al., 2003). This may simply be the result of passive attraction—the more often a male sings, the more likely a female is to hear him. However, calling song is energetically expensive and the amount of time a male spends singing is usually dependent on, for example, diet, condition and size (see Section 3). The singing effort of males in the cricket *Plebeiogryllus guttiventris*, for example, is not consistent from one night to the next (Nandi and Balakrishnan, 2016). In some species at least, those males whose songs have the most attractive features are also the ones that sing with the greatest effort (Bertram et al., 2011) so both the quality and quantity of song constitute an honest signal (Maynard Smith and Harper, 2003). In other species males may be able to compensate for unattractive features of their song by increasing their singing effort. In *G. texensis* for example (McAuley and Bertram, 2016), male calling songs at sub-optimal frequencies are less audible to females, have a smaller range, and induce reduced neural and phonotactic responses in females. However males with unattractive song frequencies can increase their singing effort and sing more loudly, though only if they are high-quality males able to invest the extra energy required.

Apart from the amount of effort a male puts into singing, and the specific features of his song, the timing of his song in relation to those of other males can influence his attractiveness to females. Such male–male interactions are considered in Section 7.

6.4 Variation in female mate preferences

Although females across a species may show a significant preference for male song with particular parameters, this does not mean that all females of that species necessarily show the same preference, or the same degree of choosiness (i.e. how selective they are in terms of accepting or rejecting a male whose song feature is different to some extent from their preferred value).

Preferences and choosiness often vary consistently between different females (e.g. Reinhold and Jacoby, 2002; Stout et al., 2010; von Helversen et al., 2004). In some species variation in choosiness can be related to the condition of the female: if she is in poor condition she cannot afford to spend a lot of energy in finding her preferred mate (e.g. Judge et al., 2014). Females infested with parasitoids (see Section 8) may also show reduced selectivity in choosing a mate, probably because they need to reproduce as quickly as possible before the parasitoids kill them (e. g. Beckers and Wagner, 2013). The strength of preference for leading songs also varies between females.

Individual females also show plasticity in their mate choice behaviour. In some species, older females (with low residual reproductive value) may be less choosy because the benefits gained from being selective no longer offset the costs (e.g. Mautz and Sakaluk, 2008; Pacheco et al., 2013), though in others there is no consistent relationship between age and choosiness (e.g. Sarmiento-Ponce et al., 2021). Female choosiness may also vary with stage of oviposition cycle (Aufderheide and Ronacher, 2019).

The female's social environment, either as a juvenile or as an adult (Swanger and Zuk, 2015), can affect both her preferences and her choosiness (see review by Rodriguez et al., 2013). In the bush-cricket *Ephippiger diurnus* for example, females reared in an environment where all the male song they hear is high quality (i.e. has the preferred number of syllables) are more selective than those that only hear low-quality male song (Rebar et al., 2019). Population density can also affect choosiness: for example in *Teleogryllus oceanicus* (Bailey and Zuk, 2008; Lierheimer and Tinghitella, 2017) and *Gryllus lineaticeps* (Atwell and Wagner, 2014), females hearing sounds that simulate a high density of males are more selective than those that hear sounds simulating a low density of males. In *T. oceanicus* females reared in the absence of male songs (i.e. in an environment where potential mates are likely to be limited) are more likely to engage in risky behaviour (i.e. perform phonotaxis when the perceived risk of predation is high), especially if they have the opportunity to mate with a male with preferred song characteristics (Ghalichi et al., 2020). In *G. lineaticeps*, Beckers and Wagner (2011) have found some evidence that a female adopts different strategies for selecting a mate depending on male density at the time. When male density is low she may use a comparison strategy whereby she continues searching as long as each new male encountered is of higher quality than the previous one; once she encounters a new male that is of lower quality than the previous one, she chooses the previous male. When male density is high, she may use a fixed threshold strategy with a 'last chance' option whereby if, after some period of time, she has not encountered a sufficiently attractive male (one that meets her quality threshold), she will mate with the next available male regardless of his quality. Bailey and Zuk (2009) also showed that female *T. oceanicus* adjust their mate choice behaviour according to their specific social experience: their response to a particular male depends on what males they have already encountered and what order they encountered them in. Females hearing a more attractive male song shortly before hearing an 'average' song are more choosy (less likely to respond to the average song) than females that had previously heard a less attractive song.

There may also be constraints on the female's ability to exercise choice. In the cricket *Teleogryllus commodus*, for example, Hall et al. (2008) showed that, after copulation, females can 'choose' males whose courtship song has particular parameters by removing the spermatophore of non-preferred males earlier than those of preferred males. The male's fitness is correlated with spermatophore attachment time and he can delay spermatophore removal by harassing the female. In the presence of the male, the female's ability to exercise choice is limited, the intensity of sexual selection on male courtship song is greatly reduced and the particular song features that are selected are different compared with when the female is able to escape from his harrassment and remove the spermatophore when she chooses. Differences in environmental conditions that make it more or less likely that the female is able to escape will therefore lead to differences in both the strength and focus of sexual selection. Such fluctuations in time and space are likely to be common (see review by Miller and Svensson, 2014).

6.5 Female song, male mate choice and role reversal

Compared with the extensive literature on female male choice, there has been relatively little research on male mate choice in Orthoptera. Where it has been demonstrated, male choice has been related to non-acoustic cues such as the female's size or age (e.g. Hare and Simmons, 2020; Jarrige et al., 2013). There is potential for male choice based on female song in those species where females sing but we have been unable to find any studies where this possibility has been investigated or which have looked for correlations between features of song and other female characteristics, such as body size or condition, that might affect the reproductive success of the male. Female song has been found to stimulate male mating activity in general. For example, in *Chorthippus biguttulus*, females sing to show they are ready to mate and males preferentially court singing females (Wirmer et al., 2010).

In duetting species, the length of the female reply can vary between and within individuals. Longer replies may provide more locational information, allowing the male to search more effectively for the female, and may indicate a preference for the singing male (Bailey, 2003). For example, in the katydid *Scudderia pistillata*, males in better condition produce more syllables per bout of singing and females respond faster and with longer songs containing more 'ticks' (Villarreal and Gilbert, 2013).

In some species, when food is scarce, females compete with each other for nutritious nuptial gifts and males become choosy. Sexual selection may act on females as a result of this role reversal. For example in *Kawanaphila nartee*, females have more sensitive hearing than males, possibly because this benefits them during intrasexual competition when they scramble for access to acoustically signalling males (Hare and Simmons, 2020). However, selection for larger ear size in females has been shown to occur only during periods of role reversal (Hare and Simmons, 2021).

6.6 Mate choice and speciation

Divergence in acoustic mating signals has been a significant driver of speciation in Orthoptera (e.g. Wilkins et al., 2013). Sexual selection has also played an important role: divergence in the male song between populations is often coupled with divergence in female song preferences (e.g. Barbosa et al., 2016; Orci and Iorgu, 2017). One of the most detailed studies has been carried out by Shaw and colleagues on the rapidly speciating Hawaiian cricket *Laupala* (e.g. Grace and Shaw, 2011; Wiley et al., 2012; Xu and Shaw, 2021). They have shown that male song and female preference coevolve in a stepwise manner and that genetic coupling underlies this signal–preference coevolution. Female mate choice thus facilitates the establishment of sexual isolation, reduced gene flow, and population differentiation, all prerequisites for speciation.

7. Competition and cooperation

The ability to hear is important to male orthopterans, even though females in most species do not sing: it allows them to detect predators (see Section 8), locate and interact acoustically with other males, and obtain information from their songs which may be important in male–male cooperation or competition.

7.1 Aggregating and spacing

As we described previously (Robinson and Hall, 2002), orthopteran species commonly form aggregated populations, either because individuals are passively attracted to an area containing resources like food or singing sites or because they are actively attracted to each other. Various possible benefits of aggregating have been suggested, such as reduced vulnerability to predation and increased attractiveness to females of a group of singing males compared with a single male (see e.g. Greenfield, 2015). The male calling song plays an important part in the distribution of individuals within a population, repelling or attracting other males as well as attracting females. Where singing males interact temporally with each other over relatively long periods of time, aggregated populations are defined as choruses (see Section 7.5).

Males may respond to another male that they can hear singing by moving away to maintain a space between them, by approaching the singer using their song to locate them (phonotactic response), or by adjusting their own singing behaviour (acoustic response). Chamorro-R et al. (2007), for example, showed experimentally that one of the functions of the calling song of the katydid *Panacanthus pallicornis* is to maintain spacing: a group of males that had been deafened tended to aggregate while a group of normal, hearing males spread out randomly. Where males use song to maintain a minimum distance with neighbours, distribution is likely to be random at low population densities, only becoming regular as density increases, as shown for example in the katydid *Sphagniana sphagnorum* (Morris et al., 2018). Distance estimation must however be relative rather than absolute because so many environmental factors, such as temperature and vegetation, affect the transmission of song (see Section 3.9). Morris et al. (2018) have suggested that *S. sphagnorum* males may estimate distance using the ratio between the perceived sound pressure levels of the audio and ultrasonic components of the calling song and Kostarakos and Römer (2018) have described a neuronal mechanism that would allow them to do so.

There may be advantages to keeping a distance from other males, at least to some extent. For example the mole cricket *Gryllotalpa major* forms leks in which males sing in close proximity. Females however prefer to visit smaller leks (Keane et al., 2018) and, within a lek, those males further from the centre and from nearest singing neighbours (Howard et al., 2011).

Males may perform phonotaxis towards other males for a variety of reasons: to join other males in an aggregation (e.g. in the bush-cricket *Conocephalus brevipennis*: Guerra and Mason, 2005), to get close to a more successful male so they can act as a satellite (Section 2.2), or to initiate an aggressive contest. For example, *Gryllus integer* males that are more able to win contests preferentially approach the song of a male that is more attractive to females (Leonard and Hedrick, 2009). They are more likely to be doing this to initiate a contest than to act as satellites because unattractive males do not behave in the same way. As *G. rubens* males get older they become less likely to perform phonotaxis and more likely to respond acoustically (Jang, 2011). This strategy may enable them to find singing sites as young adults and then swich to attracting females and maintaining spacing from other males.

7.2 Aggressive signals and fighting

7.2.1 Males

Aggressive song can play an important role in aggressive encounters between males. It may have a function before and/or after a contest. Before, it may mitigate the costs of aggression by allowing males to assess each other's competitive ability and so decide a contest without an actual fight. In Hawaii some *Teleogryllus oceanicus* males have lost the ability to sing (see Section 8.3). Contests between two silent males are much more aggressive than those in pairs where at least one male can sing (Logue et al., 2010). In *Hemideina* tree weta contests, the larger male usually wins, but if both males are of similar size contests are likely to be extended and therefore costly, and in these cases aggressive song is probably used to decide the winner (Ewers and Cowley, 2005).

If males are to assess each other based on song, that song must contain information. In some species, the energy put into the aggressive song—either the duty cycle or the singing effort—is an honest signal of competitive ability. For example, the haglid *Cyphoderris monstrosa* (Chang, 2015) is territorial and sings before a fight. The duty cycle of this aggressive song is not necessarily consistent from one fight to another but reflects their competitive ability at the time. Particular parameters of the aggressive song, such as chirp length, may also be correlated with competitive ability as for example in the crickets *Gryllus veletis* (Harrison et al., 2017) and *Acheta domesticus* (Brown et al., 2006). In *A. domesticus*, temporal components of the song are correlated with body size which in turn affects the intensity and outcome of fights.

After a contest the winner may sing aggressively as a 'victory display' (Bertram et al., 2010). This may have two functions: to 'browbeat' the loser, i.e. to show them the winner is still capable of continuing or even escalating the contest, and to advertise their success to nearby prospective mates or rivals. Information in the victory display could inform decisions made by the loser about whether or not to engage the winner again in future encounters. Such victory displays are seen in several territorial cricket species that display from, and defend, burrows or crevices, including *G. veletis*, *G. pennsylvanicus*, *G. rubens*, and *T. oceanicus* but are rarely seen in non-territorial ones like *G. fultoni* or *G. vernalis* (Jang et al., 2008).

Male–male encounters can be affected by the social environment. For example in *G. assimilis*, males are more likely to initiate or escalate fights and spend less time performing victory displays if a female is present (Montroy et al., 2016). *G. veletis* males are more aggressive and perform more victory display behaviours if another male is present (Fitzsimmons and Bertram, 2013).

Songs produced during male–male interactions may not always be directed at the rival male. In the katydid *Gnathoclita sodalis* (De Souza et al., 2011) the male produces both airborne sound and vibratory signals during contests. The 'loser' (who withdraws, usually without a fight) is usually the smaller male. The number of his vibratory signals predicts the winner of the contest, but there is no apparent relationship between winning and the male's aggressive song, which may be directed at nearby females rather than at the other male. However, it is not necessarily the case that males who are more aggressive or more likely to win contests are more attractive to females. In *G. integer*, for example, females prefer males who sing for longer but singing time is negatively correlated with aggressiveness, possibly because both require energy and there is a trade-off between the two (Hedrick and Bunting, 2014).

7.2.2 Females

Females are known to fight each other and may also produce aggressive song. For example, *Gryllus campestris* females (Rillich et al., 2009) may fight for the opportunity to mate in the presence of a courting male, though they do not produce aggressive song. Either male calling song or courtship song can induce female fighting but aggressive song rarely does. Females in the bush-crickets *Steropleurus stali* and *S. nobrei* (Hartley, 1993) may also fight in the presence of a singing male. They produce aggressive song which depresses the acoustic activity of other individuals nearby, both male and female.

7.3 Acoustic defence and attack

Males of duetting species are vulnerable to rival eavesdropping males (Section 2.2). Several strategies have been adopted among bush-crickets species to counter eavesdroppers. Most commonly, once a singing male receives a response, he reduces the interval between his songs, which enables him to exchange information with the female about her location more quickly (see e.g. Heller et al., 2020). He may also sing more quietly so he can only be heard by an individual close by, which is likely to be the female, as for example in *Anaulacomera almadaenis* and *Phaneroptera sparsa* (Heller et al., 2021). In some species, males add an element to their own calling song that masks the female reply and prevents other males from hearing it while still being able to hear it themselves (e.g. *Caedicia*: Hammond and Bailey, 2003). Others mimic a female response after their normal calling song, thereby making it more difficult for the eavesdropper to locate the actual female that is responding, for example in *Scudderia pistillata* (Villarreal and Gilbert, 2014), *Gonatoxia* (Heller and Hemp, 2017) and *A. almadaenis* (Heller et al., 2020).

Eavesdroppers may also attack singing males acoustically. *Caedicia* males for example (Bailey et al., 2006) produce a song that mimics a female response, possibly to distract the singing male away from the real female response. They also sing over the part of the male's calling song necessary for species recognition, thus reducing the likelihood that a female will respond and potentially making her available for mating with the eavesdropper.

Some species exhibit more than one of the strategies adopted by singers and eavesdroppers, giving rise to highly complex male songs. Heller and Hemp (2020) argue for the importance of sexual selection by male competition, as well as by female choice, in the evolution of male song in duetting species.

7.4 Plasticity in male behaviour

Just as individual females show plasticity in mate choice behaviour depending on their social environment (Section 6.4), individual males are able to adjust their behaviour depending on acoustic cues from other males.

It is common for males to put more effort into singing (greater singing effort or higher duty cycle) if they can hear rival males (e.g. Anichini et al., 2018; Callander et al., 2013; Mhatre and Balakrishnan, 2006; Morris et al., 2002; Noguera, 2019; Rebar et al., 2016). Males of one *Mecapoda* bush-cricket species simply start singing as soon as a rival does (Krobath et al., 2017). Males may also adjust their courtship behaviour. For example, in the cricket *Gryllodes sigillatus* males start courtship more quickly if they have been exposed to the sound of another male (Santori et al., 2020).

The amount of singing a male can hear around him is typically an indication of population density and therefore the level of sperm competition he is likely to experience, so it is unsurprising that males of several species have been found to change their investment in sperm or spermatophores if they can hear rivals singing. They may produce better quality sperm (e.g. *Teleogryllus oceanicus*: Simmons and Lovegrove, 2017) or more sperm (e.g. *Acheta domesticus* and *Gryllodes supplicans*: Gage and Barnard, 1996). Males of the bush-cricket *Ephippiger diurnus* change the size of their spermatophore depending on the quality rather than the number of songs of rival males, i.e. the perceived level of strong competition (Rebar et al., 2016).

The potential availability of mates also depends on population density and males have been shown to adjust their mating tactics in response to the amount of conspecific song they hear. For example, *T. oceanicus* males that can hear conspecifics singing are more likely to perform phonotaxis towards other males, and therefore more likely to adopt satellite behaviour (Section 2.2), compared with males kept in the absence of song (Balenger et al., 2018).

7.5 Choruses

As we described previously (Robinson and Hall, 2002), different species show different kinds of chorusing behaviour. Recent research has concentrated however on *synchronous chorusing*, in which males sing so that the repeated units (RUs) of their songs occur at almost the same time as those of other males, and *alternating chorusing* in which males sing so their RUs occur in the spaces between those of other males. Synchrony tends to be more common in species where the rate of singing is high and alternation where it is low (Hartbauer and Römer, 2016). Most of the studies of synchrony so far have been on species that sing with regular simple rhythms and the first study of synchrony in a species with a complex song (*Mecopoda* 'Two-part caller') has only been published recently (Nityananda and Balakrishnan, 2021).

Within a synchronous chorus, leader and follower males can often be identified: leaders are defined as males that habitually initiate their RU first and followers as those that initiate their RU second. It is still unclear how these types of chorus evolved but since 2002 considerable progress has been made in understanding the possible benefits of synchrony and alternation and the mechanisms by which they may arise (see for example Greenfield, 2015; Greenfield et al., 2017; Hartbauer and Römer, 2016).

7.5.1 Leaders, followers and female choice

In some species, females are more likely to respond to leading songs in a synchronous chorus. This preference appears to result from a sensory bias (Hartbauer and Römer, 2016) that causes females to attend to leading songs and ignore songs whose onsets follow the onset of the leading song within a certain time (the *preference extent*). The preference extent can be less than 70 ms in some species or as long as 2 s in others (Greenfield and Snedden, 2003).

Studies of three *Neoconocephalus* species (Murphy et al., 2017; Murphy and Schul, 2017) found no heritability of leadership ability and no correlation between leadership ability and several possible indicators of male quality such as size, condition, chirp duration, song duty cycle, or immunocompetence. But in the one species of the three where females show leadership preferences (*N. ensiger*), there is a possible direct fitness benefit for females mating with a leader: they produce offspring in better condition, possibly because leader males provide better seminal proteins.

Males with higher rates of intrinsic signalling are more likely to become leaders (e.g. *Mecopoda*: Hartbauer and Römer, 2016; Nityananda and Balakrishnan, 2007) and leaders are expected to attract more females than followers. So why should males who are likely to end up as followers take part in a chorus? It has been suggested that by cooperating to form choruses, follower males may have greater mating success than they would by singing on their own. This is considered in the next section.

However, leader and follower roles, and female preference for leaders, are not necessarily consistent or clear-cut. Males are able to adjust their signalling rate and in some species males switch between leader and follower roles relatively frequently, as in *Mecopoda* 'Chirper' (Nityananda and Balakrishnan, 2008) and *Ephippiger diurnus* (Party et al., 2014). In *M. elongata*, chirps of older males are longer than those of younger males so younger males more often take the leader role in male–male duets; given a straight choice between a young leader and an old follower, females prefer the young male (Hartbauer et al., 2015). But things are not so simple in more complex choice situations. In a small chorus of four males only 40% of females chose the leader and most instead preferred the male that had both a frequent leader role and longer chirps. As a result, older males are likely to be just as attractive as younger males even though they spend more time than younger males as followers.

The strength of the female preference for leading songs may also vary within a species. Greenfield et al. (2004), for example, found that different

E. ephippiger females differ consistently in the strength of their preference and that the preference declines with age. Mate choice is also influenced not only by whether the male is a leader but by the acoustic neighbourhood: males singing in locations with more overall singing are preferred (Berg and Greenfield, 2005).

Followers may also have strategies to improve their mating success. Nityananda and Balakrishnan (2008) found that a *Mecopoda* 'Chirper' male that is generally a follower may sing more often when leaders are not singing. Also, while leaders are typically louder than followers, some quieter males are able to distance themselves sufficiently from them to become the loudest male in their local area, which compensates for their lack of leadership. Simulations by Fertschai et al. (2007) have shown that the ability of followers to attract females may be much higher than suggested by laboratory trials. Followers may also make a trade-off between a reduced rate of mating versus reduced longevity caused by infestation with parasitoid fly larvae. *M. elongata*, for example, is subject to acoustically-orienting Ormiin parasitoids (see Section 8) but these flies are more likely to attack males with leading songs (Lee et al., 2009).

7.5.2 Evolution of choruses

Alternating choruses probably arose because they reduce acoustic interference between neighbouring males (Greenfield, 2015). Because a male cannot hear as clearly while singing himself, alternation ensures that a singer can clearly evaluate a neighbour's song and adjust his own singing and location as necessary. The same applies to the neighbour, so cooperating to create an alternating chorus benefits both males. Females also benefit from being able to discriminate more easily between the characteristics of two singing males.

Explanations for the evolution of synchrony are more diverse and not necessarily mutually exclusive (for good summaries see: Greenfield, 2015; Hartbauer and Römer, 2016). Evidence for any of them is very limited. One hypothesis is that singers that synchronize are less vulnerable to phonotactic predators and parasitoid flies, which find it more difficult to locate a single singer within the chorus. Some support for this idea has been found in *Mecapoda* (Prakash et al., 2021), where acoustically orienting bats take longer to capture an individual singing in a chorus than one singing alone.

Many species have a calling song with a rhythm or distinct sound envelope that is species specific and females only respond to the song if they recognize that rhythm or sound envelope. By cooperating to synchronize their songs, groups of males maintain the species-specific rhythm and/or distinct sound envelope. Otherwise females would hear nearly continuous sound at fairly constant intensity and would not be attracted to the group. Some support for this idea has been found in *Neoconocephalus nebrascensis* (Deily and Schul, 2009) and *Sorapagus catalaunicus* (Greenfield et al., 2017).

Synchronization could also create a 'beacon effect'—an increase in peak signal intensity which leads to an increase in the signal-to-noise-ratio and the distance at which females can detect the group signal compared with a single singer. Computer simulations of *Mecopoda* 'Chirper' chorusing behaviour (Nityananda and Balakrishnan, 2009) show that while leaders have some selective advantage, cooperation among followers to create a beacon effect is an evolutionarily stable strategy.

The above explanations are based on cooperation between synchronizing males such that followers are able to achieve greater fitness in a chorus (more mates on a per capita basis) compared with singing alone or in a non-synchronous chorus. Other explanations see synchrony as a result of competition between males. Where female preferences for leading songs occur, selection favours males that adjust their song timing to increase their incidence of leading songs. Synchrony or alternation may then be an emergent property of this competition to become leader.

7.5.3 Mechanisms of synchrony and alternation

The male's singing rhythm is regulated by a central pattern generator (oscillator) in the nervous system (Section 5.1) and each male has an intrinsic rate of singing. But the rate is not fixed and males can adjust it. In some species (e.g. *Neoconocephalus ensiger*: Murphy et al., 2016) a male modifies the rate and phase of his rhythm so that it gradually aligns with his neighbours: slow singers accelerate and fast singers decelerate until all singers are synchronized. This can lead to very precise synchrony. In other species, rather than adjusting their intrinsic rhythm, males lengthen or shorten a single song period after hearing the song of a neighbour (e.g. *Mecopoda elongata*: Hartbauer and Römer, 2016). Both these mechanisms can lead to a kind of synchrony which has been called 'type 1 synchrony' (Greenfield et al., 2017), where the group rhythm remains stable for long periods, leaders and followers can be identified most of the time, and an individual male may sing as a leader or as a follower for a high proportion of the time within a chorusing bout.

In other species such as *Ephippiger diurnus*, synchronization emerges as the result of an 'inhibitory-resetting mechanism' (Greenfield et al., 2016). In this model, which has been called 'type 2 synchrony' (Greenfield et al., 2017), a male interrupts his own song when he hears a neighbour sing, stays inhibited from singing until the neighbour's song ends, then rebounds to produce the song that was suppressed, thereafter resuming his intrinsic rhythm. When a local group of males sing with similar intrinsic rhythms, an approximate synchrony emerges if the male's rebound phase is about the same length as the free-running song period; but if the male rebounds more quickly, he will sing out of phase with his neighbour, i.e. they will alternate. Even so, there will still be some degree of synchrony because the neighbour's neighbour will also alternate with that neighbour and as a result will be in synchrony with the original male. The process therefore may create a series of small interlacing groups that chorus in a mixture of synchrony and alternation. Unlike type 1 synchrony, leader and follower roles switch frequently between males during a chorusing bout.

Greenfield et al. (2017) argue that type 1 synchronies should be favoured by selection when the overall characteristics of the chorus are adaptive, for example where it is critical to maintain a species-specific rhythm or to generate a beacon effect. Here males must cooperate if they are not to reduce the attractiveness of the group and ultimately their own mating success. On the other hand, because type 2 synchronies are much less regular, the inhibitory resetting mechanisms that generate them would not be favoured by selection if overall chorus characteristics are adaptive. Instead, they should be more likely where competition between individual males within the chorus is the key factor and overall characteristics such as synchrony are of little importance either to the males that produce the chorus or to the females who are listening. Here the synchrony probably emerges as an incidental property with negligible adaptive value.

Whether a species shows type 1 or type 2 synchrony does not seem to be related to its phylogeny and even two species from the same genus may have different types. Greenfield et al. (2017) compared *E. diurnus* (with its irregular type 2 synchrony arising from an inhibitory-resetting mechanism) with the related species *Sorapagus catalaunicus*, which has a steadier type 1 synchrony. They found that *S. catalaunicus* chorusing arises from a strikingly different mechanism from that in *E. diurnus*: males simply do not initiate songs during a 500–700 ms period after the end of a neighbour's song. This ensures a minimum period of silence between sound envelopes broadcast in the chorus. Chorusing in *S. catalaunicus* is therefore potentially the result of cooperation between males driven by a female's requirement to hear a distinct sound envelope while that in *E. diurnus* probably reflects male–male competition caused by female preference for leading songs. However, this is not the whole story because Greenfield et al. (2017) also found that females of both species must hear clear sound envelopes before they respond. Both species also strongly prefer leading songs, though in *S. catalaunicus* the preference extent is extremely short, at about 45 ms (compared with 150–600 ms in *E. diurnus*). Because of neuromuscular constraints, it is likely that an inhibitory-resetting mechanism would therefore not be able to prevent *S. catalaunicus* males producing following songs. There is also very little disadvantage to singing as a follower if the male's song occurs at least 45 ms after the leader, which will be the situation most of the time. They conclude therefore that an inhibitory-resetting mechanism would not have been favoured by selection and, in fact, males do not display it. The mechanism adopted by *S. catalaunicus* works perfectly for a species with a high duty cycle. However, in species such as *E. diurnus* that have a low duty cycle, distinct sound envelopes will occur regardless of what individual males do, making cooperation to generate synchrony unnecessary. They suggest that this may have allowed the female preference for leading songs to co-evolve with competitive male singing interactions that increased an individual male's frequency of singing as a leader, culminating in the pronounced inhibitory resetting and extended female preferences for leading songs seen in *E. diurnus*.

Support for this suggestion comes from a comparative phylogenetic analysis of 17 populations of *E. diurnus* which have diverged in male song timing and female preference for leading songs (Greenfield et al., 2016). They found that male adjustments in their song timing correlate with the preference extent, showing that male adjustments co-evolved with female preferences.

In either type of synchrony, adjustments to song timing become problematic in dense choruses. Any male trying to adjust to all of his neighbours by delaying his song each time he heard one of them could end up hardly singing himself. Males appear to solve this problem by attending only to some of their neighbours, usually the ones likely to be their most important rivals for any nearby female. Greenfield and Snedden (2003) found evidence for such selective attention in two alternating species: the grasshopper *Ligurotettix planum* and the bush-cricket *E. ephippiger*. In the synchronizing bush-cricket *Neoconocephalus spiza*, however, they found that 40% of all males show no selective attention at all and suggested that selective attention is less necessary for synchronizing species: because many of his neighbours sing synchronously, a male does not hear most of them and is therefore unlikely to be 'reset' by them. However some synchronizing species do show selective attention, such as *Mecopoda* and *E. diurnus*.

Fertschai et al. (2007) looked at whether *Mecopoda* males restrict their attention to louder chirps neurophysiologically, behaviourally or through spacing. They found that louder leading chirps are preferentially represented in the omega neuron (Section 5.2) but the representation of softer following chirps is not completely abolished. Following chirps that are 20 dB louder than leading chirps are better represented than leading chirps. Males synchronize with leading chirps even when the following chirps are 20 dB louder. They do not pay attention only to louder chirps but are instead affected by all chirps above a particular threshold. Within the chorus males on average have only one or two neighbours whose songs are above this threshold. Selective attention is thus achieved in this bush-cricket through spacing rather than neurophysiological filtering of softer signals.

Marin-Cudraz and Greenfield (2016) showed that *E. diurnus* shows selective attention to just a few of their loudest neighbours. They also found that attention is most restricted—sometimes to only one neighbour—in populations with the highest duty cycle. In these populations males need to be more selective because indiscriminate attention would seriously interfere with their singing. Thus choruses appear to be finely tuned and controlled by feedback loops in which individual singers generate a collective chorus that, in turn, influences the singing behaviour of those same individuals.

7.5.4 Multispecies choruses

It is common for several species of orthopteran to contribute to the acoustic environment in a particular area: Diwakar and Balakrishnan (2007b) studied a community of 20 orthopteran species in an Indian evergreen forest, all singing over the same 12-h time period each night and showing considerable overlap in both frequency and temporal patterning of song features.

In such multispecies choruses, therefore, acoustic overlap between species has considerable potential to reduce the efficiency of signal transmission between conspecifics. The amount of overlap depends on several factors, including song structure and intensity, the degree of signal attenuation caused by the habitat, receiver tuning, singer density and the species composition of the chorus. How individuals of a particular species within the chorus manage to communicate with each other despite the potential problems has been the subject of an increasing amount of research in recent years.

Balakrishnan et al. (2014) used simulations to look at the effects of the relative abundance of species and singer density on the levels of acoustic overlap among five orthopteran species from the same Indian evergreen forest mentioned above. There is greater acoustic overlap if species are equally abundant compared with when one species is dominant. Acoustic overlap also increases with singer density but the amount of increase depends on both the song structures of the species and on the relative abundance of species. Differences in acoustic overlap between species pairs are not predictable simply from song structures.

Acoustic interference may be reduced if different species occupy different acoustic spaces. Tiwari and Diwakar (2019) looked at six conehead katydid species from two areas in India and found that sympatric species with similar songs never sing at the same time or from the same location. Some forest multispecies choruses have been found to show vertical stratification in singing height, whereby different species tend to sing at different heights between the ground and the top of the canopy (e.g. Diwakar and Balakrishnan, 2007a; Schmidt et al., 2013).

Schmidt et al. (2016), however, found no evidence in a tropical forest multispecies chorus to suggest that selection pressures resulting from acoustic competition have led to divergence in songs so as to avoid masking interference: different species that sing at the same time, or are close in space, do not have less similar songs than those that sing at different times or at a distance from each other. They argue that selection pressures on sensory/neuronal mechanisms may have been more important in maintaining reliable communication. They have shown, for example, that some of these forest species have a more sharply tuned song frequency than field crickets, which do not need to compete with heterospecific singers.

Similarly, Jain et al. (2014) looked at temporal, spectral and active space overlap between species in a rain-forest assemblage of orthopterans. Although they found that species that sing over the same time period each day may avoid singing together within that time period, there was no relationship between spectral overlap and singing activity and there was little evidence that species actively use spacing to minimize acoustic overlap. A simulation of acoustic spaces integrating temporal, spectral and amplitudinal characteristics of multiple signals showed that increasing song intensity and tuning receivers could be powerful ways to minimize effective acoustic overlap and that for most individuals in the chorus this overlap may actually be close to zero. Thus, song temporal structure, intensity and frequency together provide sufficient acoustic space for several species to sing together yet communicate effectively with other conspecifics with little interference, despite the apparent cacophony of a rain-forest chorus.

8. Predators and parasites

8.1 Natural enemies: Risks and defences

As we described in Robinson and Hall (2002), orthopterans are attacked by a large variety of natural enemies and have evolved various defences against them. Many of these enemies, in particular bats (e.g. Jones et al., 2011) and parasitoid tachinid flies from the Ormiini tribe (Lehmann, 2003), eavesdrop on a prey's acoustic signals and use them to home in on its position. Non-singing females approaching a singing male are also vulnerable: predators can locate them due to the noise and vibrations they make while moving or they may be attacked by parasitoids when close to a singing male. The risks associated with singing for the male have been shown to be equivalent to risks associated with phonotaxis for the female in the tree cricket *Oecanthus henryi* (Torsekar et al., 2019). However, in some species at least, movement may actually increase the risk of bat predation for both males and females more than singing does (Geipel et al., 2020; Raghuram et al., 2015).

Each predator or parasitoid species usually attacks a variety of orthopteran and other species, but may prefer some prey species over others or find them easier to locate (Falk et al., 2015; Grant and Samways, 2015). For a particular orthopteran species, therefore, the selection pressures affecting them will be determined not only by which species of predators and parasitoids are present but by which other species of orthopteran they mix with (Goodale et al., 2019; Trillo et al., 2019).

Orthopterans in turn can detect approaching predators and parasitoids using the noises and vibrations they make (e.g. Kilmer et al., 2010) and the air disturbances they create (Lewkiewicz and Zuk, 2004). Filiform hairs located on the cerci can detect the faintest movements of the air (Jacob and Hedwig, 2015; Magal et al., 2006). Some species have auditory units tuned specifically to frequencies appropriate for detecting predators, for example the echolocation calls of bats (Section 5.2).

Orthopteran species have evolved various strategies for avoiding detection by predators and parasitoids, including reducing the time spent singing, reducing the duty cycle, avoiding singing when predators or parasitoids are active (e.g. Bertram et al., 2004), singing at a frequency that the predator or parasitoid cannot hear well (e.g. Dobbs et al., 2020), singing more quietly, switching to vibrational instead of sound signals (e.g. Römer et al., 2010), and/or singing at the same time as other conspecifics (e.g. Trillo et al., 2019). Once a predator has been detected, orthopterans may react defensively to try and avoid being captured, such as switching to low-intensity song so only a nearby mate can hear, stopping singing completely for a time (e.g. ter Hofstede et al., 2010), and/or producing a defensive sound (a noise directed at a predator that improves the chances of survival; for a review of defensive sounds see Low et al., 2021). Haglids in the genus *Cyphoderris* have an unusual stridulatory method of producing a defensive sound, probably directed at ground predators, using Ander's organ (Woodrow et al., 2021). Ceasing singing has been shown to be a successful strategy in *Neoconocephalus ensiger* (ter Hofstede et al., 2008). But attempts to avoid detection and capture are not always successful, and acoustically-orienting predators also use other methods of finding prey such as echolocation (e.g. ter Hofstede et al., 2017).

Some orthopteran species are predatory themselves and may use sound to capture their own prey. The katydid *Chlorobalius leucoviridis*, for example, preys on cicadas and can attract male cicadas by imitating the species-specific wing-flick replies of sexually receptive female cicadas (Marshall and Hill, 2009). This aggressive mimicry is accomplished both acoustically, with tegminal clicks, and visually, with synchronized body jerks.

8.2 Variation in response

Just because a species is able to hear echolocating bats does not necessarily mean that it will react defensively if a bat is detected. Symes et al. (2020) found that many of the species of neotropical katydids on Barro Colorado Island do not stop singing when exposed to bat echolocation sounds, even though they are able to hear them. Probably fewer than 10 of the 76 bat species on the island eavesdrop on singing katydids, and the other bats pose no threat. Since the katydids are unlikely to be able to discriminate between the sounds of different bat species, stopping singing every time they hear a bat would severely disrupt their ability to attract a mate. Instead most of the katydid species show 'proactive' defence in the form of very reduced signalling rates, making them difficult for eavesdropping bats to locate. Symes et al. (2020) argue that even when they are capable of detecting predators, proactive defence has greater benefits than reactive defence when predator cues are not strongly correlated with the risk of actual attack.

Defence strategies that reduce the duty cycle, the overall amount of singing, or the distance at which the calling song can be heard, or which change the frequency or characteristics of the calling song can be advantageous in terms of reducing the selection pressure imposed by predators and parasitoids. However, they are disadvantageous both for males in terms of attracting mates (e.g. Dobbs et al., 2020) and for females in terms of perceiving and discriminating a conspecific male song against background noise (Römer et al., 2011). The balance between natural selection (predator pressure) and sexual selection (mating success) will vary between species, and between populations of the same species, depending on the particular enemy species that are present and the local environmental conditions that affect signal transmission (see Section 3.9). Differences in signalling behaviour between populations has been recorded in numerous species, for example the bladder grasshopper, *Bullacris unicolor* (Couldridge and Gordon, 2015). Hawaiian populations of *Teleogryllus oceanicus* vary in the level of risk from parasitoids and the level of risk correlates with the length of time males cease singing after a potential parasitoid is detected (Lewkiewicz and Zuk, 2004).

On the other hand, some species have been found to show no differences between populations in the total time spent singing, or the temporal pattern of singing, regardless of the level of risk, for example *Gryllus lineaticeps* (Beckers and Wagner, 2012). And some species respond to high risk by actually increasing 'risky' behaviour. Beckers and Wagner (2018) found that in *G. lineaticeps*, both females and parasitoids prefer faster chirp rates, yet males in populations with a higher risk of parasitoid attack actually produce faster, i.e. riskier, chirp rates. They suggest that, because the parasitoids attack late in the breeding season, males have increased investment in singing so as to maximize their mating success before the advent of parasitoid activity. Lehmann and Lakes-Harlan (2019) argue that, in some species of the genus *Poecilimon*, mating in aggregation might select for increased, not reduced male singing, despite the risk of attack from parasitoids. This is because, once a male has mated, he drops out of the chorus, being unable to mate for a few days while he produces another spermatophore. During this refractory period he is not vulnerable to parasitoid attack. Males preferred by females are also preferred by parasitoids but quick and risky singing might be advantageous if better singers are able to attract females before parasitoids can find them.

In some species, there are consistent differences between individuals in their behaviour in relation to predators. Wilson et al. (2010), for example, identified a 'boldness' syndrome of behaviour in *Acheta domesticus*, in which bolder individuals explore more and are more willing to emerge after a predation event. Bolder males also sing more and bolder females are more willing to perform phonotaxis.

Female mate choice behaviour (choosiness and preferences) and male singing behaviour may also vary with predation risk. In a Florida population of *G. rubens*, adults emerging in spring are not subject to parasitoids whereas those emerging in autumn are. Autumn females are less likely to perform phonotaxis and autumn males are less likely to sing and more likely to switch to alternative strategies such as silent searching (Velez and Brockmann, 2006).

Individuals may also vary their behaviour depending on their local circumstances, for example in relation to temporal differences in predator pressure, such as in the katydid *Docidocercus gigliotosi* (Römer et al., 2010; see Section 2.3) and the cricket *Acanthoplus speiseri* (Bateman and Fleming, 2013). The likelihood of a male switching to silent searching or satellite behaviour instead of singing may vary depending on the perceived level of risk from predators, as shown in the tree cricket *Oecanthus henryiand* (Torsekar and Balakrishnan, 2020). The two phases of the locust *Schistocerca gregaria* differ in their hearing ability: solitary locusts, which are at greater risk of predation from bats because they fly at night, have hearing that is more responsive to the higher frequencies of bat echolocation calls than is the hearing of gregarious locusts (Gordon et al., 2014).

8.3 A cricket and a parasitoid in Hawaii: Evolution in real time

The Australian cricket *Teleogryllus oceanicus* colonized the Hawaiian islands between 150 and 2500 years ago; it shares its range on some of the islands with the eavesdropping parasitoid *Ormia ochracea* which was introduced some time before 1989 and is most common on the islands of Kaua'i, O'ahu and the 'Big Island' of Hawai'i (Tinghitella et al., 2021). The Hawaiian islands are the only place where the two species occur together (Lehmann, 2003).

In the early 2000s a sex-linked mutation ('flatwing') arose in the *T. oceanicus* population on Kaua'i (Zuk et al., 2006). This mutation reduces or eliminates sound-producing structures by feminizing wing venation and flatwing males are unable to sing. The wings of females, which do not sing, are unaffected by the mutation. The mutation has other costs for males apart from not being able to attract females with a calling song. It appears to act as a master regulatory switch during early development and has pleiotropic effects including feminization of the cuticular hydrocarbon profile (Pascoal et al., 2020). Cuticular hydrocarbons (CHCs) are sexually dimorphic in *T. oceanicus* and function as sexual signals during courtship

(Section 2.3); CHCs of flatwing males are less attractive than those of normal males (Gray et al., 2014). The lack of courtship song is also a problem, since it is important in releasing female mounting and in mate choice and as a result females prefer not to mate with flatwings when they encounter them (Kota et al., 2020). Flatwing males also continue to make stridulatory movements and maintain the same degree of 'singing' effort as normal males but cannot actually produce any sound, so they still pay the energetic price of singing without the benefit of attracting females (Rayner et al., 2020).

A different mutation producing a similar flatwing phenotype was found on O'ahu in 2005; O'ahu flatwings show a less drastic loss of secondary wing structures than Kaua'i flatwings, tending to retain more of their harp and cross veins, and often having a scraper, but are still unable to sing (Bailey et al., 2019; Pascoal et al., 2014). Since 2010 the occasional flatwing has also been found on Big Island (Zuk et al., 2018). Recently two other mutations have been found that lead to song loss: 'curlywing' on O'ahu and Kaua'i and 'small-wing' on Big Island (Rayner et al., 2019).

Silent males (i.e. males with these various song-loss mutations) are rarely attacked by *O. ochracea* but cannot attract mates by singing (Zuk et al., 2018). However, silent males are able to compensate for this inability to attract females: they are much more likely than normal males to search silently for females (Balenger and Zuk, 2015) or to show satellite behaviour (Zuk et al., 2018). The effects on male reproductive success of adopting satellite behaviour versus singing when the chances of attack by parasitoids are high have been modelled by Rotenberry and Zuk (2016). They found that satellite behaviour can compensate for the negative effects of parasitoids, though the effect is slight. However, flatwings also show a strong preference for calling songs with characteristics preferred by females, whereas normal males do not prefer any particular calling song type. Flatwings therefore tend to act as satellites to the most attractive normal males, thus maximizing their chances of intercepting females and simultaneously reducing the mating success of those singing males (Olzer and Zuk, 2018). In addition, flatwings sire more offspring per mating than do singers (Heinen-Kay et al., 2019).

Tinghitella and Zuk (2009) have shown that females from ancestral, unparasitized populations, as well as parasitized Hawaiian populations, will mate with silent flatwing males, so this behaviour must pre-date song loss. Ancestral Australian females discriminate against flatwing males more than Hawaiian females. Tinghitella and Zuk suggest that island colonization favoured females with relaxed mating requirements, which facilitated the rapid evolution of song loss in Hawaii.

However, there is still strong sexual selection favouring singing males. Both flatwing and normal genotype females prefer normal males and are more likely to mate in the presence of courtship song; normal females are especially likely to mate in the presence of song (Heinen-Kay et al., 2020). Females reared in silent conditions mimicking the population on Kaua'i are less choosy and more responsive with respect to male calling song compared with females that hear song during rearing (Zuk et al., 2018). This suggests they may be able to compensate for the reduced availability of singers by increasing their responsiveness to the few that are left. Tanner et al. (2019) also found that females are predominantly homozygous normal-wing and their offspring are sired disproportionately by singing males. Yet, overall, the benefits of song loss in terms of avoiding parasitoids must outweigh the costs because the spread of the song-loss mutations has been extremely rapid: more than 90% of males on Kaua'i and about 50% on O'ahu are now flatwings (Zuk et al., 2018). The persistence of singing males in each population reflects the large advantage they have over silent males in terms of mating success, which increases the fewer of them there are.

Very recently, yet another *T. oceanicus* variant has emerged and started to spread. Since 2017 Tinghitella et al. (2021) have increasingly been finding males that produce a novel and extremely variable song. This song, which they call purring, appears to be heritable. It is spectrally unlike the normal song and is produced using structurally different wing morphology. They first found purring males on the island of Moloka'i, in a population which otherwise only had silent males, but since then they have found five more populations on three islands, all of which also have either silent males or a mixture of silent and normal males. *Ormia ochracea* does not respond to purring song but *T. oceanicus* females do. Although females still much prefer normal males, this puts purring males at an advantage over silent males.

8.4 The role of predation in the evolution of communication systems and population divergence

Predation pressure has played an important role in the evolution of communication systems in Orthoptera. In lebinthine crickets, for example, ter Hofstede et al. (2015) argue that the duet between male calling song and female vibrational response (see Section 2.3) probably originated soon after the evolution of bat echolocation. They suggest that the frequency of the male's calling song increased until it was able to release the female's vibrational startle response, normally triggered by bat echolocation sounds. By then performing vibrotaxis towards the female, the male was able to locate her without having to depend on her choosing him and, by remaining still, the female reduced her risk of predation. So both sexes benefited.

Evolutionary loss of traits is common in diverse taxa over evolutionary time. The selection pressures affecting sexual signals and other non-signalling traits are different. For sexual signals, persistence is favoured by sexual selection because they contribute to mating success for both males and females but loss will be favoured by natural selection if they are exploited by natural enemies. Heinen-Kay and Zuk (2019) argue that most examples of sexual signal loss come from phylogenetic studies, so it is difficult to ascertain the context and key factors responsible for their loss. Since song loss in *T. oceanicus* in the Hawaiian islands has been documented in real time over the last 20 years, they were able to use this species to identify factors that facilitated this evolution and found that losses of non-signalling traits in other taxa show important parallels. In general, a significant cost from the environment, weak selection for persistence, and alternative ways of achieving the purpose of the trait appear to be critical in the evolutionary loss of both sexual signals and non-signal traits.

Selection pressures from parasitoids may also have played an important role in the divergence of populations and ultimately, in speciation. For example, five distinct song types are found in various parapatric and sympatric populations of *Mecopoda elongata* in South India. Dutta et al. (2019) looked at three of the sympatric song types and found that there are significant differences between song types in the probability of being infected with parasitoid larvae. Parasitoid infection rates are extremely high, and Dutta et al. (2019) argue that this creates a strong selection pressure likely to drive any changes in acoustic signals that reduce the probability of detection by the flies, especially as parasitoids show preferences for some song types over others (Sakaguchi and Gray, 2011).

Bailey et al. (2019) argue that the release of cryptic variation following sexual signal loss resulting from predator pressure could have played an important role in speciation. In Hawaiian *T. oceanicus*, the wings of flatwings are morphologically very variable and produce a broad range of acoustic resonances well outside the species' normal range. This could facilitate the rapid evolution of novel songs with frequencies far removed from the narrow range of around 5 kHz typical for the species. Receiver structures and physiology would need to coevolve, but on Moloka'i females already prefer to respond to signals produced by some flatwings. Female responses to novel frequencies may therefore be less of a barrier to signal evolution than are the biomechanical constraints on sound production. Widespread secondary trait losses have been inferred from fossil and phylogenetic evidence across many taxa, and Bailey et al. (2019) suggest that such reversals could have played a role in shaping historical patterns of speciation.

9. Conclusions and future work

Despite enormous advances in research on orthopteran acoustics over the last 20 years, there are still many outstanding questions. For example, to what extent is synchronous singing in a chorus adaptive, especially for followers, versus simply being an emergent property, and how do ecological and physiological factors interact to determine the structure of a chorus in the field?

Research since 2002 has also begun to move into some relatively new areas. It is becoming increasingly apparent that behaviour, such as that shown in mate choice, male signalling, and responses to predators and parasites, can be extremely plastic depending on the particular developmental, physiological, social, and environmental circumstances experienced by a species or an individual. Possibly as a result of such plasticity, field researchers or those who try to create more naturalistic protocols for their laboratory experiments are finding that there can be major differences in behaviour in simple laboratory trials compared with the real world. Similarly, it has been recognized that sound signalling cannot be considered in isolation and the interactions between acoustics and other signalling modalities must be considered if a complete picture of orthopertan communications is to be revealed. Studies in all these areas are likely to become increasingly important in the future.

There is also a growing literature on the generally adverse effects on orthopteran communication and behaviour caused by anthropogenic factors such as climate change (Cusano et al., 2016), radiation resulting from nuclear accidents (Fuciarelli and Rollo, 2021), artificial light at night (Levy et al., 2021) and urban and traffic noise (e.g. Bent et al., 2021; Gurule-Small and Tinghitella, 2019; Schmidt et al., 2014). Some species are able to adapt to anthropogenic noise to some extent at least (Gallego-Abenza et al., 2020; Lampe et al., 2014; Sathyan and Couldridge, 2021; Tan, 2020). Others may be relatively unaffected (e.g. Costello and Symes, 2014) and the effects in some cases may even be beneficial (Phillips et al., 2019). The COVID-19 pandemic has provided opportunities for natural experiments to learn how human activities affect orthopteran behaviour (Tan and Robillard, 2021a).

Studies of song loss in the Hawaiian populations of *Teleogryllus oceanicus* that have been taking place since the early 2000s have also dramatically improved our understanding of the evolution of communication systems in orthopterans, and further study of this species and its parasitoid are likely to provide invaluable data for many years to come.

Another relatively recent development is the emergence of automated methods that utilize orthopteran acoustic signalling to survey or monitor population size or density (e.g. Wang et al., 2014), species diversity (e.g. Thompson et al., 2020; van der Mescht et al., 2021), responses to environmental change (e.g. Gibb et al., 2019), or environmental health (e.g. Hill et al., 2019). With the accelerating pace of climate change, these methods can only become more important.

Orthopterans, especially crickets, have long provided model organisms for work on auditory processing in the nervous system and substantial research in this area has continued over the last 20 years. These studies are not only useful for understanding insect systems but are applicable to auditory systems in general and as such will probably continue to be widely relevant.

There are however still some continuing gaps in research. For example, there has been very little work on female song and, in particular, on questions such as whether female song contains any information or whether males choose their mates based on female song. Given the increased interest in male mate choice in recent studies, it is to be hoped that this topic will soon be addressed.

Acknowledgements

We would like to thank the Open University for their financial and academic support, and Professor David Gowing, Roger Lowry and Patricia Ash for their invaluable help and support. The photograph of *Gryllus bimaculatus* in Fig. 7 was provided courtesy of Professor Axel Hochkirch of Trier University and we would like to thank him for allowing us to use it.

References

Acosta, R.C., Timm, V.F., Szinwelski, N., Matiotti Da Costa, M.K., Zefa, E., 2020. Mating behavior and acoustic communication of the long-legged cricket *Endecous (Notendecous) onthophagus* (Berg, 1891) from Southern Brazil (Orthoptera: Grylloidea: Phalangopsidae). Zootaxa 4743, 427–437. https://doi.org/10.11646/zootaxa.4743.3.10.

Andrade, M.C.B., Mason, A.C., 2000. Male condition, female choice, and extreme variation in repeated mating in a scaly cricket, *Ornebius aperta* (Orthoptera: Gryllidae: Mogoplistinae). J. Insect Behav. 13, 483–497.

Anichini, M., Frommolt, K.-H., Lehmann, G.U.C., 2018. To compete or not to compete: bushcricket song plasticity reveals male body condition and rival distance. Anim. Behav. 142, 59–68. https://doi.org/10.1016/j.anbehav.2018.05.022.

Atwell, A., Wagner Jr., W.E., 2014. Female mate choice plasticity is affected by the interaction between male density and female age in a field cricket. Anim. Behav. 98, 177–183. https://doi.org/10.1016/j.anbehav.2014.10.007.

Aufderheide, J., Ronacher, B., 2019. The impact of age and egg-laying cycle on female grasshoppers' preference functions for acoustic signals. J. Comp. Physiol. A 205, 103–111. https://doi.org/10.1007/s00359-018-1307-8.

Bailey, W.J., 2003. Insect duets: underlying mechanisms and their evolution. Physiol. Entomol. 28, 157–174.

Bailey, N.W., Zuk, M., 2008. Acoustic experience shapes female mate choice in field crickets. Proc. R. Soc. Lond. Ser. B Biol. Sci. 275, 2645–2650.

Bailey, N.W., Zuk, M., 2009. Field crickets change mating preferences using remembered social information. Biol. Lett. 5, 449–451. https://doi.org/10.1098/rsbl.2009.0112.

Bailey, W.J., Ager, E.I., O'Brien, E.K., Watson, D.L., 2003. Searching by visual and acoustic cues among bushcrickets (Orthoptera: Tettigoniidae): will females remain faithful to a male who stops calling? Physiol. Entomol. 28, 209–214. https://doi.org/10.1046/j.1365-3032.2003.00334.x.

Bailey, W., MacLeay, C., Gordon, T., 2006. Acoustic mimicry and disruptive alternative calling tactics in an Australian bushcricket (*Caedicia*; Phaneropterinae; Tettigoniidae; Orthoptera): does mating influence male calling tactic? Physiol. Entomol. 31, 201–210. https://doi.org/10.1111/j.1365-3032.2006.00501.x.

Bailey, N.W., Gray, B., Zuk, M., 2010. Acoustic experience shapes alternative mating tactics and reproductive investment in male field crickets. Curr. Biol. 20, 845–849. https://doi.org/10.1016/j.cub.2010.02.063.

Bailey, N.W., Gray, B., Zuk, M., 2011. Exposure to sexual signals during rearing increases immune defence in adult field crickets. Biol. Lett. 7, 217–220. https://doi.org/10.1098/rsbl.2010.0659.

Bailey, N.W., Moran, P.A., Hennig, R.M., 2017. Divergent mechanisms of acoustic mate recognition between closely related field cricket species (*Teleogryllus* spp.). Anim. Behav. 130, 17–25. https://doi.org/10.1016/j.anbehav.2017.06.007.

Bailey, N.W., Pascoal, S., Montealegre-Z, F., 2019. Testing the role of trait reversal in evolutionary diversification using song loss in wild crickets. Proc. Natl. Acad. Sci. U. S. A. 116, 8941–8949. https://doi.org/10.1073/pnas.1818998116.

Baker, E., Chesmore, D., 2020. Standardisation of bioacoustic terminology for insects. Biodivers. Data J. 8, E54222. https://doi.org/10.3897/BDJ.8.e54222.

Baker, A., Sarria-S, F., Morris, G., Jonsson, T., Montealegre-Z, F., 2017. Wing resonances in a new dead-leaf-mimic katydid (Tettigoniidae: Pterochrozinae) from the Andean cloud forests. Zool. Anz. 270, 60–70. https://doi.org/10.1016/j.jcz.2017.10.001.

Baker, A.A., Jonsson, T., Aldridge, S., Montealegre-Z, F., 2019. Complex wing motion during stridulation in the katydid *Nastonotus foreli* (Orthoptera: Tettigoniidae: Pseudophyllinae). J. Insect Physiol. 114, 100–108. https://doi.org/10.1016/j.jinsphys.2019.03.005.

Balakrishnan, R., Pollack, G.S., 1997. The role of antennal sensory cues in female responses to courting males in the cricket *Teleogryllus oceanicus*. J. Exp. Biol. 200, 511–522.

Balakrishnan, R., Bahuleyan, J., Nandi, D., Jain, M., 2014. Modelling the effects of chorus species composition and caller density on acoustic masking interference in multispecies choruses of crickets and katydids. Ecol. Inform. 21, 50–58. https://doi.org/10.1016/j.ecoinf.2013.11.006.

Balenger, S.L., Zuk, M., 2014. Testing the Hamilton–Zuk hypothesis: past, present, and future. Integr. Comp. Biol. 54, 601–613. https://doi.org/10.1093/icb/icu059.

Balenger, S.L., Zuk, M., 2015. Roaming Romeos: male crickets evolving in silence show increased locomotor behaviours. Anim. Behav. 101, 213–219. https://doi.org/10.1016/j.anbehav.2014.12.023.

Balenger, S.L., Bastiaans, E., Zuk, M., 2018. Is it the song or the singers? Acoustic and social experiences shape adult reproductive tactics and condition. J. Insect Behav. 31, 552–568. https://doi.org/10.1007/s10905-018-9698-5.

Barbosa, F., Rebar, D., Greenfield, M.D., 2016. Female preference functions drive interpopulation divergence in male signalling: call diversity in the bushcricket *Ephippiger diurnus*. J. Evol. Biol. 29, 2219–2228. https://doi.org/10.1111/jeb.12940.

Bateman, P.W., Fleming, P.A., 2013. Signaling or not-signaling: variation in vulnerability and defense tactics of armored ground crickets (*Acanthoplus Speiseri*: Orthoptera, Tettigoniidae, Hetrodinae). J. Insect Behav. 26, 14–22. https://doi.org/10.1007/s10905-012-9329-5.

Beckers, O.M., 2020. Phenotypic plasticity related to temperature induces song variation in the field cricket *Gryllus rubens*. Ethology 126, 781–790. https://doi.org/10.1111/eth.13035.

Beckers, O.M., Wagner Jr., W.E., 2011. Mate sampling strategy in a field cricket: evidence for a fixed threshold strategy with last chance option. Anim. Behav. 81, 519–527. https://doi.org/10.1016/j.anbehav.2010.11.022.

Beckers, O.M., Wagner Jr., W.E., 2012. Eavesdropping parasitoids do not cause the evolution of less conspicuous signalling behaviour in a field cricket. Anim. Behav. 84, 1457–1462. https://doi.org/10.1016/j.anbehav.2012.09.016.

Beckers, O.M., Wagner Jr., W.E., 2013. Parasitoid infestation changes female mating preferences. Anim. Behav. 85, 791–796. https://doi.org/10.1016/j.anbehav.2013.01.025.

Beckers, O.M., Wagner Jr., W.E., 2018. Males and females evolve riskier traits in populations with eavesdropping parasitoids. Behav. Ecol. Sociobiol. 72, 174. https://doi.org/10.1007/s00265-018-2588-1.

Bell, P.D., 1980. Multimodal communication by the black-horned tree cricket, *Oecanthus nigricornis* (Walker) (Orthoptera, Gryllidae). Can. J. Zool. 58, 1861–1868. https://doi.org/10.1139/z80-254.

Benavides-Lopez, J.L., ter Hofstede, H., Robillard, T., 2020. Novel system of communication in crickets originated at the same time as bat echolocation and includes male–male multimodal communication. Sci. Nat. 107, 9. https://doi.org/10.1007/s00114-020-1666-1.

Benediktov, A.A., 2012. Acoustic signals of the pygmy mole cricket *Xya variegata* (Latreille, 1809) (Orthoptera: Tridactyloidea). Russ. Entomol. J. 21, 307–308.

Benediktov, A., 2015. Vibro-acoustical signals of the locust *Trilophidia annulata* (Thunb.) (Orthoptera, Acrididae, Oedipodinae). Mosc. Univ. Biol. Sci. Bull. 70, 35–38. https://doi.org/10.3103/S0096392515010022.

Benediktov, A., 2018. Mixed sound-vibrational communication in Orthoptera. Metaleptea 38, 43.

Bennet-Clark, H.C., 1970. The mechanism and efficiency of sound production in mole crickets. J. Exp. Biol. 52, 619–652.

Bennet-Clark, H.C., 1999. Resonators in insect sound production: how insects produce loud pure-tone songs. J. Exp. Biol. 202, 3347–3357.

Bennet-Clark, H.C., 2003. Wing resonances in the Australian field cricket *Teleogryllus oceanicus*. J. Exp. Biol. 206, 1479–1496. https://doi.org/10.1242/jeb.00281.

Bennet-Clark, H.C., Bailey, W.J., 2002. Ticking of the clockwork cricket: the role of the escapement mechanism. J. Exp. Biol. 205, 613–625. https://doi.org/10.1242/jeb.205.5.613.

Bent, A.M., Ings, T.C., Mowles, S.L., 2021. Anthropogenic noise disrupts mate choice behaviors in female *Gryllus bimaculatus*. Behav. Ecol. 32, 201–210. https://doi.org/10.1093/beheco/araa124.

Benton, T., 2012. Grasshoppers and Crickets. HarperCollins, London.

Bentsen, C.L., Hunt, J., Jennions, M.D., Brooks, R., 2006. Complex multivariate sexual selection on male acoustic signaling in a wild population of *Teleogryllus commodus*. Am. Nat. 167, E102–E116. https://doi.org/10.1086/501376.

Berg, A., Greenfield, M.D., 2005. Sexual selection in insect choruses: influences of call power and relative timing. J. Insect Behav. 18, 59–75. https://doi.org/10.1007/s10905-005-9347-7.

Berger, D., Gottsberger, B., 2010. Analysis of the courtship of *Myrmeleotettix antennatus* (Fieber, 1853)—with general remarks on multimodal courtship behaviour in gomphocerine grasshoppers. Articulata 25, 1–21.

Bertram, S.M., Bowen, M., 2006. Field cricket species differences in the temporal patterns of long-distance mate attraction signals. Ethology 112, 850–857. https://doi.org/10.1111/j.1439-0310.2006.01234.x.

Bertram, S.M., Rook, V., 2011. Jamaican field cricket mate attraction signals provide age cues. Ethology 117, 1050–1055. https://doi.org/10.1111/j.1439-0310.2011.01958.x.

Bertram, S.M., Rook, V., 2012. Relationship between condition, aggression, signaling, courtship, and egg laying in the field cricket, *Gryllus assimilis*. Ethology 118, 360–372. https://doi.org/10.1111/j.1439-0310.2011.02019.x.

Bertram, S.M., Orozco, S.X., Bellani, R., 2004. Temporal shifts in conspicuousness: mate attraction displays of the Texas field cricket, *Gryllus texensis*. Ethology 110, 963–975.

Bertram, S.M., Schade, J.D., Elser, J.J., 2006. Signalling and phosphorus: correlations between mate signalling effort and body elemental composition in crickets. Anim. Behav. 72, 899–907. https://doi.org/10.1016/j.anbehav.2006.02.012.

Bertram, S.M., Rook, V.L.M., Fitzsimmons, L.P., 2010. Strutting their stuff: victory displays in the spring field cricket, *Gryllus veletis*. Behaviour 147, 1249–1266. https://doi.org/10.1163/000579510x514535.

Bertram, S.M., Fitzsimmons, L.P., McAuley, E.M., Rundle, H.D., Gorelick, R., 2011. Phenotypic covariance structure and its divergence for acoustic mate attraction signals among four cricket species. Ecol. Evol. 2, 181–195. https://doi.org/10.1002/ece3.76.

Bertram, S.M., Harrison, S.J., Thomson, I.R., Fitzsimmons, L.P., 2013. Adaptive plasticity in wild field cricket's acoustic signaling. PLoS One 8, e69247. https://doi.org/10.1371/journal.pone.0069247.

Bertram, S.M., Loranger, M.J., Thomson, I.R., Harrison, S.J., Ferguson, G.L., Reifer, M.L., Corlett, D.H., Gowaty, P.A., 2016. Linking mating preferences to sexually selected traits and offspring viability: good versus complementary genes hypotheses. Anim. Behav. 119, 75–86. https://doi.org/10.1016/j.anbehav.2016.06.003.

Broughton, W.B., 1963. Method in bio-acoustic terminology. In: Busnel, R.-G. (Ed.), Acoustic Behaviour of Animals. Elsevier, Amsterdam, pp. 3–24.

Broughton, W.B., 1976. Proposal for a new term 'echeme' to replace 'chirp' in animal acoustics. Physiol. Entomol. 1, 103–106.

Brown, W.D., 2016. Mating behavior of the endemic Hawaiian cricket *Leptogryllus elongatus* (Orthoptera: Gryllidae: Oecanthinae). J. Insect Behav. 29, 449–458. https://doi.org/10.1007/s10905-016-9574-0.

Brown, W.D., Smith, A.T., Moskalik, B., Gabriel, J., 2006. Aggressive contests in house crickets: size, motivation and the information content of aggressive songs. Anim. Behav. 72, 225–233. https://doi.org/10.1016/j.anbehav.2006.01.012.

Brumm, H., Slabbekoorn, H., 2005. Acoustic communication in noise. Adv. Stud. Behav. 35, 151–209.

Bull, C.M., 1979. Function of complexity in the courtship of the grasshopper *Myrmeleotettix maculatus*. Behaviour 69, 201–216. https://doi.org/10.1163/156853979x00476.

Burrows, M., 2003. Froghopper insects leap to new heights. Nature 424, 509. https://doi.org/10.1038/424509a.

Busnel, R.-G., Pasquinelly, F., Dumortier, B., 2019. Body tremulations and their transmission as vibrations for short distance information transfer between *Ephippiger* male and female. In: Hill, P., Lakes-Harlan, R., Mazzoni, V., Narins, P., Virant-Doberlet, M., Wessel, A. (Eds.), Biotremology: Studying Vibrational Behavior. Animal Signals and Communication, Springer, Cham, Switzerland, pp. 43–49.

Callander, S., Kahn, A.T., Hunt, J., Backwell, P.R.Y., Jennions, M.D., 2013. The effect of competitors on calling effort and life span in male field crickets. Behav. Ecol. 24, 1251–1259. https://doi.org/10.1093/beheco/art059.

Candolin, U., 2003. The use of multiple cues in mate choice. Biol. Rev. 78, 575–595. https://doi.org/10.1017/s1464793103006158.

Celiker, E., Jonsson, T., Montealegre-Z, F., 2020. The auditory mechanics of the outer ear of the bush cricket: a numerical approach. Biophys. J. 118, 464–475. https://doi.org/10.1016/j.bpj.2019.11.3394.

Centeno, E., Zefa, E., 2019. The complex communication signals in the mating behavior of the tropical cricket *Cranistus colliurides* Stål, 1861 (Orthoptera: Grylloidea; Trigonidiidae: Phylloscyrtini). Zootaxa 4623, 571–576. https://doi.org/10.11646/zootaxa.4623.3.9.

Centeno, E., Calixto, E.S., Zefa, E., 2020. Acoustic repertoire of the sword-tail cricket *Cranistus colliurides* Stål, 1861 (Orthoptera: Grylloidea, Trigonidiidae: Phylloscyrtini). Bioacoustics. https://doi.org/10.1080/09524622.2020.1838325.

Chamorro-R, J., Montealegre-Z, F., Gonzalez-O, R., 2007. Determinants of male spacing behaviour in *Panacanthus pallicornis* (Orthoptera: Tettigoniidae). Ethology 113, 1158–1172. https://doi.org/10.1111/j.1439-0310.2007.01428.x.

Champagnon, J., del Castillo, R.C., 2008. Female mate choice, calling song and genetic variance in the cricket, *Gryllodes sigillatus*. Ethology 114, 223–230. https://doi.org/10.1111/j.1439-0310.2007.01470.x.

Chang, T. T., 2015. Effect of Acoustic Signaling, Metabolic Rate, and Size on Territoriality in Male *Cyphoderris monstrosa* (Orthoptera: Haglidae). MSc Thesis. University of Toronto: Available online http://hdl.handle.net/1807/70254.

Chivers, B.D., Jonsson, T., Soulsbury, C.D., Montealegre-Z, F., 2017. Structural biomechanics determine spectral purity of bush-cricket calls. Biol. Lett. 13, 20170573. https://doi.org/10.1098/rsbl.2017.0573.

Cigliano, M. M., Braun, H., Eades, D. C. & Otte, D., 2021. Orthoptera Species File. Version 5.0/5.0. Available online: http://Orthoptera.SpeciesFile.org [Accessed 25 March 2021].

Clemens, J., Aufderheide, J., Ronacher, B., 2017. Relative weighting of acoustic information during mating decisions in grasshoppers indicates signatures of sexual selection. J. Comp. Physiol. A 203, 891–901. https://doi.org/10.1007/s00359-017-1200-x.

Clemens, J., Ronacher, B., Reichert, M.S., 2021. Sex-specific speed-accuracy trade-offs shape neural processing of acoustic signals in a grasshopper. Proc. R. Soc. Lond. Ser. B: Biol. Sci. 288, 20210005. https://doi.org/10.1098/rspb.2021.0005.

Cole, J.A., 2010. Clinal variation explains taxonomic discrepancy in the calling songs of shield-back katydids (Orthoptera: Tettigoniidae: Tettigoniinae: *Aglaothorax*). Biol. J. Linn. Soc. 101, 910–921. https://doi.org/10.1111/j.1095-8312.2010.01532.x.

Copeland, E.K., Fedorka, K.M., 2012. The influence of male age and simulated pathogenic infection on producing a dishonest sexual signal. Proc. R. Soc. Lond. Ser. B: Biol. Sci. 279, 4740–4746. https://doi.org/10.1098/rspb.2012.1914.

Costello, R.A., Symes, L.B., 2014. Effects of anthropogenic noise on male signalling behaviour and female phonotaxis in *Oecanthus* tree crickets. Anim. Behav. 95, 15–22. https://doi.org/10.1016/j.anbehav.2014.05.009.

Couldridge, V.C.K., Gordon, M.L., 2015. Diel variation in signalling and signal transmission in the bladder grasshopper, *Bullacris unicolor* (Orthoptera; Pneumoridae). Behaviour 152, 1706–1723. https://doi.org/10.1163/1568539x-00003300.

Couldridge, V.C.K., van Staaden, M.J., 2004. Habitat-dependent transmission of male advertisement calls in bladder grasshoppers (Orthoptera; Pneumoridae). J. Exp. Biol. 207, 2777–2786. https://doi.org/10.1242/jeb.01092.

Couldridge, V.C.K., van Staaden, M.J., 2006. Female preferences for male calling songs in the bladder grasshopper *Bullacris membracioides*. Behaviour 143, 1439–1456. https://doi.org/10.1163/156853906779367035.

Creutzig, F., Wohlgemuth, S., Stumpner, A., Benda, J., Ronacher, B., Herz, A.V.M., 2009. Timescale-invariant representation of acoustic communication signals by a bursting neuron. J. Neurosci. 29, 2575–2580. https://doi.org/10.1523/jneurosci.0599-08.2009.

Cusano, D.A., Matthews, L.P., Grapstein, E., Parks, S.E., 2016. Effects of increasing temperature on acoustic advertisement in the Tettigoniidae. J. Orthoptera Res. 25, 39–47.

De Graaf, J., Schoeman, A.S., Brandenburg, R.L., 2005. Stridulation of *Gryllotalpa africana* (Orthoptera: Gryllotalpidae) on turf grass in South Africa. Fla. Entomol. 88, 292–299. https://doi.org/10.1653/0015-4040(2005)088[0292:sogaog]2.0.co;2.

De Souza, L.R., Kasumovic, M.M., Judge, K.A., Morris, G.K., 2011. Communicating male size by tremulatory vibration in a Columbian rainforest katydid, *Gnathoclita sodalis* (Orthoptera, Tettigoniidae). Behaviour 148, 341–357. https://doi.org/10.1163/000579511x559418.

Deb, R., Modak, S., Balakrishnan, R., 2020. Baffling: a condition-dependent alternative mate attraction strategy using self-made tools in tree crickets. Proc. R. Soc. Lond. Ser. B: Biol. Sci. 287, 20202229. https://doi.org/10.1098/rspb.2020.2229.

deCarvalho, T.N., Shaw, K.L., 2010. Divergence of courtship and mating behaviors among endemic Hawaiian swordtail crickets. Behaviour 147, 479–504. https://doi.org/10.1163/000579509x12586249027577.

Deily, J.A., Schul, J., 2009. Selective phonotaxis in *Neoconocephalus nebrascensis* (Orthoptera: Tettigoniidae): call recognition at two temporal scales. J. Comp. Physiol. A 195, 31–37. https://doi.org/10.1007/s00359-008-0379-2.

Diwakar, S., Balakrishnan, R., 2006. Male and female stridulation in an Indian weta (Orthoptera: Anostostomatidae). Bioacoustics 16, 75–85. https://doi.org/10.1080/09524622.2006.9753565.

Diwakar, S., Balakrishnan, R., 2007a. Vertical stratification in an acoustically communicating ensiferan assemblage of a tropical evergreen forest in southern India. J. Trop. Ecol. 23, 479–486. https://doi.org/10.1017/s0266467407004208.

Diwakar, S., Balakrishnan, R., 2007b. The assemblage of acoustically communicating crickets of a tropical evergreen forest in southern India: call diversity and diel calling patterns. Bioacoustics 16, 113–135. https://doi.org/10.1080/09524622.2007.9753571.

Dobbs, O.L., Talavera, J.B., Rossi, S.M., Menjivar, S., Gray, D.A., 2020. Signaler–receiver–eavesdropper: risks and rewards of variation in the dominant frequency of male cricket calls. Ecol. Evol. 10, 12364–12371. https://doi.org/10.1002/ece3.6866.

Donelson, N.C., van Staaden, M.J., 2005. Alternate tactics in male bladder grasshoppers *Bullacris membracioides* (Orthoptera: Pneumoridae). Behaviour 142, 761–778.

Donelson, N.C., Smith, A.R., van Staaden, M.J., 2008. Variation in adult longevity in a polymorphic grasshopper species. J. Orthoptera Res. 17, 279–282. https://doi.org/10.1665/1082-6467-17.2.279.

Doubell, M., Grant, P.B.C., Esterhuizen, N., Bazelet, C.S., Addison, P., Terblanche, J.S., 2017. The metabolic costs of sexual signalling in the chirping katydid *Plangia graminea* (Serville) (Orthoptera: Tettigoniidae) are context dependent: cumulative costs add up fast. J. Exp. Biol. 220, 4440–4449. https://doi.org/10.1242/jeb.160036.

Dutta, R., Balakrishnan, R., Tregenza, T., 2018. Divergence in potential contact pheromones and genital morphology among sympatric song types of the bush cricket *Mecopoda elongata*. Front. Ecol. Evol. 6, 158. https://doi.org/10.3389/fevo.2018.00158.

Dutta, R., Reddy, M., Tregenza, T., 2019. Discovery of an acoustically locating parasitoid with a potential role in divergence of song types among sympatric populations of the bush cricket *Mecopoda elongata*. J. Orthoptera Res. 28, 181–186.

Elaeva, N.F., Korsunovskaya, O.S., 2012. Ecology and sound signaling of relict katydid *Deracantha onos* (Pallas, 1772) (Orthoptera, Tettigoniidae, Bradyporinae). Mosc. Univ. Biol. Sci. Bull. 67, 111–116. https://doi.org/10.3103/S0096392512020046.

Elias-Quevedo, A., Cueva del Castillo, R., 2020. On the discovery of a male morph with a novelty alternative mating tactic in the neotropical cricket *Macroanaxipha macilenta* (Orthoptera: Gryllidae). J. Ethol. 38, 3–11. https://doi.org/10.1007/s10164-019-00614-5.

Elliott, C.J.H., Koch, U.T., 1985. The clockwork cricket. Naturwissenschaften 72, 150–153.

Elsner, N., 1974. Neuroethology of sound production in gomphocerine grasshoppers I. Song pattern and stridulatory movements. Z. Vergl. Physiol. 88, 67–102. https://doi.org/10.1007/BF00695923.

Elsner, N., Wasser, G., 1994. Leg and wing stridulation in various populations of the gomphocerine grasshopper *Stenobothrus rubicundus* (Germar 1817). 1. Sound patterns and singing movements. Zool. Anal. Complex Syst. 98, 179–190.

Ely, S.O., Mahamat, H., Njagi, P.G.N., Bashir, M.O., El-Amin, S.E.-T., Hassanali, A., 2006. Mate location mechanism and phase-related mate preferences in solitarius desert locust, *Schistocerca gregaria*. J. Chem. Ecol. 32, 1057–1069. https://doi.org/10.1007/s10886-006-9045-8.

Endler, J.A., 2000. Evolutionary implications of the interaction between animal signals and the environment. In: Espmark, Y., Amundsen, T., Rosenqvist, G. (Eds.), Animal Signals: Signalling and Signal Design in Animal Communication. Tapir Academic Press, Trondheim, Norway, pp. 11–46.

Erregger, B., Schmidt, A.K.D., 2018. Anthropogenic calling sites boost the sound amplitude of advertisement calls produced by a tropical cricket. Anim. Behav. 142, 31–38. https://doi.org/10.1016/j.anbehav.2018.05.021.

Erregger, B., Kovac, H., Stabentheiner, A., Hartbauer, M., Römer, H., Schmidt, A.K.D., 2017. Cranking up the heat: relationships between energetically costly song features and the increase in thorax temperature in male crickets and katydids. J. Exp. Biol. 220, 2635–2644. https://doi.org/10.1242/jeb.155846.

Ewers, R.M., Cowley, G., 2005. The role of sound production in determining dominance in agonistic interactions between male tree wetas (*Hemideina crassidens*, Orthoptera: Anostostomatidae). N. Z. Nat. Sci. 30, 11–17.

Falk, J.J., ter Hofstede, H.M., Jones, P.L., Dixon, M.M., Faure, P.A., Kalko, E.K.V., Page, R.A., 2015. Sensory-based niche partitioning in a multiple predator–multiple prey community. Proc. R. Soc. Lond. Ser. B: Biol. Sci. 282, 20150520. https://doi.org/10.1098/rspb.2015.0520.

Faulkes, Z., Pollack, G.S., 2000. Effects of inhibitory timing on contrast enhancement in auditory circuits in crickets (*Teleogryllus oceanicus*). J. Neurophysiol. 84, 1247–1255.

Faulkes, Z., Pollack, G.S., 2001. Mechanisms of frequency-specific responses of omega neuron 1 in crickets (*Teleogryllus oceanicus*): a polysynaptic pathway for song? J. Exp. Biol. 204, 1295–1305.

Fedorka, K.M., Mousseau, T.A., 2007. Immune system activation affects male sexual signal and reproductive potential in crickets. Behav. Ecol. 18, 231–235. https://doi.org/10.1093/beheco/arl067.

Fergus, D.J., Shaw, K.L., 2013. Circadian rhythms and period expression in the Hawaiian cricket genus *Laupala*. Behav. Genet. 43, 241–253. https://doi.org/10.1007/s10519-012-9576-4.

Fertschai, I., Stradner, J., Römer, H., 2007. Neuroethology of female preference in the synchronously singing bushcricket *Mecopoda elongata* (Tettigoniidae; Orthoptera): why do followers call at all? J. Exp. Biol. 210, 465–476. https://doi.org/10.1242/jeb.02655.

Fianco, M., de Souza-Dias, P.G.B., de Farias-Martins, F., Magro, S., Mateus Prasniewski, V., Ricci, J., Zefa, E., Szinwelski, N., 2018. Ethology of the cricket *Endecous* (*Endecous*) *chape* Souza-Dias & de Mello, 2017 (Orthoptera: Grylloidea: Phalangopsidae) I: agonistic and reproductive behavior. J. Orthoptera Res. 27, 193–201.

Field, L.H., Bailey, W.J., 1997. Sound production in primitive Orthoptera from Western Australia: sounds used in defence and social communication in *Ametrus* sp. and *Hadrogryllacris* sp. (Gryllacrididae: Orthoptera). J. Nat. Hist. 31, 1127–1141.

Finck, J., Ronacher, B., 2017. Components of reproductive isolation between the closely related grasshopper species *Chorthippus biguttulus* and *C. mollis*. Behav. Ecol. Sociobiol. 71, 70. https://doi.org/10.1007/s00265-017-2295-3.

Finck, J., Kuntze, J., Ronacher, B., 2016. Chemical cues from females trigger male courtship behaviour in grasshoppers. J. Comp. Physiol. A 202, 337–345. https://doi.org/10.1007/s00359-016-1081-4.

Fitzsimmons, L.P., Bertram, S.M., 2011. The calling songs of male spring field crickets (*Gryllus veletis*) change as males age. Behaviour 148, 1045–1065. https://doi.org/10.1163/000579511x588812.

Fitzsimmons, L.P., Bertram, S.M., 2013. Playing to an audience: the social environment influences aggression and victory displays. Biol. Lett. 9, 20130449. https://doi.org/10.1098/rsbl.2013.0449.

Forrest, T.G., 1982. Acoustic communication and baffling behaviors of crickets. Fla. Entomol. 65, 33–44. https://doi.org/10.2307/3494144.

Franklin, M., Droege, S., Dawson, D., Royle, J.A., 2009. Nightly and seasonal patterns of calling in common true katydids (Orthoptera: Tettigoniidae: *Pterophylla camellifolia*). J. Orthoptera Res. 18, 15–18. https://doi.org/10.1665/034.018.0108.

Franzke, A., Reinhold, K., 2012. Limited condition dependence of male acoustic signals in the grasshopper *Chorthippus biguttulus*. Ecol. Evol. 2, 1914–1921. https://doi.org/10.1002/ece3.309.

Fuciarelli, T.M., Rollo, C.D., 2021. Ionizing radiation alters male *Acheta domesticus* courtship songs that are critical for mating success. Anim. Behav. 178, 209–216. https://doi.org/10.1016/j.anbehav.2021.06.018.

Funk, D.H., 2016. Mating behavior of *Phyllopalpus pulchellus* Uhler (Orthoptera: Gryllidae: Trigonidiinae). J. Orthoptera Res. 25, 7–13.

Gabel, E., Hennig, R.M., 2016. Evidence for comparative decision making in female crickets. Behav. Ecol. 27, 1216–1222. https://doi.org/10.1093/beheco/arw030.

Gage, A.R., Barnard, C.J., 1996. Male crickets increase sperm number in relation to competition and female size. Behav. Ecol. Sociobiol. 38, 349–353.

Gallego-Abenza, M., Mathevon, N., Wheatcroft, D., 2020. Experience modulates an insect's response to anthropogenic noise. Behav. Ecol. 31, 90–96. https://doi.org/10.1093/beheco/arz159.

Garcia, M.D., Gomez, R., Clemente, M.E., Presa, J.J., 2014. Sound production in the genus *Acinipe* Rambur, 1832 (Orthoptera: Pamphagidae). Ital. J. Zool. 81, 264–270. https://doi.org/10.1080/11250003.2014.895059.

Garcia, M.D., Clemente, M.E., Presa, J.J., 2015. The courtship behaviour of *Calliptamus wattenwylianus* Pantel, 1896 (Orthoptera, Catantopinae). Articulata 30, 33–51.

Geipel, I., Kernan, C.E., Litterer, A.S., Carter, G.G., Page, R.A., ter Hofstede, H.M., 2020. Predation risks of signalling and searching: bats prefer moving katydids. Biol. Lett. 16, 20190837. https://doi.org/10.1098/rsbl.2019.0837.

Ghalichi, N.S., Heinen-Kay, J.L., Zuk, M., 2020. Acoustic experience interacts with perceived risk of predation in shaping female response in crickets. J. Insect Behav. 33, 38–47. https://doi.org/10.1007/s10905-020-09744-y.

Gibb, R., Browning, E., Glover-Kapfer, P., Jones, K.E., 2019. Emerging opportunities and challenges for passive acoustics in ecological assessment and monitoring. Methods Ecol. Evol. 10, 169–185. https://doi.org/10.1111/2041-210x.13101.

Glen-Worrall, G. M., 2003. Repeat and Multiple Mating in an Australian Scaly Cricket, *Ornebius aperta* (Orthoptera: Mogoplistidae). MSc Thesis. University of Toronto: Available online https://www.bac-lac.gc.ca/eng/services/theses/Pages/item.aspx?idNumber=57061477 [Accessed April 2021].

Golov, Y., Rillichi, J., Harari, A., Ayali, A., 2018a. Precopulatory behavior and sexual conflict in the desert locust. PeerJ 6, e4356. https://doi.org/10.7717/peerj.4356.

Golov, Y., Rillich, J., Douek, M., Harari, A.R., Ayali, A., 2018b. Sexual behavior of the desert locust during intra- and inter-phase interactions. J. Insect Behav. 31, 629–641. https://doi.org/10.1007/s10905-018-9703-z.

Goodale, E., Ruxton, G.D., Beauchamp, G., 2019. Predator eavesdropping in a mixed-species environment: how prey species may use grouping, confusion, and the cocktail party effect to reduce predator detection. Front. Ecol. Evol. 7, 141. https://doi.org/10.3389/fevo.2019.00141.

Gordon, S.D., Jackson, J.C., Rogers, S.M., Windmill, J.F.C., 2014. Listening to the environment: hearing differences from an epigenetic effect in solitarious and gregarious locusts. Proc. R. Soc. Lond. Ser. B Biol. Sci. 281, 20141693. https://doi.org/10.1098/rspb.2014.1693.

Grace, J.L., Shaw, K.L., 2011. Coevolution of male mating signal and female preference during early lineage divergence of the Hawaiian cricket, *Laupala cerasina*. Evolution 65, 2184–2196. https://doi.org/10.1111/j.1558-5646.2011.01278.x.

Grant, P.B.C., Samways, M.J., 2015. Acoustic prey and a listening predator: interaction between calling katydids and the bat-eared fox. Bioacoustics 24, 49–61. https://doi.org/10.1080/09524622.2014.927789.

Gray, D.A., Cade, W.H., 2000. Sexual selection and speciation in field crickets. Proc. Natl. Acad. Sci. U. S. A. 97, 14449–14454. https://doi.org/10.1073/pnas.97.26.14449.

Gray, B., Bailey, N.W., Poon, M., Zuk, M., 2014. Multimodal signal compensation: do field crickets shift sexual signal modality after the loss of acoustic communication? Anim. Behav. 93, 243–248. https://doi.org/10.1016/j.anbehav.2014.04.033.

Gray, D.A., Gabel, E., Blankers, T., Hennig, R.M., 2016. Multivariate female preference tests reveal latent perceptual biases. Proc. R. Soc. Lond. Ser. B: Biol. Sci. 283, 20161972. https://doi.org/10.1098/rspb.2016.1972.

Greenfield, M.D., 2002. Signalers and Receivers: Mechanisms of Evolution of Arthropod Communication. Oxford University Press, Oxford.

Greenfield, M.D., 2015. Signal interactions and interference in insect choruses: singing and listening in the social environment. J. Comp. Physiol. A 201, 143–154. https://doi.org/10.1007/s00359-014-0938-7.

Greenfield, M.D., Snedden, W.A., 2003. Selective attention and the spatio-temporal structure of orthopteran choruses. Behaviour 140, 1–26. https://doi.org/10.1163/156853903763999863.

Greenfield, M.D., Siegfreid, E., Snedden, W.A., 2004. Variation and repeatability of female choice in a chorusing katydid, *Ephippiger ephippiger*: an experimental exploration of the precedence effect. Ethology 110, 287–299.

Greenfield, M.D., Esquer-Garrigos, Y., Streiff, R., Party, V., 2016. Animal choruses emerge from receiver psychology. Sci. Rep. 6, 34369. https://doi.org/10.1038/srep34369.

Greenfield, M.D., Marin-Cudraz, T., Party, V., 2017. Evolution of synchronies in insect choruses. Biol. J. Linn. Soc. 122, 487–504.

Guerra, P.A., Mason, A.C., 2005. Male competition and aggregative behaviour are mediated by acoustic cues within a temporally unstructured aggregation. Behaviour 142, 71–90. https://doi.org/10.1163/1568539053627730.

Gurule-Small, G.A., Tinghitella, R.M., 2019. Life history consequences of developing in anthropogenic noise. Global Change Biol. 25, 1957–1966. https://doi.org/10.1111/gcb.14610.

Gwynne, D.T., 2004. Reproductive behavior of ground weta (Orthoptera: Anostostomatidae): drumming behavior, nuptial feeding, post-copulatory guarding and maternal care. J. Kans. Entomol. Soc. 77, 414–428. https://doi.org/10.2317/e-34.1.

Hall, M., Robinson, D., 2020. Mating in the speckled bush-cricket, *Leptophyes punctatissima*. Antenna 44, 173–177.

Hall, M.D., Bussiere, L.F., Hunt, J., Brooks, R., 2008. Experimental evidence that sexual conflict influences the opportunity, form and intensity of sexual selection. Evolution 62, 2305–2315. https://doi.org/10.1111/j.1558-5646.2008.00436.x.

Hammond, T.J., Bailey, W.J., 2003. Eavesdropping and defensive auditory masking in an Australian bushcricket, *Caedicia* (Phaneropterinae: Tettigoniidae: Orthoptera). Behaviour 140, 79–95.

Hare, R.M., Simmons, L.W., 2020. Ecological determinants of sex roles and female sexual selection. Adv. Stud. Behav. 52, 1–28. https://doi.org/10.1016/bs.asb.2019.11.001.

Hare, R.M., Simmons, L.W., 2021. Sexual selection maintains a female-specific character in a species with dynamic sex roles. Behav. Ecol. 32, 609–616. https://doi.org/10.1093/beheco/arab005.

Harness, N.C.E., Campbell, T., 2021. Acoustic ranging in meadow katydids: female preference for attenuated calls. Bioacoustics. https://doi.org/10.1080/09524622.2021.1879681.

Harrison, S.J., Thomson, I.R., Grant, C.M., Bertram, S.M., 2013. Calling, courtship, and condition in the fall field cricket, *Gryllus pennsylvanicus*. PLoS One 8, e60356. https://doi.org/10.1371/journal.pone.0060356.

Harrison, S.J., Godin, J.-G.J., Bertram, S.M., 2017. Influence of dietary nutrient balance on aggression and signalling in male field crickets. Anim. Behav. 134, 123–134. https://doi.org/10.1016/j.anbehav.2017.10.006.

Hartbauer, M., Römer, H., 2016. Rhythm generation and rhythm perception in insects: the evolution of synchronous choruses. Front. Neurosci. 10, 223. https://doi.org/10.3389/fnins.2016.00223.

Hartbauer, M., Siegert, M.E., Römer, H., 2015. Male age and female mate choice in a synchronizing katydid. J. Comp. Physiol. A 201, 763–772. https://doi.org/10.1007/s00359-015-1012-9.

Hartley, J.C., 1993. Acoustic behaviour and phonotaxis in the duetting ephippigerines, *Steropleurus nobrei* and *Steropleurus stali* (Tettigoniidae). Zool. J. Linn. Soc. 107, 155–167.

Head, M.L., Hunt, J., Jennions, M.D., Brooks, R., 2005. The indirect benefits of mating with attractive males outweigh the direct costs. PLoS Biol. 3, 289–294. https://doi.org/10.1371/journal.pbio.0030033.

Hebets, E.A., Papaj, D.R., 2005. Complex signal function: developing a framework of testable hypotheses. Behav. Ecol. Sociobiol. 57, 197–214. https://doi.org/10.1007/s00265-004-0865-7.

Hedrick, A., Bunting, J., 2014. An attractive male trait and aggressiveness are negatively correlated in wild field crickets, but uncorrelated in lab-reared crickets. Behav. Ecol. Sociobiol. 68, 233–238. https://doi.org/10.1007/s00265-013-1638-y.

Hedwig, B., 1986. On the role in stridulation of plurisegmental interneurons of the acridid grasshopper *Omocestus viridulus* L. 1. Anatomy and physiology of descending cephalothoracic interneurons. J. Comp. Physiol. A 158, 413–427. https://doi.org/10.1007/BF00603625.

Hedwig, B., 2000. Control of cricket stridulation by a command neuron: efficacy depends on the behavioral state. J. Neurophysiol. 83, 712–722.
Hedwig, B., 2001. Singing and hearing: neuronal mechanisms of acoustic communication in Orthopterans. Zool. Anal. Complex Syst. 103, 140–149.
Hedwig, B.G., 2016. Sequential filtering processes shape feature detection in crickets: a framework for song pattern recognition. Front. Physiol. 7, 46. https://doi.org/10.3389/fphys.2016.00046.
Heinen-Kay, J.L., Zuk, M., 2019. When does sexual signal exploitation lead to signal loss? Front. Ecol. Evol. 7, 255. https://doi.org/10.3389/fevo.2019.00255.
Heinen-Kay, J.L., Urquhart, E.M., Zuk, M., 2019. Obligately silent males sire more offspring than singers in a rapidly evolving cricket population. Biol. Lett. 15, 20190198. https://doi.org/10.1098/rsbl.2019.0198.
Heinen-Kay, J.L., Nichols, R.E., Zuk, M., 2020. Sexual signal loss, pleiotropy, and maintenance of a male reproductive polymorphism in crickets. Evolution 74, 1002–1009. https://doi.org/10.1111/evo.13952.
Heinrich, R., Elsner, N., 1997. Central nervous control of hindleg coordination in stridulating grasshoppers. J. Comp. Physiol., A 180, 257–269. https://doi.org/10.1007/s003590050046.
Heinrich, R., Kunst, M., Wirmer, A., 2012. Reproduction-related sound production of grasshoppers regulated by internal state and actual sensory environment. Front. Neurosci. 6, 89. https://doi.org/10.3389/fnins.2012.00089.
Heller, K.-G., 1984. Zur Bioacustik und Phylogenie der Gattung *Poecilimon* (Orthoptera: Tettigoniidae: Phaneropterinae). Zool. Jahrb. Syst. 111, 69–117.
Heller, K.-G., Hemp, C., 2017. Context specific signaling with different frequencies—directed to different receivers? A case study in *Gonatoxia* katydids (Orthoptera, Phaneropteridae). J. Insect Behav. 30, 420–431. https://doi.org/10.1007/s10905-017-9628-y.
Heller, K.-G., Hemp, C., 2020. Hyperdiverse songs, duetting, and the roles of intra- and intersexual selection in the acoustic communication of the genus *Eurycorypha* (Orthoptera: Tettigonioidea, Phaneropterinae). Org. Divers. Evol. 20, 597–617. https://doi.org/10.1007/s13127-020-00452-1.
Heller, K.G., Willemse, L., Odé, B., Volleth, M., Feist, R., Reinhold, K., 2011. Bioacoustics and systematics of the *Poecilimon hamatus* group (Tettigonioidea: Phaneropteridae: *Poecilimon: Hamatopoecilimon* n. subg.). J. Orthoptera Res. 20, 81–95. https://doi.org/10.1665/034.020.0108.
Heller, K.-G., Hemp, C., Ingrisch, S., Liu, C., 2015. Acoustic communication in Phaneropterinae (Tettigonioidea)—a global review with some new data. J. Orthoptera Res. 24, 7–18. https://doi.org/10.1665/034.024.0103.
Heller, K.-G., Korsunovskaya, O., Massa, B., Iorgu, I.S., 2018. High-speed duetting—latency times of the female acoustic response within the bush-cricket genera *Leptophyes* and *Andreiniimon* (Orthoptera, Phaneropteridae). Zookeys 750, 45–58. https://doi.org/10.3897/zookeys.750.23874.
Heller, K.-G., Volleth, M., Vedenina, V., Maryańska-Nadachowska, A., Warchałowska-Šliwa, E., 2020. A perfect duet? The acoustic behaviour of *Anaulacomera almadaenis* sp. nov., a species with an unusual chromosome complement, discovered in the footsteps of the explorers Spix and Martius in Brazil. Spixiana 43, 105–118.
Heller, K.G., Heller, M., Volleth, M., Samietz, J., Hemp, C., 2021. Similar songs, but different mate localization strategies of the three species of Phaneroptera occurring in Western Europe (Orthoptera: Phaneropteridae). Eur. J. Entomol. 118, 111–122. https://doi.org/10.14411/eje.2021.012.
Hershberger, W.L., 2021. Calling and courtship songs of the rare, robust ground cricket, *Allonemobius walkeri*. J. Orthoptera Res. 30, 81–85.

Hill, P.S.M., 2000. Elements of the acoustic repertoire of the prairie mole cricket (Orthoptera: Gryllotalpidae: *Gryllotalpa major* Suassure). J. Kans. Entomol. Soc. 73, 95–102.

Hill, P.S.M., Shadley, J.R., 1997. Substrate vibration as a component of a calling song. Naturwissenschaften 84, 460–463.

Hill, A.P., Prince, P., Snaddon, J.L., Doncaster, C.P., Rogers, A., 2019. AudioMoth: a low-cost acoustic device for monitoring biodiversity and the environment. Hardwarex 6, e00073. https://doi.org/10.1016/j.ohx.2019.e00073.

Hirtenlehner, S., Kueng, S., Kainz, F., Römer, H., 2013. Asymmetry in cricket song: female preference and proximate mechanism of discrimination. J. Exp. Biol. 216, 2046–2054. https://doi.org/10.1242/jeb.083774.

Holzer, B., Jacot, A., Brinkhof, M.W.G., 2003. Condition-dependent signaling affects male sexual attractiveness in field crickets, *Gryllus campestris*. Behav. Ecol. 14, 353–359.

Houslay, T.M., Houslay, K.F., Rapkin, J., Hunt, J., Bussiere, L.F., 2017. Mating opportunities and energetic constraints drive variation in age-dependent sexual signalling. Funct. Ecol. 31, 728–741. https://doi.org/10.1111/1365-2435.12766.

Howard, D.R., Hill, P.S.M., 2006. Morphology and calling song characteristics in *Gryllotalpa major* Saussure (Orthoptera: Gryllotalpidae). J. Orthoptera Res. 15, 53–57. https://doi.org/10.1665/1082-6467(2006)15[53:macsci]2.0.co;2.

Howard, D.R., Lee, N., Hall, C.L., Mason, A.C., 2011. Are centrally displaying males always the centre of female attention? Acoustic display position and female choice in a lek mating subterranean insect. Ethology 117, 199–207. https://doi.org/10.1111/j.1439-0310.2010.01858.x.

Howard, D.R., Schmidt, A.P., Hall, C.L., Mason, A.C., 2018. Substrate-borne vibration mediates intrasexual agonism in the New Zealand Cook Strait giant weta (*Deinacrida rugosa*). J. Insect Behav. 31, 599–615. https://doi.org/10.1007/s10905-018-9700-2.

Huber, F., 1960. Untersuchungen über die Funktion des Zentralnervensystems und insbesondere des Gehirnes bei der Fortbewegung und der Lauterzeugung der Grillen. Z. Vergl. Physiol. 44, 60–132.

Hunt, J., Brooks, R., Jennions, M.D., Smith, M.J., Bentsen, C.L., Bussiere, L.F., 2004. High-quality male field crickets invest heavily in sexual display but die young. Nature 432, 1024–1027. https://doi.org/10.1038/nature03084.

Imaizumi, K., Pollack, G.S., 1999. Neural coding of sound frequency by cricket auditory receptors. J. Neurosci. 19, 1508–1516.

Imaizumi, K., Pollack, G.S., 2001. Neural representation of sound amplitude by functionally different auditory receptors in crickets. J. Acoust. Soc. Am. 109, 1247–1260.

Imaizumi, K., Pollack, G.S., 2005. Central projections of auditory receptor neurons of crickets. J. Comp. Neurol. 493, 439–447. https://doi.org/10.1002/cne.20756.

Ingrisch, S., Rentz, D.C.F., 2009. Orthoptera: grasshoppers, locusts, katydids, crickets. In: Resh, V.H., Cardé, R.T. (Eds.), Encyclopedia of Insects. second ed., Academic Press, San Diego, pp. 732–743.

Ivkovic, S., Iorgu, I.S., Horvat, L., Chobanov, D., Korsunovskaya, O., Heller, K.-G., 2017. New data on the bush-cricket *Montana medvedevi* (Orthoptera: Tettigoniidae), critically endangered in Europe (EU 28), and a comparison of its song with all known song patterns within the genus. Zootaxa 4263, 527–542. https://doi.org/10.11646/zootaxa.4263.3.5.

Jacob, P.F., Hedwig, B., 2015. Impact of cercal air currents on singing motor pattern generation in the cricket (*Gryllus bimaculatus* DeGeer). J. Neurophysiol. 114, 2649–2660. https://doi.org/10.1152/jn.00669.2015.

Jacob, P.F., Hedwig, B., 2016. Acoustic signalling for mate attraction in crickets: abdominal ganglia control the timing of the calling song pattern. Behav. Brain Res. 309, 51–66. https://doi.org/10.1016/j.bbr.2016.04.025.

Jacob, P.F., Hedwig, B., 2019. Structure, activity and function of a singing CPG interneuron controlling cricket species-specific acoustic signaling. J. Neurosci. 39, 96–111. https://doi.org/10.1523/jneurosci.1109-18.2018.

Jacob, P.F., Hedwig, B., 2020. Modular timer networks: abdominal interneurons controlling the chirp and pulse pattern in a cricket calling song. J. Comp. Physiol. A 206, 921–938. https://doi.org/10.1007/s00359-020-01448-0.

Jacobs, K., Otte, B., Lakes-Harlan, R., 1999. Tympanal receptor cells of *Schistocerca gregaria*: correlation of soma positions and dendrite attachment sites, central projections and physiologies. J. Exp. Zool. 283, 270–285.

Jacot, A., Scheuber, H., Brinkhof, M.W.G., 2004. Costs of an induced immune response on sexual display and longevity in field crickets. Evolution 58, 2280–2286.

Jacot, A., Scheuber, H., Brinkhof, M.W.G., 2007. The effect of age on a sexually selected acoustic display. Ethology 113, 615–620. https://doi.org/10.1111/j.1439-0310.2007.01360.x.

Jain, M., Diwakar, S., Bahuleyan, J., Deb, R., Balakrishnan, R., 2014. A rain forest dusk chorus: cacophony or sounds of silence? Evol. Ecol. 28, 1–22. https://doi.org/10.1007/s10682-013-9658-7.

Jang, Y., 2011. Male responses to conspecific advertisement signals in the field cricket *Gryllus rubens* (Orthoptera: Gryllidae). PLoS One 6, e16063. https://doi.org/10.1371/journal.pone.0016063.

Jang, Y., Gerhardt, H.C., Choe, J.C., 2008. A comparative study of aggressiveness in eastern North American field cricket species (genus *Gryllus*). Behav. Ecol. Sociobiol. 62, 1397–1407. https://doi.org/10.1007/s00265-008-0568-6.

Jarrige, A., Greenfield, M.D., Goubault, M., 2013. Male song as a predictor of the nuptial gift in bushcrickets: on the confounding influence of male choice. Anim. Behav. 85, 1427–1434. https://doi.org/10.1016/j.anbehav.2013.03.039.

Jones, P.L., Page, R.A., Hartbauer, M., Siemers, B.M., 2011. Behavioral evidence for eavesdropping on prey song in two Palearctic sibling bat species. Behav. Ecol. Sociobiol. 65, 333–340. https://doi.org/10.1007/s00265-010-1050-9.

Jonsson, T., Montealegre-Z, F., Soulsbury, C.D., Brown, K.A.R., Robert, D., 2016. Auditory mechanics in a bush-cricket: direct evidence of dual sound inputs in the pressure difference receiver. J. R. Soc. Interface 13, 20160560. https://doi.org/10.1098/rsif.2016.0560.

Judge, K.A., 2011. Do male field crickets, *Gryllus pennsylvanicus*, signal their age? Anim. Behav. 81, 185–194. https://doi.org/10.1016/j.anbehav.2010.09.032.

Judge, K.A., Ting, J.J., Gwynne, D.T., 2008. Condition dependence of male life span and calling effort in a field cricket. Evolution 62, 868–878. https://doi.org/10.1111/j.1558-5646.2008.00318.x.

Judge, K.A., Ting, J.J., Gwynne, D.T., 2014. Condition dependence of female choosiness in a field cricket. J. Evol. Biol. 27, 2529–2540. https://doi.org/10.1111/jeb.12509.

Keane, K.T., Booth, W., Howard, D.R., Golden, T.M.J., Hill, P.S.M., 2018. Competition decreases with relatedness and lek size in mole crickets: a role for kin selection? Behav. Ecol. 29, 1067–1074. https://doi.org/10.1093/beheco/ary081.

Kelly, C.D., Gwynne, D.T., 2016. The effect of condition on mate searching speed and copulation frequency in the Cook Strait giant weta. Behav. Ecol. Sociobiol. 70, 1403–1409. https://doi.org/10.1007/s00265-016-2150-y.

Kilmer, M.K., Barrus, B.B., Schul, J., 2010. Ultrasound avoidance behaviors in two species of *Neoconocephalus* (Orthoptera, Tettigoniidae). J. Orthoptera Res. 19, 81–88. https://doi.org/10.1665/034.019.0114.

Klappert, K., Reinhold, K., 2007. Indirect benefits for choosy female grasshoppers (*Chorthippus biguttulus*)? Zool. Anal. Complex Syst. 110, 354–359. https://doi.org/10.1016/j.zool.2007.07.002.

Koch, U.T., Elliott, C.J.H., Schäffner, K.H., Kleindienst, H.U., 1988. The mechanics of stridulation of the cricket *Gryllus campestris*. J. Comp. Physiol. A 162, 213–223. https://doi.org/10.1007/BF00606086.

Korsunovskaya, O., 2008. Acoustic signals in katydids (Orthoptera, Tettigonidae). Communication I. Entomol. Rev. 88, 1032–1050. https://doi.org/10.1134/S0013873808090029.

Korsunovskaya, O., 2009. Acoustic signals in katydids (Orthoptera, Tettigonidae). Communication 2. Entomol. Rev. 89, 16–20. https://doi.org/10.1134/S0013873809010035.

Korsunovskaya, O.S., Zhantiev, R.D., 2007. Effect of temperature on auditory receptor functions in crickets (Orthoptera, Tettigoniodea). J. Evol. Biochem. Physiol. 43 (3), 327–334. https://doi.org/10.1134/s0022093007030076.

Korsunovskaya, O., Berezin, M., Heller, K.-G., Tkacheva, E., Kompantseva, T., Zhantiev, R., 2020. Biology, sounds and vibratory signals of hooded katydids (Orthoptera: Tettigoniidae: Phyllophorinae). Zootaxa 4852, 309–322. https://doi.org/10.11646/zootaxa.4852.3.3.

Kostarakos, K., Hedwig, B., 2012. Calling song recognition in female crickets: temporal tuning of identified brain neurons matches behavior. J. Neurosci. 32, 9601–9612. https://doi.org/10.1523/jneurosci.1170-12.2012.

Kostarakos, K., Römer, H., 2018. Listening in the bog: II. Neural correlates for acoustic interactions and spacing between *Sphagniana sphagnorum* males. J. Comp. Physiol. A 204, 353–367. https://doi.org/10.1007/s00359-018-1251-7.

Kostarakos, K., Hartbauer, M., Römer, H., 2008. Matched filters, mate choice and the evolution of sexually selected traits. PLoS One 3, e3005. https://doi.org/10.1371/journal.pone.0003005.

Kota, M.V., Urquhart, E.M., Zuk, M., 2020. Spermatophore retention may accommodate sexual signal loss in Pacific field crickets. Behav. Ecol. Sociobiol. 74, 95. https://doi.org/10.1007/s00265-020-02850-z.

Kowalski, K., Lakes-Harlan, R., 2011. Temporal patterns of intra- and interspecific acoustic signals differ in two closely related species of *Acanthoplus* (Orthoptera: Tettigoniidae: Hetrodinae). Zool. Anal. Complex Syst. 114, 29–35. https://doi.org/10.1016/j.zool.2010.09.002.

Kowalski, K., Lakes-Harlan, R., 2013. The acoustic communication system in six species of armoured ground crickets (Orthoptera: Tettigoniidae: Hetrodinae): commonalities and species-specific differences. Zool. Anz. 252, 204–216. https://doi.org/10.1016/j.jcz.2012.05.005.

Krobath, I., Römer, H., Hartbauer, M., 2017. Plasticity of signaling and mate choice in a trilling species of the *Mecopoda* complex (Orthoptera: Tettigoniidae). Behav. Ecol. Sociobiol. 71, 164. https://doi.org/10.1007/s00265-017-2381-6.

Lampe, U., Reinhold, K., Schmoll, T., 2014. How grasshoppers respond to road noise: developmental plasticity and population differentiation in acoustic signalling. Funct. Ecol. 28, 660–668. https://doi.org/10.1111/1365-2435.12215.

Larrosa, E., Garcia, M.-D., Clemente, M.E., Presa, J.J., 2008. Sound production in *Calliptamus barbarus* Costa 1836 (Orthoptera: Acrididae: Catantopinae). Ann. Soc. Entomol. Fr. 44, 129–138. https://doi.org/10.1080/00379271.2008.10697549.

Larsen, O.N., 1981. Mechanical time resolution in some insect ears. J. Comp. Physiol. A 143, 297–304. https://doi.org/10.1007/BF00611165.

Lee, N., Elias, D.O., Mason, A.C., 2009. A precedence effect resolves phantom sound source illusions in the parasitoid fly *Ormia ochracea*. Proc. Natl. Acad. Sci. U. S. A. 106, 6357–6362. https://doi.org/10.1073/pnas.0809886106.

Lehmann, G.U.C., 2003. Review of biogeography, host range and evolution of acoustic hunting in *Ormiini* (Insecta, Diptera, Tachinidae), parasitoids of night-calling bushcrickets and crickets (Insecta, Orthoptera, Ensifera). Zool. Anz. 242, 107–120. https://doi.org/10.1078/0044-5231-00091.

Lehmann, G.U.C., 2012. Weighing costs and benefits of mating in bushcrickets (Insecta: Orthoptera: Tettigoniidae), with an emphasis on nuptial gifts, protandry and mate density. Front. Zool. 9, 19. https://doi.org/10.1186/1742-9994-9-19.

Lehmann, G.U.C., Lakes-Harlan, R., 2019. Adaptive strategies in life-history of bushcrickets (Orthoptera) and cicadas (Homoptera) to parasitoids pressure on their acoustic communication systems—a case for sociality? Front. Ecol. Evol. 7, 295. https://doi.org/10.3389/fevo.2019.00295.

Lehmann, G., Lehmann, A., 2008. Bushcricket song as a clue for spermatophore size? Behav. Ecol. Sociobiol. 62, 569–578.

Leonard, A.S., Hedrick, A.V., 2009. Male and female crickets use different decision rules in response to mating signals. Behav. Ecol. 20, 1175–1184. https://doi.org/10.1093/beheco/arp115.

Leonard, A.S., Hedrick, A.V., 2010. Long-distance signals influence assessment of close range mating displays in the field cricket, *Gryllus integer*. Biol. J. Linn. Soc. 100, 856–865. https://doi.org/10.1111/j.1095-8312.2010.01472.x.

Levy, K., Wegrzyn, Y., Efronny, R., Barnea, A., Ayali, A., 2021. Lifelong exposure to artificial light at night impacts stridulation and locomotion activity patterns in the cricket *Gryllus bimaculatus*. Proc. R. Soc. Lond. Ser. B: Biol. Sci. 288, 20211626. https://doi.org/10.1098/rspb.2021.1626.

Lewkiewicz, D.A., Zuk, M., 2004. Latency to resume calling after disturbance in the field cricket, *Teleogryllus oceanicus*, corresponds to population-level differences in parasitism risk. Behav. Ecol. Sociobiol. 55, 569–573.

Lierheimer, V.F., Tinghitella, R.M., 2017. Quantity and quality of available mates alters female responsiveness but not investment in the Pacific field cricket, *Teleogryllus oceanicus*. Behav. Ecol. Sociobiol. 71, 80. https://doi.org/10.1007/s00265-017-2298-0.

Lin, C.-C., Hedwig, B., 2021a. Lesions of abdominal connectives reveal a conserved organization of the calling song central pattern generator (CPG) network in different cricket species. J. Comp. Physiol. A 207, 533–552. https://doi.org/10.1007/s00359-021-01495-1.

Lin, C.-C., Hedwig, B., 2021b. Wing movements underlying sound production in calling, rivalry, and courtship songs of the cricket *Gryllus bimaculatus* (DeGeer). J. Insect Physiol. 134, 104299. https://doi.org/10.1016/j.jinsphys.2021.104299.

Liu, C.-X., Heller, K.-G., Wang, X.-S., Yang, Z., Wu, C., Liu, F., Zhang, T., 2020. Taxonomy of a katydid genus *Mecopoda* Servile (Orthoptera: Tettigoniidae, Mecopodinae) from East Asia. Zootaxa 4758, 296–310. https://doi.org/10.11646/zootaxa.4758.2.5.

Logue, D.M., Abiola, I.O., Rains, D., Bailey, N.W., Zuk, M., Cade, W.H., 2010. Does signalling mitigate the cost of agonistic interactions? A test in a cricket that has lost its song. Proc. R. Soc. Lond. Ser. B: Biol. Sci. 277, 2571–2575. https://doi.org/10.1098/rspb.2010.0421.

Lopez, H., Garcia, M.D., Clemente, E., Presa, J.J., Oromi, P., 2008. Sound production mechanism in pamphagid grasshoppers (Orthoptera). J. Zool. 275, 1–8. https://doi.org/10.1111/j.1469-7998.2007.00394.x.

Loranger, M.J., Bertram, S.M., 2016. The effect of sire dominance and aggression on fitness measures in a field cricket (*Gryllus assimilis*). Anim. Behav. 119, 135–142. https://doi.org/10.1016/j.anbehav.2016.06.020.

Low, M.L., Naranjo, M., Yack, J.E., 2021. Survival sounds in insects: diversity, function, and evolution. Front. Ecol. Evol. 9, 641740. https://doi.org/10.3389/fevo.2021.641740.

Magal, C., Dangles, O., Caparroy, P., Casas, J., 2006. Hair canopy of cricket sensory system tuned to predator signals. J. Theor. Biol. 241, 459–466.

Makowicz, A.M., Daniel, M.J., Jones, B.C., Rivers, P.R., Dye, M., Kuzel, M.R., Guerrera, A.G., Kettelkamp, S., Whitcher, C., DuVal, E.H., 2020. Foundations and frontiers in mate choice. Review of: Rosenthal, G. 2017. Mate Choice: The Evolution of Sexual Decision Making from Microbes to Humans. Evolution 74, 1575–1583. https://doi.org/10.1111/evo.14018.

Marin-Cudraz, T., Greenfield, M.D., 2016. Finely tuned choruses: bush crickets adjust attention to neighboring singers in relation to the acoustic environment they create. Behav. Ecol. Sociobiol. 70, 1581–1589. https://doi.org/10.1007/s00265-016-2166-3.

Marsat, G., Pollack, G.S., 2005. Effect of the temporal pattern of contralateral inhibition on sound localization cues. J. Neurosci. 25, 6137–6144. https://doi.org/10.1523/jneurosci.0646-05.2005.

Marsat, G., Pollack, G., 2012. Bursting neurons and ultrasound avoidance in crickets. Front. Neurosci. 6, 95. https://doi.org/10.3389/fnins.2012.00095.

Marshall, D.C., Hill, K.B.R., 2009. Versatile aggressive mimicry of cicadas by an Australian predatory katydid. PLoS One 4, e4185. https://doi.org/10.1371/journal.pone.0004185.

Martin, C.M., Wagner Jr., W.E., 2010. Female field crickets incur increased parasitism risk when near preferred song. PLoS One 5, e9592. https://doi.org/10.1371/journal.pone.0009592.

Massa, B., 2012. The role of the Krauss's organ in sound production in Pamphagidae (Caelifera: Orthoptera). Ital. J. Zool. 79, 441–449. https://doi.org/10.1080/11250003.2012.667158.

Massa, B., 2020. Remarks on some interesting African Pamphagidae and Acrididae (Insecta: Orthoptera: Acridoidea). Zootaxa 4751, 29–54. https://doi.org/10.11646/zootaxa.4751.1.2.

Mautz, B.S., Sakaluk, S.K., 2008. The effects of age and previous mating experience on pre- and post-copulatory mate choice in female house crickets (*Acheta domesticus* L.). J. Insect Behav. 21, 203–212.

Maynard Smith, J., Harper, D., 2003. Animal Signals. Oxford University Press, Oxford.

Mays, D.L., 1971. Mating behavior of nemobiine crickets: *Hygronemobius*, *Nemobius*, and *Pteronemobius* (Orthoptera: Gryllidae). Fla. Entomol. 54, 113–126. https://doi.org/10.2307/3493557.

McAuley, E.M., Bertram, S.M., 2016. Field crickets compensate for unattractive static long-distance call components by increasing dynamic signalling effort. PLoS One 11, e0167311. https://doi.org/10.1371/journal.pone.0167311.

McCartney, J., Kokko, H., Heller, K.G., Gwynne, D.T., 2012. The evolution of sex differences in mate searching when females benefit: new theory and a comparative test. Proc. R. Soc. Lond. Ser. B: Biol. Sci. 279, 1225–1232. https://doi.org/10.1098/rspb.2011.1505.

Meckenhauser, G., Kramer, S., Farkhooi, F., Ronacher, B., Nawrot, M.P., 2014. Neural representation of calling songs and their behavioral relevance in the grasshopper auditory system. Front. Syst. Neurosci. 8, 183. https://doi.org/10.3389/fnsys.2014.00183.

Mhatre, N., Balakrishnan, R., 2006. Male spacing behaviour and acoustic interactions in a field cricket: implications for female mate choice. Anim. Behav. 72, 1045–1058.

Mhatre, N., Montealegre-Z, F., Balakrishnan, R., Robert, D., 2012. Changing resonator geometry to boost sound power decouples size and song frequency in a small insect. Proc. Natl. Acad. Sci. U. S. A. 109, E1444–E1452. https://doi.org/10.1073/pnas.1200192109.

Mhatre, N., Pollack, G., Mason, A., 2016. Stay tuned: active amplification tunes tree cricket ears to track temperature-dependent song frequency. Biol. Lett. 12, 20160016. https://doi.org/10.1098/rsbl.2016.0016.

Michelsen, A., Larsen, O.N., 2008. Pressure difference receiving ears. Bioinspir. Biomim. 3, 011001. https://doi.org/10.1088/1748-3182/3/1/011001.

Michelsen, A., Popov, A.V., Lewis, B., 1994. Physics of directional hearing in the cricket *Gryllus bimaculatus*. J. Comp. Physiol. A 175, 153–164.

Miller, C.W., Svensson, E.I., 2014. Sexual selection in complex environments. Annu. Rev. Entomol. 59, 427–445. https://doi.org/10.1146/annurev-ento-011613-162044.

Mitra, C., Wagner Jr., W.E., Zera, A.J., Tolle, A.E., 2011. Variation in singing behaviour among morphs of the sand field cricket, *Gryllus firmus*. Ecol. Entomol. 36, 152–160. https://doi.org/10.1111/j.1365-2311.2010.01254.x.

Montealegre-Z, F., 2012. Reverse stridulatory wing motion produces highly resonant calls in a neotropical katydid (Orthoptera: Tettigoniidae: Pseudophyllinae). J. Insect Physiol. 58, 116–124. https://doi.org/10.1016/j.jinsphys.2011.10.006.

Montealegre-Z, F., Mason, A.C., 2005. The mechanics of sound production in *Panacanthus pallicornis* (Orthoptera: Tettigoniidae: Conocephalinae): the stridulatory motor patterns. J. Exp. Biol. 208, 1219–1237. https://doi.org/10.1242/jeb.01526.

Montealegre-Z, F., Morris, G.K., 2004. The spiny devil katydids, *Panacanthus* Walker (Orthoptera: Tettigoniidae): an evolutionary study of acoustic behaviour and morphological traits. Syst. Entomol. 29, 21–57. https://doi.org/10.1111/j.1365-3113.2004.00223.x.

Montealegre-Z, F., Postles, M., 2010. Resonant sound production in *Copiphora gorgonensis* (Tettigoniidae: Copiphorini), an endemic species from Parque Nacional Natural Gorgona. Colombia. J. Orthoptera Res. 19, 347–355. https://doi.org/10.1665/034.019.0223.

Montealegre-Z, F., Morris, G.K., Mason, A.C., 2006. Generation of extreme ultrasonics in rainforest katydids. J. Exp. Biol. 209, 4923–4937. https://doi.org/10.1242/jeb.02608.

Montealegre-Z, F., Windmill, J.F.C., Morris, G.K., Robert, D., 2009. Mechanical phase shifters for coherent acoustic radiation in the stridulating wings of crickets: the plectrum mechanism. J. Exp. Biol. 212, 257–269. https://doi.org/10.1242/jeb.022731.

Montealegre-Z, F., Jonsson, T., Robert, D., 2011a. Sound radiation and wing mechanics in stridulating field crickets (Orthoptera: Gryllidae). J. Exp. Biol. 214, 2105–2117. https://doi.org/10.1242/jeb.056283.

Montealegre-Z, F., Morris, G.K., Sarria-S, F.A., Mason, A.C., 2011b. Quality calls: phylogeny and biogeography of a new genus of neotropical katydid (Orthoptera: Tettigoniidae) with ultra pure-tone ultrasonics. Syst. Biodivers. 9, 77–94. https://doi.org/10.1080/14772000.2011.560209.

Montealegre-Z, F., Ogden, J., Jonsson, T., Soulsbury, C.D., 2017. Morphological determinants of signal carrier frequency in katydids (Orthoptera): a comparative analysis using biophysical evidence of wing vibration. J. Evol. Biol. 30, 2068–2078. https://doi.org/10.1111/jeb.13179.

Montroy, K., Loranger, M.J., Bertram, S.M., 2016. Male crickets adjust their aggressive behavior when a female is present. Behav. Processes 124, 108–114. https://doi.org/10.1016/j.beproc.2015.11.003.

Moradian, N.R., Walker, S.E., 2008. Relationships between body size and sound-producing structures in crickets: do large males have large harps? Invertebr. Biol. 127, 444–451. https://doi.org/10.1111/j.1744-7410.2008.00142.x.

Moran, P.A., Hunt, J., Mitchell, C., Ritchie, M.G., Bailey, N.W., 2019. Behavioural mechanisms of sexual isolation involving multiple modalities and their inheritance. J. Evol. Biol. 32, 243–258. https://doi.org/10.1111/jeb.13408.

Morris, G.K., 1980. Calling display and mating behaviour of *Copiphora rhinoceros* Pictet (Orthoptera: Tettigoniidae). Anim. Behav. 28, 42–51.

Morris, G.K., DeLuca, P.A., Norton, M., Mason, A.C., 2002. Calling-song function in male haglids (Orthoptera: Haglidae, *Cyphoderris*). Can. J. Zool. 80, 271–285. https://doi.org/10.1139/z02-003.

Morris, G.K., Hall, A.M., Römer, H., 2018. Listening in the bog: I. Acoustic interactions and spacing between males of *Sphagniana sphagnorum*. J. Comp. Physiol. A 204, 339–351. https://doi.org/10.1007/s00359-018-1250-8.

Mowles, S.L., 2014. The physiological cost of courtship: field cricket song results in anaerobic metabolism. Anim. Behav. 89, 39–43. https://doi.org/10.1016/j.anbehav.2013.12.014.

Mullen, S.P., Mendelson, T.C., Schal, C., Shaw, K.L., 2007. Rapid evolution of cuticular hydrocarbons in a species radiation of acoustically diverse Hawaiian crickets (Gryllidae: Trigonidiinae: *Laupala*). Evolution 61, 223–231. https://doi.org/10.1111/j.1558-5646.2007.00019.x.

Mulloney, B., Smarandache, C., 2010. Fifty years of CPGs: two neuroethological papers that shaped the course of neuroscience. Front. Behav. Neurosci. 4, 45. https://doi.org/10.3389/fnbeh.2010.00045.

Murphy, M.A., Schul, J., 2017. Does leadership indicate male quality in *Neoconocephalus* katydids? Behav. Ecol. Sociobiol. 71, 22. https://doi.org/10.1007/s00265-016-2253-5.

Murphy, M.A., Thompson, N.L., Schul, J., 2016. Keeping up with the neighbor: a novel mechanism of call synchrony in *Neoconocephalus ensiger* katydids. J. Comp. Physiol. A 202, 225–234. https://doi.org/10.1007/s00359-016-1068-1.

Murphy, M.A., Gerhardt, H.C., Schul, J., 2017. Leader preference in *Neoconocephalus ensiger* katydids: a female preference for a nonheritable male trait. J. Evol. Biol. 30, 2222–2229. https://doi.org/10.1111/jeb.13188.

Nandi, D., Balakrishnan, R., 2013. Call intensity is a repeatable and dominant acoustic feature determining male call attractiveness in a field cricket. Anim. Behav. 86, 1003–1012. https://doi.org/10.1016/j.anbehav.2013.09.003.

Nandi, D., Balakrishnan, R., 2016. Spatio-temporal dynamics of field cricket calling behaviour: implications for female mate search and mate choice. PLoS One 11, e0165807. https://doi.org/10.1371/journal.pone.0165807.

Narvaez, A., Robillard, T., 2012. The reproductive behaviour of the cricket *Lebinthus santoensis* Robillard, 2009 (Grylloidea, Eneopterinae, Lebinthini). Zoosystema 34, 279–286. https://doi.org/10.5252/z2012n2a6.

Nityananda, V., Balakrishnan, R., 2007. Synchrony during acoustic interactions in the bushcricket *Mecopoda* 'Chirper' (Tettigoniidae: Orthoptera) is generated by a combination of chirp-by-chirp resetting and change in intrinsic chirp rate. J. Comp. Physiol. A 193, 51–65. https://doi.org/10.1007/s00359-006-0170-1.

Nityananda, V., Balakrishnan, R., 2008. Leaders and followers in katydid choruses in the field: call intensity, spacing and consistency. Anim. Behav. 76, 723–735. https://doi.org/10.1016/j.anbehav.2008.04.015.

Nityananda, V., Balakrishnan, R., 2009. Modeling the role of competition and cooperation in the evolution of katydid acoustic synchrony. Behav. Ecol. 20, 484–489.

Nityananda, V., Balakrishnan, R., 2021. Synchrony of complex signals in an acoustically communicating katydid. J. Exp. Biol. 224, jeb241877. https://doi.org/10.1242/jeb.241877.

Noguera, J.C., 2019. Crickets increase sexual signalling and sperm protection but live shorter in the presence of rivals. J. Evol. Biol. 32, 49–57. https://doi.org/10.1111/jeb.13390.

Olzer, R., Zuk, M., 2018. Obligate, but not facultative, satellite males prefer the same male sexual signal characteristics as females. Anim. Behav. 144, 37–43. https://doi.org/10.1016/j.anbehav.2018.07.014.

Orci, K.M., 2007. Female preferences for male song characters in the bush-cricket *Isophya camptoxypha* (Orthoptera, Tettigonioidea). J. Insect Behav. 20, 503–513. https://doi.org/10.1007/s10905-007-9096-x.

Orci, K.M., Iorgu, I.S., 2017. Subspecies-specific song preferences and diverged heterospecific discrimination in females of the bush-cricket *Isophya kraussii* (Orthoptera: Phaneropterinae). Biol. J. Linn. Soc. 122, 505–513.

Ostrowski, T.D., Sradnick, J., Stumpner, A., Elsner, N., 2009. The elaborate courtship behavior of *Stenobothrus clavatus* Willemse, 1979 (Acrididae: Gomphocerinae). J. Orthoptera Res. 18, 171–182. https://doi.org/10.1665/034.018.0206.

Ower, G.D., Judge, K.A., Steiger, S., Caron, K.J., Smith, R.A., Hunt, J., Sakaluk, S.K., 2013. Multivariate sexual selection on male song structure in wild populations of sagebrush crickets, *Cyphoderris strepitans* (Orthoptera: Haglidae). Ecol. Evol. 3, 3590–3603. https://doi.org/10.1002/ece3.736.

Ower, G.D., Hunt, J., Sakaluk, S.K., 2017. Multivariate sexual selection on male tegmina in wild populations of sagebrush crickets, *Cyphoderris strepitans* (Orthoptera: Haglidae). J. Evol. Biol. 30, 338–351. https://doi.org/10.1111/jeb.13008.

Pacheco, K., Dawson, J.W., Jutting, M., Bertram, S.M., 2013. How age influences phonotaxis in virgin female Jamaican field crickets (*Gryllus assimilis*). PeerJ 1, e130. https://doi.org/10.7717/peerj.130.

Party, V., Brunel-Pons, O., Greenfield, M.D., 2014. Priority of precedence: receiver psychology, female preference for leading calls and sexual selection in insect choruses. Anim. Behav. 87, 175–185. https://doi.org/10.1016/j.anbehav.2013.10.029.

Pascoal, S., Cezard, T., Eik-Nes, A., Gharbi, K., Majewska, J., Payne, E., Ritchie, M.G., Zuk, M., Bailey, N.W., 2014. Rapid convergent evolution in wild crickets. Curr. Biol. 24, 1369–1374. https://doi.org/10.1016/j.cub.2014.04.053.

Pascoal, S., Risse, J.E., Zhang, X., Blaxter, M., Cezard, T., Challis, R.J., Gharbi, K., Hunt, J., Kumar, S., Langan, E., Liu, X., Rayner, J.G., Ritchie, M.G., Snoek, B.L., Trivedi, U., Bailey, N.W., 2020. Field cricket genome reveals the footprint of recent, abrupt adaptation in the wild. Evol. Lett. 4, 19–33. https://doi.org/10.1002/evl3.148.

Pfau, H.K., Pfau, B., 2015. On the bioacoustics of the Iberian Ephippigerini (Orthoptera, Tettigoniidae, Bradyporinae). Mitt. Schweiz. Entomol. Ges. 88, 105–154.

Phillips, J.N., Ruef, S.K., Garvin, C.M., Le, M.-L.T., Francis, C.D., 2019. Background noise disrupts host–parasitoid interactions. R. Soc. Open Sci. 6, 190867. https://doi.org/10.1098/rsos.190867.

Pollack, G.S., Hedwig, B., 2017. The cricket auditory pathway: neural processing of acoustic signals. In: Horch, H.W., Mito, T., Popadić, A., Ohuchi, H., Noji, S. (Eds.), The Cricket as a Model Organism: Development. Regeneration, and Behavior, Springer Japan, Tokyo, pp. 155–167.

Pollack, G.S., Kim, J.S., 2013. Selective phonotaxis to high sound-pulse rate in the cricket *Gryllus assimilis*. J. Comp. Physiol. A 199, 285–293. https://doi.org/10.1007/s00359-013-0792-z.

Prakash, H., Greif, S., Yovel, Y., Balakrishnan, R., 2021. Acoustically eavesdropping bat predators take longer to capture katydid prey signalling in aggregation. J. Exp. Biol. 224, jeb233262. https://doi.org/10.1242/jeb.233262.

Prestwich, K.N., 1994. The energetics of acoustic signaling in anurans and insects. Am. Zool. 34, 625–643.

Prestwich, K.N., O'Sullivan, K., 2005. Simultaneous measurement of metabolic and acoustic power and the efficiency of sound production in two mole cricket species (Orthoptera: Gryllotalpidae). J. Exp. Biol. 208, 1495–1512. https://doi.org/10.1242/jeb.01550.

Ragge, D.R., Reynolds, W.J., 1998. The Songs of the Grasshoppers and Crickets of Western Europe. Harley Books, Colchester.

Raghuram, H., Deb, R., Nandi, D., Balakrishnan, R., 2015. Silent katydid females are at higher risk of bat predation than acoustically signalling katydid males. Proc. R. Soc. Lond. Ser. B Biol. Sci. 282, 20142319. https://doi.org/10.1098/rspb.2014.2319.

Rajaraman, K., Godthi, V., Pratap, R., Balakrishnan, R., 2015. A novel acoustic–vibratory multimodal duet. J. Exp. Biol. 218, 3042–3050. https://doi.org/10.1242/jeb.122911.

Rajaraman, K., Nair, A., Dey, A., Balakrishnan, R., 2018. Response mode choice in a multimodally duetting paleotropical pseudophylline bushcricket. Front. Ecol. Evol. 6, 172. https://doi.org/10.3389/fevo.2018.00172.

Rantala, M.J., Kortet, R., 2003. Courtship song and immune function in the field cricket *Gryllus bimaculatus*. Biol. J. Linn. Soc. 79, 503–510.

Rayner, J.G., Aldridge, S., Montealegre-Z, F., Bailey, N.W., 2019. A silent orchestra: convergent song loss in Hawaiian crickets is repeated, morphologically varied, and widespread. Ecology 100, e02694. https://doi.org/10.1002/ecy.2694.

Rayner, J.G., Schneider, W.T., Bailey, N.W., 2020. Can behaviour impede evolution? Persistence of singing effort after morphological song loss in crickets. Biol. Lett. 16, 20190931. https://doi.org/10.1098/rsbl.2019.0931.

Rebar, D., Bailey, N.W., Zuk, M., 2009. Courtship song's role during female mate choice in the field cricket *Teleogryllus oceanicus*. Behav. Ecol. 20, 1307–1314. https://doi.org/10.1093/beheco/arp143.

Rebar, D., Barbosa, F., Greenfield, M.D., 2016. Acoustic experience influences male and female pre- and postcopulatory behaviors in a bushcricket. Behav. Ecol. 27, 434–443. https://doi.org/10.1093/beheco/arv171.

Rebar, D., Barbosa, F., Greenfield, M.D., 2019. Female reproductive plasticity to the social environment and its impact on male reproductive success. Behav. Ecol. Sociobiol. 73, 48. https://doi.org/10.1007/s00265-019-2661-4.

Reichert, M.S., Finck, J., Ronacher, B., 2017. Exploring the hidden landscape of female preferences for complex signals. Evolution 71, 1009–1024. https://doi.org/10.1111/evo.13202.

Reinhold, K., 1999. Paternal investment in *Poecilimon veluchianus* bushcrickets: beneficial effects of nuptial feeding on offspring viability. Behav. Ecol. Sociobiol. 45, 293–299.

Reinhold, K., Jacoby, K.J., 2002. Dissecting the repeatability of female choice in the grasshopper *Chorthippus biguttulus*. Anim. Behav. 64, 245–250. https://doi.org/10.1006/anbe.2002.3061.

Rheinlaender, J., Shen, J.-X., Römer, H., 2006. Auditory lateralization in bushcrickets: a new dichotic paradigm. J. Comp. Physiol. A 192, 389–397. https://doi.org/10.1007/s00359-005-0078-1.

Riede, K., 1987. A comparative study of mating behaviour in some neotropical grasshoppers (Acridoidea). Ethology 76, 265–296.

Rillich, J., Buhl, E., Schildberger, K., Stevenson, P.A., 2009. Female crickets are driven to fight by the male courting and calling songs. Anim. Behav. 77, 737–742.

Robillard, T., Montealegre-Z, F., Desutter-Grandcolas, L., Grandcolas, P., Robert, D., 2013. Mechanisms of high-frequency song generation in brachypterous crickets and the role of ghost frequencies. J. Exp. Biol. 216, 2001–2011. https://doi.org/10.1242/jeb.083964.

Robinson, D., 1990. Acoustic communication between the sexes in bushcrickets. In: Bailey, W.J., Rentz, D.C.F. (Eds.), The Tettigoniidae: Biology. Systematics and Evolution, Crawford House Press, Bathurst, Australia, pp. 112–129.

Robinson, D.J., Hall, M.J., 2002. Sound signalling in Orthoptera. Adv. Insect Physiol. 29, 151–278. https://doi.org/10.1016/S0065-2806(02)29003-7.

Rodriguez, R.L., Rebar, D., Fowler-Finn, K.D., 2013. The evolution and evolutionary consequences of social plasticity in mate preferences. Anim. Behav. 85, 1041–1047. https://doi.org/10.1016/j.anbehav.2013.01.006.

Rodriguez-Munoz, R., Hopwood, P., Fisher, D., Skicko, I., Tucker, R., Woodcock, K., Slate, J., Walling, C., Tregenza, T., 2019. Older males attract more females but get fewer matings in a wild field cricket. Anim. Behav. 153, 1–14. https://doi.org/10.1016/j.anbehav.2019.04.011.

Roemschied, F.A., Eberhard, M.J.B., Schleimer, J.-H., Ronacher, B., Schreiber, S., 2014. Cell-intrinsic mechanisms of temperature compensation in a grasshopper sensory receptor neuron. eLife 3, e02078. https://doi.org/10.7554/eLife.02078.

Römer, H., Lang, A., Hartbauer, M., 2010. The signaller's dilemma: a cost–benefit analysis of public and private communication. PLoS One 5, e13325. https://doi.org/10.1371/journal.pone.0013325.

Römer, H., Lang, A., Pfeiffer, M., Hartbauer, M., 2011. A cost–benefit analysis of public and private communication. Commun. Integr. Biol. 4, 106–108.

Ronacher, B., 1989. Stridulation of acridid grasshoppers after hemisection of thoracic ganglia: evidence for hemiganglionic oscillators. J. Comp. Physiol. A 164, 723–736.

Ronacher, B., 2019. Innate releasing mechanisms and fixed action patterns: basic ethological concepts as drivers for neuroethological studies on acoustic communication in Orthoptera. J. Comp. Physiol. A 205, 33–50. https://doi.org/10.1007/s00359-018-01311-3.

Ronacher, B., Stange, N., 2013. Processing of acoustic signals in grasshoppers—a neuroethological approach towards female choice. J. Physiol. (Paris) 107, 41–50. https://doi.org/10.1016/j.jphysparis.2012.05.005.

Rosenthal, G.G., 2017. Mate Choice: The Evolution of Sexual Decision Making From Microbes to Humans. Princeton University Press, Princeton, NJ.

Rössler, W., Jatho, M., Kalmring, K., 2006. The auditory–vibratory sensory system in bushcrickets. In: Drosopoulos, S., Claridge, M.F. (Eds.), Insect Sounds and Communication: Physiology, Behaviour, Ecology and Evolution. CRC Press/Taylor & Francis Group, Boca Raton, FL, USA, pp. 35–69.

Rotenberry, J.T., Zuk, M., 2016. Alternative reproductive tactics in context: how demography, ecology, and behavior affect male mating success. Am. Nat. 188, 582–588. https://doi.org/10.1086/688507.

Rotenberry, J.T., Swanger, E., Zuk, M., 2015. Alternative reproductive tactics arising from a continuous behavioral trait: callers versus satellites in field crickets. Am. Nat. 185, 469–490. https://doi.org/10.1086/680219.

Ryan, K.M., Sakaluk, S.K., 2009. Dulling the senses: the role of the antennae in mate recognition, copulation and mate guarding in decorated crickets. Anim. Behav. 77, 1345–1350. https://doi.org/10.1016/j.anbehav.2009.02.011.

Ryder, J.J., Siva-Jothy, M.T., 2000. Male calling song provides a reliable signal of immune function in a cricket. Proc. R. Soc. Lond. Ser. B Biol. Sci. 267, 1171–1175.

Safi, K., Heinzle, J., Reinhold, K., 2006. Species recognition influences female mate preferences in the common european grasshopper (*Chorthippus biguttulus* Linnaeus, 1758). Ethology 112, 1225–1230. https://doi.org/10.1111/j.1439-0310.2006.01282.x.

Saidy, J., Gray, D.A., 2004. Observations of the acoustic behavior of *Hoplosphyrum boreale* (Scudder): a common scaly cricket of southern California. Bull. Southern California Acad. Sci. 103, 34–43.

Sakaguchi, K.M., Gray, D.A., 2011. Host song selection by an acoustically orienting parasitoid fly exploiting a multispecies assemblage of cricket hosts. Anim. Behav. 81, 851–858. https://doi.org/10.1016/j.anbehav.2011.01.024.

Samietz, J., Schumacher, J., Reinhardt, K., 2014. Comparison of the mating behaviour of a bush cricket in the laboratory and the field: calling activity and mating frequency of a long-winged species, *Phaneroptera falcata* (Ensifera: Tettigoniidae). Eur. J. Entomol. 111, 189–197. https://doi.org/10.14411/eje.2014.019.

Samuel, L., Stumpner, A., Atkins, G., Stout, J., 2013. Processing of model calling songs by the prothoracic AN2 neurone and phonotaxis are significantly correlated in individual female *Gryllus bimaculatus*. Physiol. Entomol. 38, 344–354. https://doi.org/10.1111/phen.12040.

Santori, C., Bussière, L.F., Houslay, T.M., 2020. Heightened perception of competition hastens courtship. Behav. Ecol. 31, 239–246. https://doi.org/10.1093/beheco/arz183.

Sarmiento-Ponce, E.J., Rogers, S., Hedwig, B., 2021. Does the choosiness of female crickets change as they age? J. Exp. Biol. 224, jeb241802. https://doi.org/10.1242/jeb.241802.

Sarria-S, F.A., Morris, G.K., Windmill, J.F.C., Jackson, J., Montealegre, F., 2014. Shrinking wings for ultrasonic pitch production: hyperintense ultra-short-wavelength calls in a new genus of neotropical katydids (Orthoptera: Tettigoniidae). PLoS One 9, e98708. https://doi.org/10.1371/journal.pone.0098708.

Sarria-S, F.A., Buxton, K., Jonsson, T., Montealegre, Z.F., 2016. Wing mechanics, vibrational and acoustic communication in a new bush-cricket species of the genus *Copiphora* (Orthoptera: Tettigoniidae) from Colombia. Zool. Anz. 263, 55–65. https://doi.org/10.1016/j.jcz.2016.04.008.

Sathyan, R., Couldridge, V., 2021. The effect of anthropogenic noise and weather conditions on male calls in the bladder grasshopper *Bullacris unicolor*. Bioacoustics 30, 110–123. https://doi.org/10.1080/09524622.2019.1703818.

Scherberich, J., Hummel, J., Schöneich, S., Nowotny, M., 2016. Auditory fovea in the ear of a duetting katydid nows male-specific adaptation to the female call. Curr. Biol. 26 (23), R1222–R1223. https://doi.org/10.1016/j.cub.2016.10.035.

Scherberich, J., Hummel, J., Schöneich, S., Nowotny, M., 2017. Functional basis of the sexual dimorphism in the auditory fovea of the duetting bushcricket *Ancylecha fenestrata*. Proc. R. Soc. Lond. Ser. B: Biol. Sci. 284, 20171426. https://doi.org/10.1098/rspb.2017.1426.

Scheuber, H., Jacot, A., Brinkhof, M.W.G., 2004. Female preference for multiple condition-dependent components of a sexually selected signal. Proc. R. Soc. Lond. Ser. B Biol. Sci. 271, 2453–2457. https://doi.org/10.1098/rspb.2004.2907.

Schmidt, A.K.D., Römer, H., 2013. Diversity of acoustic tracheal system and its role for directional hearing in crickets. Front. Zool. 10, 61. https://doi.org/10.1186/1742-9994-10-61.

Schmidt, A., Ronacher, B., Hennig, R.M., 2008. The role of frequency, phase and time for processing of amplitude modulated signals by grasshoppers. J. Comp. Physiol. A 194, 221–233. https://doi.org/10.1007/s00359-007-0295-x.

Schmidt, A.K.D., Römer, H., Riede, K., 2013. Spectral niche segregation and community organization in a tropical cricket assemblage. Behav. Ecol. 24, 470–480. https://doi.org/10.1093/beheco/ars187.

Schmidt, R., Morrison, A., Kunc, H.P., 2014. Sexy voices—no choices: male song in noise fails to attract females. Anim. Behav. 94, 55–59. https://doi.org/10.1016/j.anbehav.2014.05.018.

Schmidt, A.K.D., Riede, K., Römer, H., 2016. No phenotypic signature of acoustic competition in songs of a tropical cricket assemblage. Behav. Ecol. 27, 211–218. https://doi.org/10.1093/beheco/arv141.

Scholtz, C., Bazelet, C.S., de Klerk, H., 2018. Gryllacrididae (Orthoptera: Ensifera) in southern Africa. J. Orthoptera Res. 27, 183–186. https://doi.org/10.3897/jor.27.29645.

Schöneich, S., 2020. Neuroethology of acoustic communication in field crickets—from signal generation to song recognition in an insect brain. Prog. Neurobiol. 194, 101882. https://doi.org/10.1016/j.pneurobio.2020.101882.

Schöneich, S., Hedwig, B., 2010. Hyperacute directional hearing and phonotactic steering in the cricket (*Gryllus bimaculatus* deGeer). PLoS One 5, e15141. https://doi.org/10.1371/journal.pone.0015141.

Schöneich, S., Hedwig, B., 2011. Neural basis of singing in crickets: central pattern generation in abdominal ganglia. Naturwissenschaften 98, 1069–1073. https://doi.org/10.1007/s00114-011-0857-1.

Schöneich, S., Hedwig, B., 2012. Cellular basis for singing motor pattern generation in the field cricket (*Gryllus bimaculatus* DeGeer). Brain Behav. 2, 707–725. https://doi.org/10.1002/brb3.89.

Schöneich, S., Hedwig, B., 2017. Neurons and networks underlying singing behaviour. In: Horch, H.W., Mito, T., Popadić, A., Ohuchi, H., Noji, S. (Eds.), The Cricket as a

Model Organism: Development, Regeneration, and Behavior. Springer Japan, Tokyo, pp. 141–153.
Schöneich, S., Kostarakos, K., Hedwig, B., 2015. An auditory feature detection circuit for sound pattern recognition. Sci. Adv. 1, e1500325. https://doi.org/10.1126/sciadv.1500325.
Senter, P., 2008. Voices of the past: a review of Paleozoic and Mesozoic animal sounds. Hist. Biol. 20, 255–287. https://doi.org/10.1080/08912960903033327.
Shestakov, L.S., Vedenina, V.Y., 2015. Broad selectivity for courtship song in the cricket *Gryllus bimaculatus*. Ethology 121, 966–976. https://doi.org/10.1111/eth.12409.
Simmons, L.W., Bailey, W.J., 1993. Agonistic communication between males of a zaprochiline katydid (Orthoptera: Tettigoniidae). Behav. Ecol. 4, 364–368.
Simmons, L.W., Lovegrove, M., 2017. Socially cued seminal fluid gene expression mediates responses in ejaculate quality to sperm competition risk. Proc. R. Soc. Lond. Ser. B: Biol. Sci. (284), 20171486. https://doi.org/10.1098/rspb.2017.1486.
Simmons, L.W., Ritchie, M.G., 1996. Symmetry in the songs of crickets. Proc. R. Soc. Lond. Ser. B Biol. Sci. 263, 1305–1311.
Simmons, L.W., Zuk, M., Rotenberry, J.T., 2005. Immune function reflected in calling song characteristics in a natural population of the cricket *Teleogryllus commodus*. Anim. Behav. 69, 1235–1241. https://doi.org/10.1016/j.anbehav.2004.09.011.
Simmons, L.W., Thomas, M.L., Simmons, F.W., Zuk, M., 2013. Female preferences for acoustic and olfactory signals during courtship: male crickets send multiple messages. Behav. Ecol. 24, 1099–1107. https://doi.org/10.1093/beheco/art036.
Singh, R., Jain, M., 2020. Variation in call types, calling activity patterns and relationship between call frequency and body size in a field cricket, *Acanthogryllus asiaticus*. Bioacoustics 30, 284–302. https://doi.org/10.1080/09524622.2020.1720817.
Singh, R., Prathibha, P., Jain, M., 2020. Effect of temperature on life-history traits and mating calls of a field cricket, *Acanthogryllus asiaticus*. J. Therm. Biol. 93, 102740. https://doi.org/10.1016/j.jtherbio.2020.102740.
Song, H., 2018. Biodiversity of Orthoptera. In: Foottit, R.G., Adler, P.H. (Eds.), Insect Biodiversity: Science and Society. Wiley-Blackwell, Chichester, pp. 245–279.
Song, H., Bethoux, O., Shin, S., Donath, A., Letsch, H., Liu, S., McKenna, D.D., Meng, G., Misof, B., Podsiadlowski, L., Zhou, X., Wipfler, B., Simon, S., 2020. Phylogenomic analysis sheds light on the evolutionary pathways towards acoustic communication in Orthoptera. Nat. Commun. 11, 4939. https://doi.org/10.1038/s41467-020-18739-4.
Spooner, J.D., 1964. Texas bush katydid—its sounds and their significance. Anim. Behav. 12, 235–244. https://doi.org/10.1016/0003-3472(64)90007-7.
Spooner, J.D., 1968. Pair-forming acoustic systems of phaneropterine katydids (Orthoptera Tettigoniidae). Anim. Behav. 16, 197–212.
Stiedl, O., Stumpner, A., Mbungu, D.N., Atkins, G., Stout, J.F., 1997. Morphology and physiology of local auditory interneurons in the prothoracic ganglion of the cricket *Acheta domesticus*. J. Exp. Zool. 279, 43–53.
Stoffer, B., Walker, S.E., 2012. The use of multimodal communication in mate choice decisions by female house crickets, *Acheta domesticus*. Anim. Behav. 83, 1131–1138. https://doi.org/10.1016/j.anbehav.2012.02.006.
Stölting, H., Stumpner, A., 1998. Tonotopic organization of auditory receptors of the bushcricket *Pholidoptera griseoaptera* (Tettigoniidae, Decticinae). Cell Tissue Res. 294, 377–386.
Stout, J., Navia, B., Jeffery, J., Samuel, L., Hartwig, L., Butlin, A., Chung, M., Wilson, J., Dashner, E., Atkins, G., 2010. Plasticity of the phonotactic selectiveness of four species of chirping crickets (Gryllidae): implications for call recognition. Physiol. Entomol. 35, 99–116. https://doi.org/10.1111/j.1365-3032.2009.00713.x.

Stout, J., Stumpner, A., Jeffery, J., Samuel, L., Atkins, G., 2011. Response properties of the prothoracic AN2 auditory interneurone to model calling songs in the cricket *Gryllus bimaculatus*. Physiol. Entomol. 36, 343–359. https://doi.org/10.1111/j.1365-3032.2011.00803.x.

Strauß, J., 2019. What determines the number of auditory sensilla in the tympanal hearing organs of Tettigoniidae? Perspectives from comparative neuroanatomy and evolutionary forces. J. Orthoptera Res. 28, 205–219.

Strauß, J., Lehmann, G.U.C., Lehmann, A.W., Lakes-Harlan, R., 2012. Spatial organization of tettigoniid auditory receptors: insights from neuronal tracing. J. Morphol. 273, 1280–1290. https://doi.org/10.1002/jmor.20058.

Strauß, J., Lomas, K., Field, L.H., 2017a. The complex tibial organ of the New Zealand ground weta: sensory adaptations for vibrational signal detection. Sci. Rep. 7, 2031. https://doi.org/10.1038/s41598-017-02132-1.

Strauß, J., Alt, J.A., Ekschmitt, K., Schul, J., Lakes-Harlan, R., 2017b. Evolutionary diversification of the auditory organ sensilla in *Neoconocephalus* katydids (Orthoptera: Tettigoniidae) correlates with acoustic signal diversification over phylogenetic relatedness and life history. J. Evol. Biol. 30, 1094–1109. https://doi.org/10.1111/jeb.13066.

Stritih-Peljhan, N., Virant-Doberlet, M., 2021. Vibrational signalling, an underappreciated mode in cricket communication. Sci. Nat. 108, 41. https://doi.org/10.1007/s00114-021-01749-7.

Stumpner, A., Dann, A., Schink, M., Gubert, S., Hugel, S., 2013. True katydids (Pseudophyllinae) from Guadeloupe: acoustic signals and functional considerations of song production. J. Insect Sci. 13, 157. https://doi.org/10.1673/031.013.1.

Swanger, E., Zuk, M., 2015. Cricket responses to sexual signals are influenced more by adult than juvenile experiences. J. Insect Behav. 28, 328–337. https://doi.org/10.1007/s10905-015-9504-6.

Symes, L.B., 2018. Spatial and temporal variation in three call traits and preferences of the tree cricket *Oecanthus forbesi*. Behav. Ecol. Sociobiol. 72, 35. https://doi.org/10.1007/s00265-018-2442-5.

Symes, L.B., Page, R.A., ter Hofstede, H.M., 2016. Effects of acoustic environment on male calling activity and timing in neotropical forest katydids. Behav. Ecol. Sociobiol. 70, 1485–1495.

Symes, L.B., Martinson, S.J., Kernan, C.E., ter Hofstede, H.M., 2020. Sheep in wolves' clothing: prey rely on proactive defences when predator and non-predator cues are similar. Proc. R. Soc. Lond. Ser. B: Biol. Sci. 287, 20201212. https://doi.org/10.1098/rspb.2020.1212.

Symes, L.B., Robillard, T., Martinson, S.J., Dong, J., Kernan, C.E., Miller, C.R., ter Hofstede, H.M., 2021. Daily signaling rate and the duration of sound per signal are negatively related in neotropical forest katydids. Integr. Comp. Biol. 61, 887–899. https://doi.org/10.1093/icb/icab138.

Tan, M.K., 2020. Soundscape of urban-tolerant crickets (Orthoptera: Gryllidae, Trigonidiidae) in a tropical Southeast Asia city, Singapore. Bioacoustics 30, 469–486. https://doi.org/10.1080/09524622.2020.1813627.

Tan, M.K., Ingrisch, S., Wahab, R.B.A., Japir, R., Chung, A.Y.C., 2020. Ultrasonic bioacoustics and stridulum morphology reveal cryptic species among *Lipotactes* big-eyed katydids (Orthoptera: Tettigoniidae: Lipotactinae) from Borneo. Syst. Biodivers. 18, 510–524. https://doi.org/10.1080/14772000.2020.1769223.

Tan, M.K., Robillard, T., 2021a. Population divergence in the acoustic properties of crickets during the COVID-19 pandemic. Ecology 102, e03323. https://doi.org/10.1002/ecy.3323.

Tan, M.K., Robillard, T., 2021b. Highly diversified circadian rhythms in the calling activity of eneopterine crickets (Orthoptera: Grylloidea: Gryllidae) from Southeast Asia. Bioacoustics. https://doi.org/10.1080/09524622.2021.1973562.

Tanner, J.C., Swanger, E., Zuk, M., 2019. Sexual signal loss in field crickets maintained despite strong sexual selection favoring singing males. Evolution 73, 1482–1489. https://doi.org/10.1111/evo.13761.

ter Hofstede, H.M., Ratcliffe, J.M., Fullard, J.H., 2008. The effectiveness of katydid (*Neoconocephalus ensiger*) song cessation as antipredator defence against the gleaning bat *Myotis septentrionalis*. Behav. Ecol. Sociobiol. 63, 217–226.

ter Hofstede, H.M., Kalko, E.K.V., Fullard, J.H., 2010. Auditory-based defence against gleaning bats in neotropical katydids (Orthoptera: Tettigoniidae). J. Comp. Physiol. A 196, 349–358. https://doi.org/10.1007/s00359-010-0518-4.

ter Hofstede, H.M., Schöneich, S., Robillard, T., Hedwig, B., 2015. Evolution of a communication system by sensory exploitation of startle behavior. Curr. Biol. 25, 3245–3252. https://doi.org/10.1016/j.cub.2015.10.064.

ter Hofstede, H., Voigt-Heucke, S., Lang, A., Römer, H., Page, R., Faure, P., Dechmann, D., 2017. Revisiting adaptations of neotropical katydids (Orthoptera: Tettigoniidae) to gleaning bat predation. Neotrop. Biodivers. 3, 41–49.

ter Hofstede, H.M., Symes, L.B., Martinson, S.J., Robillard, T., Faure, P., Madhusudhana, S., Page, R.A., 2020. Calling songs of neotropical katydids (Orthoptera: Tettigoniidae) from Panama. J. Orthoptera Res. 29, 137–201. https://doi.org/10.3897/jor.29.46371.

Thomas, M.L., Gray, B., Simmons, L.W., 2011. Male crickets alter the relative expression of cuticular hydrocarbons when exposed to different acoustic environments. Anim. Behav. 82, 49–53. https://doi.org/10.1016/j.anbehav.2011.03.023.

Thompson, A.C., Samways, M.J., Bazelet, C.S., 2020. Biosphere reserve zones are equal in terms of katydid ecoacoustics. Bioacoustics 29, 249–265. https://doi.org/10.1080/09524622.2019.1595147.

Thomson, I.R., Bertram, S.M., 2014. Spring field crickets (*Gryllus veletis*) use two different pulse types when signaling to attract mates. J. Insect. Behav. 27, 217–227. https://doi.org/10.1007/s10905-013-9421-5.

Tian, D., Shen, C.-Z., Chen, L., Chen, G.-Y., Zhang, T., Li, K., He, Z.-Q., 2019. An integrative taxonomy of *Vescelia pieli pieli* species complex based on morphology, genes and songs from China (Orthoptera: Grylloidea: Phalangopsidae: Phaloriinae). Zootaxa 4695, 67–75. https://doi.org/10.11646/zootaxa.4695.1.5.

Ting, J.J., Judge, K.A., Gwynne, D.T., 2017. Listening to male song induces female field crickets to differentially allocate reproductive resources. J. Orthoptera Res. 26, 205–210.

Tinghitella, R.M., Zuk, M., 2009. Asymmetric mating preferences accommodated the rapid evolutionary loss of a sexual signal. Evolution 63, 2087–2098. https://doi.org/10.1111/j.1558-5646.2009.00698.x.

Tinghitella, R.M., Broder, E.D., Gallagher, J.H., Wikle, A.W., Zonana, D.M., 2021. Responses of intended and unintended receivers to a novel sexual signal suggest clandestine communication. Nat. Commun. 12, 797. https://doi.org/10.1038/s41467-021-20971-5.

Tiwari, C., Diwakar, S., 2019. Singers in the grass: call description of conehead katydids (family: Tettigoniidae) and observations on avoidance of acoustic overlap. Bioacoustics 28, 522–538. https://doi.org/10.1080/09524622.2018.1499553.

Tomar, M., Diwakar, S., 2020. Investigating host plant association, calling activity, and sexual dimorphism in Indian *Gryllacropsis* sp. (Orthoptera: Anostostomatidae). Ecol. Evol. 10, 12364–12371. https://doi.org/10.1002/ece3.6819.

Torsekar, V.R., Balakrishnan, R., 2020. Sex differences in alternative reproductive tactics in response to predation risk in tree crickets. Funct. Ecol. 34, 2326–2337. https://doi.org/10.1111/1365-2435.13652.

Torsekar, V.R., Isvaran, K., Balakrishnan, R., 2019. Is the predation risk of mate-searching different between the sexes? Evol. Ecol. 33, 329–343. https://doi.org/10.1007/s10682-019-09982-3.

Tregenza, T., Simmons, L.W., Wedell, N., Zuk, M., 2006. Female preference for male courtship song and its role as a signal of immune function and condition. Anim. Behav. 72, 809–818. https://doi.org/10.1016/j.anbehav.2006.01.019.

Trillo, P.A., Benson, C.S., Caldwell, M.S., Lam, T.L., Pickering, O.H., Logue, D.M., 2019. The influence of signaling conspecific and heterospecific neighbors on eavesdropper pressure. Front. Ecol. Evol. 7, 292. https://doi.org/10.3389/fevo.2019.00292.

Trobe, D., Schuster, R., Römer, H., 2011. Fast and reliable decisions for a dynamic song parameter in field crickets. J. Comp. Physiol. A 197, 131–135. https://doi.org/10.1007/s00359-010-0589-2.

Ulagaraj, S.M., 1976. Sound production in mole crickets (Orthoptera: Gryllotalpidae: *Scapteriscus*). Ann. Entomol. Soc. Am. 69, 299–306.

Vahed, K., 1998. The function of nuptial feeding in insects: a review of empirical studies. Biol. Rev. 73, 43–78.

van der Mescht, A.C., Pryke, J.S., Gaigher, R., Samways, M.J., 2021. Ecological and acoustic responses of bush crickets to anthropogenic and natural ecotones. Biodivers. Conserv. https://doi.org/10.1007/s10531-021-02278-5.

van Staaden, M.J., Römer, H., 1997. Sexual signalling in bladder grasshoppers: tactical design for maximizing calling range. J. Exp. Biol. 200, 2597–2608.

van Staaden, M.J., Römer, H., 1998. Evolutionary transition from stretch to hearing organs in ancient grasshoppers. Nature 394, 773–776.

Vedenina, V., Sevastianov, N., Tarasova, T., 2020. Contributions to the study of the grasshopper (Orthoptera: Acrididae: Gomphocerinae) courtship songs from Kazakhstan and adjacent territories. Zootaxa 4895, 505–527. https://doi.org/10.11646/zootaxa.4895.4.3.

Velez, M.J., Brockmann, H.J., 2006. Seasonal variation in female response to male calling song in the field cricket, *Gryllus rubens*. Ethology 112, 1041–1049. https://doi.org/10.1111/j.1439-0310.2006.01262.x.

Verburgt, L., Ferguson, J.W.H., 2010. Mate choice in field crickets: can females acoustically detect male body size? J. Ethol. 28, 141–151. https://doi.org/10.1007/s10164-009-0166-8.

Verburgt, L., Ferreira, M., Ferguson, J.W.H., 2011. Male field cricket song reflects age, allowing females to prefer young males. Anim. Behav. 81, 19–29. https://doi.org/10.1016/j.anbehav.2010.09.010.

Villarreal, S.M., Gilbert, C., 2013. Female acoustic reply to variation in the male call in a duetting katydid, *Scudderia pistillata*. Behaviour 150, 525–546. https://doi.org/10.1163/1568539x-00003065.

Villarreal, S.M., Gilbert, C., 2014. Male *Scudderia pistillata* katydids defend their acoustic duet against eavesdroppers. Behav. Ecol. Sociobiol. 68, 1669–1675. https://doi.org/10.1007/s00265-014-1775-y.

von Helversen, D., Wendler, G., 2000. Coupling of visual to auditory cues during phonotactic approach in the phaneropterine bushcricket *Poecilimon affinis*. J. Comp. Physiol. A 186, 729–736.

von Helversen, D., Balakrishnan, R., von Helversen, O., 2004. Acoustic communication in a duetting grasshopper: receiver response variability, male strategies and signal design. Anim. Behav. 68, 131–144.

von Helversen, D., von Helversen, O., Heller, K.-G., 2012. When to give up responding acoustically in *Poecilimon* bush-crickets: a clue to population density. Articulata 27, 57–66.

von Helversen, O., von Helversen, D., Rohrseitz, K., Koopmann, I., Stumpner, A., Heller, K.-G., 2015. Signals and sex-specific active space in a duetting insect (Orthoptera: Phaneropteridae: *Poecilimon affinis*). J. Orthoptera Res. 24, 19–28. https://doi.org/10.1665/034.024.0104.

Wagner Jr., W.E., 2005. Male field crickets that provide reproductive benefits to females incur higher costs. Ecol. Entomol. 30, 350–357.

Wagner Jr., W.E., Harper, C.J., 2003. Female life span and fertility are increased by the ejaculates of preferred males. Evolution 57, 2054–2066.

Wagner Jr., W.E., Reiser, M.G., 2000. The importance of calling song and courtship song in female mate choice in the variable field cricket. Anim. Behav. 59, 1219–1226.

Wagner Jr., W.E., 2011. Direct benefits and the evolution of female mating preferences: conceptual problems, potential solutions, and a field cricket. Adv. Stud. Behav. 43, 273–319. https://doi.org/10.1016/b978-0-12-380896-7.00006-x.

Walker, S., Cade, W., 2003. The effects of temperature and age on calling song in a field cricket with a complex calling song, *Teleogryllus oceanicus* (Orthoptera: Gryllidae). Can. J. Zool. 81, 1414–1420.

Walker, T.J., Figg, D.E., 1990. Song and acoustic burrow of the prairie mole cricket, *Gryllotalpa major* (Orthoptera, Gryllidae). J. Kans. Entomol. Soc. 63, 237–242.

Wang, Y., Huang, Q., Zhang, X., Ren, B., 2014. Application of insect songs in monitoring population density level of *Locusta migratoria migratoria* (Orthoptera: Acrididae). Zool. Stud. 53, 55. https://doi.org/10.1186/s40555-014-0055-x.

Watts, C., Thornburrow, D., 2011. Habitat use, behavior and movement patterns of a threatened New Zealand giant weta, *Deinacrida heteracantha* (Anostostomatidae: Orthoptera). J. Orthoptera Res. 20, 127–135.

Wedell, N., Ritchie, M.G., 2004. Male age, mating status and nuptial gift quality in a bushcricket. Anim. Behav. 67, 1059–1065.

Wedell, N., Tregenza, T., 1999. Successful fathers sire successful sons. Evolution 53, 620–625.

Wegehaupt, A.K., Wagner Jr., W.E., 2017. Females can solve the problem of low signal reliability by assessing multiple male traits. Biol. Lett. 13, 20170386. https://doi.org/10.1098/rsbl.2017.0386.

Weissman, D.B., Vandergast, A.G., Song, H.J., Shin, S., McKenna, D.D., Ueshima, N., 2021. Generic relationships of New World Jerusalem crickets (Orthoptera: Stenopelmatoidea: Stenopelmatinae), including all known species of *Stenopelmatus*. Zootaxa 4917, 1–122. https://doi.org/10.11646/zootaxa.4917.1.1.

White, C.R., Matthews, P.G.D., Seymour, R.S., 2008. In situ measurement of calling metabolic rate in an Australian mole cricket, *Gryllotalpa monanka*. Comp. Biochem. Physiol. A Mol. Integr. Physiol. 150, 217–221. https://doi.org/10.1016/j.cbpa.2006.08.030.

Whitman, D.W., 2008. The significance of body size in the Orthoptera: a review. J. Orthoptera Res. 17, 117–134. https://doi.org/10.1665/1082-6467-17.2.117.

Wiley, C., Ellison, C.K., Shaw, K.L., 2012. Widespread genetic linkage of mating signals and preferences in the Hawaiian cricket *Laupala*. Proc. R. Soc. Lond. Ser. B Biol. Sci. 279, 1203–1209. https://doi.org/10.1098/rspb.2011.1740.

Wilkins, M.R., Seddon, N., Safran, R.J., 2013. Evolutionary divergence in acoustic signals: causes and consequences. Trends Ecol. Evol. 28, 156–166. https://doi.org/10.1016/j.tree.2012.10.002.

Wilson, K.M.Z., Walker, S.E., 2019. Age at mating and male quality influence female patterns of reproductive investment and survival. Ecol. Evol. 9, 5440–5449. https://doi.org/10.1002/ece3.5137.

Wilson, A.D.M., Whattam, E.M., Bennett, R., Visanuvimol, L., Lauzon, C., Bertram, S.M., 2010. Behavioral correlations across activity, mating, exploration, aggression, and antipredator contexts in the European house cricket, *Acheta domesticus*. Behav. Ecol. Sociobiol. 64, 703–715. https://doi.org/10.1007/s00265-009-0888-1.

Wirmer, A., Faustmann, M., Heinrich, R., 2010. Reproductive behaviour of female *Chorthippus biguttulus* grasshoppers. J. Insect Physiol. 56, 745–753. https://doi.org/10.1016/j.jinsphys.2010.01.006.

Wittmann, J.P., Kolss, M., Reinhold, K., 2011. A neural network-based analysis of acoustic courtship signals and female responses in *Chorthippus biguttulus* grasshoppers. J. Comput. Neurosci. 31, 105–115. https://doi.org/10.1007/s10827-010-0299-3.

Woo, B., 2020. First natural history observations of the canyon pygmy mole cricket, *Ellipes monticolus* (Orthoptera: Tridactylidae). J. Orthoptera Res. 29, 1–7. https://doi.org/10.3897/jor.29.33413.

Woodrow, C., Judge, K.A., Pulver, C., Jonsson, T., Montealegre-Z, F., 2021. The Ander's organ: a mechanism for anti-predator ultrasound in a relict orthopteran. J. Exp. Biol. 224, jeb237289. https://doi.org/10.1242/jeb.237289.

Xu, M., Shaw, K.L., 2021. Extensive linkage and genetic coupling of song and preference loci underlying rapid speciation in *Laupala* crickets. J. Hered. 112, 204–213. https://doi.org/10.1093/jhered/esab001.

Yack, J.E., 2004. The structure and function of auditory chordotonal organs in insects. Microsc. Res. Tech. 63, 315–337. https://doi.org/10.1002/jemt.20051.

Zefa, E., Martins, L.D., Szinwelski, N., 2008. Complex mating behavior in *Adelosgryllus rubricephalus* (Orthoptera, Phalangopsidae, Grylloidea). Iheringia. Sér. Zool. 98, 325–328. https://doi.org/10.1590/s0073-47212008000300006.

Zefa, E., Oliveira, G.L., Redu, D.R., Martins, L.P., 2013. Calling song of two sympatric species of cricket *Phylloscyrtini* (Orthoptera Gryllidae Trigonidiinae). Ethol. Ecol. Evol. 25, 21–27. https://doi.org/10.1080/03949370.2012.685764.

Zhang, X., Hedwig, B., 2019. Bilateral auditory processing studied by selective cold-deactivation of cricket hearing organs. J. Exp. Biol. 222, jeb210799. https://doi.org/10.1242/jeb.210799.

Zhantiev, R.D., Korsunovskaya, O.S., 2019. Acoustic signals of katydids of the tribe Gampsocleidini (Orthoptera, Tettigoniidae) in Russia. Zool. Zhurnal 98, 773–777. https://doi.org/10.1134/s004451341905012x.

Zhantiev, R.D., Korsunovskaya, O.S., Sorokin, N.N., Chukanov, V.S., 2003. Sound signals in mole crickets (Orthoptera, Gryllotalpidae) from Eastern Europe. Zool. Zhurnal 82, 1339–1346.

Zhantiev, R., Korsunovskaya, O., Benediktov, A., 2017. Acoustic signals of the bush-crickets Isophya (Orthoptera: Phaneropteridae) from Eastern Europe, Caucasus and adjacent territories. Eur. J. Entomol. 114, 301–311. https://doi.org/10.14411/eje.2017.037.

Zhao, L.-Q., Chai, H.-L., Wu, H.-J., Zhu, D.-H., 2017. Mating advantage of short-winged over long-winged adult males in the cricket *Velarifictorus ornatus* (Orthoptera: Gryllidae). Eur. J. Entomol. 114, 267–274. https://doi.org/10.14411/eje.2017.032.

Zhemchuzhnikov, M.K., Kutcherov, D.A., Kymre, J.H., Knyazev, A.N., 2017. Louder songs can enhance attractiveness of old male crickets (*Gryllus bimaculatus*). J. Insect Behav. 30, 211–219. https://doi.org/10.1007/s10905-017-9611-7.

Zorovic, M., Hedwig, B., 2011. Processing of species-specific auditory patterns in the cricket brain by ascending, local, and descending neurons during standing and walking. J. Neurophysiol. 105, 2181–2194. https://doi.org/10.1152/jn.00416.2010.

Zuk, M., Rotenberry, J.T., Tinghitella, R.M., 2006. Silent night: adaptive disappearance of a sexual signal in a parasitized population of field crickets. Biol. Lett. 2, 521–524. https://doi.org/10.1098/rsbl.2006.0539.

Zuk, M., Rebar, D., Scott, S.P., 2008. Courtship song is more variable than calling song in the field cricket *Teleogryllus oceanicus*. Anim. Behav. 76, 1065–1071.

Zuk, M., Bailey, N.W., Gray, B., Rotenberry, J.T., 2018. Sexual signal loss: the link between behaviour and rapid evolutionary dynamics in a field cricket. J. Anim. Ecol. 87, 623–633. https://doi.org/10.1111/1365-2656.12806.

CHAPTER TWO

Sound production and hearing in insects

Thomas R. Neil and Marc W. Holderied
School of Biological Sciences, University of Bristol, Bristol, United Kingdom

Contents

Abstract

Among the oldest land animals, insects have roamed the planet for over 400 million years. Acoustics plays a vital role in the ecology of many insects across the globe, facilitating mate finding, defence and information transfer. Many insect sounds are readily perceived by human ears, with the sound of singing cicadas or grasshoppers familiar to many. Yet how insects produce sounds, and what they use them for is not always evident, and they might not even be perceptible to human ears. The Belgian painter Erik

Advances in Insect Physiology, Volume 61
ISSN 0065-2806
https://doi.org/10.1016/bs.aiip.2021.10.001

Pevernagie encouraged acoustic awareness: 'Let us drop our tin ear and listen to the sounds of the real world veiled beyond our inattention'. His appeal can be elegantly applied to insect bioacoustics, as while singing cicadas epitomise the searing heat of a southern summer to most, there is a vast hidden world of insect sounds, which is both wondrous and complex. Generations of scientists have tasked themselves with uncovering its secrets, with new chapters written still today. Here we will explore and summarise the many fascinating ways insects both produce and perceive sounds, as well as look into how they exploit sound to their benefit in an ecological context. We look to bring together both classic and modern research, while looking to the future to speculate on what the next steps in understanding the acoustic world of insects might be.

1. Bioacoustics

The influence of sound on insect ecology is profound and diverse, with both sound production and hearing evolving numerous times across many different taxa. The following introduces some biologically relevant aspects of sound signals. Sound is simply any form of vibration that is transmitted through a medium, such as air, water or a solid. The transmission of vibration through the medium causes localised particle displacement, which travels through the medium as a wave. Soundwaves are characterised by their frequency, wavelength, amplitude, and propagation speed. The oscillating motion of the individual medium particles means that they experience their own velocity, a quantity that is different from the propagation speed of sound itself.

Sound pressure level (SPL) is a measure of the 'loudness' of a sound and is based on the local pressure oscillations within a medium as a sound wave is transmitted within it. The unit of sound pressure is the pascal (Pa), equivalent to 1 N per square metre. Biologically relevant changes in pressure cover many orders of magnitude, so a logarithmic scale is more commonly used, with SPL being given in decibels (dB), as the sound pressure relative to a reference (0 dB SPL equals 20 μPa for airborne sounds and 1 μPa in water). 0 dB SPL was anthropocentrically selected as a measure of the lowest sound pressure humans can hear at their best frequency.

The sound field surrounding an object is typically divided into near field and far field. The near field describes the region immediately adjacent to the object, typically within one wavelength of the sound that is interacting with it. Beyond this distance the region is known as the far field. Within the far field of a sound source, sound spreads spherically and hence its pressure attenuates at a rate of 6 dB per doubling of distance, providing sound propagation is not influenced by external factors such as reflecting surfaces.

Acoustic impedance describes the resistance of a medium to the flow of acoustic energy. Its importance for bioacoustics is that the acoustic impedances of structures receiving and transmitting sound need to be well matched such that they can transfer this information to the ear. As such insect hearing structures are often small and light so that they may be susceptible to changes in pressure and medium particle velocity brought about by sound.

The mechanics required to detect a sound by its changes in pressure or its particle oscillations are fundamentally different, and many examples of both pressure and particle motion receivers are found in insects. To detect the pressure component of sound a receiver must oscillate in response to the associated pressure changes. Insects have evolved an organ called the tympanum, a section of cuticle or cuticular structures that underwent some form of localised thinning, which enables the cuticle to move in response to the pressure oscillations acting upon its surfaces. Each tympanum is accompanied by a mechanosensory system, which is activated by the movement of the tympanum. Sensory cells convert the movement into neuronal signals to be interpretated by the animal's nervous system.

Unlike tympanal ears that pick up changes in sound pressure, sensors that are able to detect the oscillations of medium particles must be capable of being actuated by particle motion. Structures capable of following the oscillations of the medium particles are typically thin and slender, such as individual hairs or antennae. Often, such particle motion sensors operate best at specific frequencies.

2. Evolution of hearing and sound production in insects

The ecological function of insect acoustics falls into two broad contexts—intraspecific communication and predator avoidance. The evolution of communication among insects is something that has intrigued scientists for decades. For two insects to communicate, the sender must produce some form of signal that is matched by the receiver's ability to perceive and interpret that signal.

The type of insect communication likely under most intense selection pressure is sexual communication—the ability of an insect to communicate with and localise a partner. One key question in the evolution of insect communication is whether signal generation and perception evolved jointly, or one preceded the other. The two possibilities here can be summarised as either the Fisherian mechanism (Lande, 1981), which support coevolution, or the sensory-bias mechanism (Endler and Basolo, 1998), which stipulates that the ability to perceive signals is ancestral. There is evidence for both theories.

To understand the evolution of acoustic signalling in insects, researchers have looked to couple the limited fossil record (Gu et al., 2012) of acoustic structures with phylogenomics (Kawahara et al., 2019; Regier et al., 2013; Schubnel et al., 2021). This has enabled a fairly comprehensive overview of possible origins and development of both signalling and sound perception in insects. Evidence for acoustic communication in insects is relatively poorly represented in the fossil record and is phylogenetically less comprehensively represented compared to the plethora of information available on the evolution of communication in vertebrates. Among insects, only species in nine of the 32 recognised orders have been shown to possess ears capable of detecting airborne sounds. But within these taxa acoustic communication has evolved independently at least 24–29 times (Yager, 1999). In almost every instance of insect communication, its evolution revolves around some aspect of mating, such as advertising, courting or competition. It is often the males of the species which produce sound signals incurring the associated predation risks and waiting for females to travel towards them. There are examples of females responding with their own calls forcing males to suffer increased predation pressure both through signalling and phonotaxis (Heller and von Helversen, 1986). An exception to this is found in mosquitoes, where it is the female that initiates contact between the two partners (Cator et al., 2009). In social insects, airborne acoustic signalling is rare. However, there is one confirmed case in honeybees: their famous waggle dance to recruit fellow foragers includes an acoustic component, which can be detected by conspecifics. This signal is also accompanied by substrate vibrations and tactile stimulation, and it seems that this combination of various signals informs the nestmates of fruitful foraging sites (Hunt and Richard, 2013).

The evolution of ears must have preceded the rise of acoustic communication in insects and so it is important to consider what selection pressures to evolve sensors for airborne sounds preceded communication. One of the biggest and most intriguing selection pressures favouring the evolution of hearing organs in insects is the emergence of echolocating bats. Bats hunt using ultrasound, and there are numerous examples of insect ears that are specifically tuned to the ultrasonic frequencies of bat biosonar across four insect orders (Mantodea, Neuroptera, Coleoptera, Lepidoptera), having evolved independently at least 14 times (Hoy, 1992). Molecular analysis estimates the evolution of ultrasound sensitive ears to be around 65 MYA, suggesting a coevolutionary response to the echolocation signals used by bats. It is among the Lepidoptera that there exists the strongest evidence in support of the coevolution of hearing and bat predation, giving rise to one of the most studied and well understood evolutionary arms races. Almost all the origins of hearing in this order occur at the time of the emergence of echolocating bats, with hearing in three of the groups being widespread throughout a large superfamily (Pyraloidea, Geometroidea, and Noctuoidea; Regier et al., 2013). Examples of a loss of hearing in species in these groups generally occurs in ecological regions that lack echolocating insectivorous bats or in which their ecological niches no longer overlap (Fullard, 1994), and there is evidence that local moth populations have frequency sensitivity that matches the frequencies used by the local bat populations (ter Hofstede et al., 2013).

In insects, sound signalling is rarer than hearing capabilities, with only four orders known to include species that produce airborne acoustic signals. One example of this asymmetry can be seen at a suborder level in Caelifera of the Orthoptera. Hearing in this suborder is present in three of the eight extant superfamilies (Acridoidea, Pyrgomorphoidea and Pneumoroidea) whereas sound signalling is only present in one (Acridoidea). This suggests the ability to hear was basal in this suborder and that sound signalling evolved later. Hearing may have originally functioned as a means of detecting the sounds of terrestrial vertebrates, before going on to being involved in courtship. Within the other orthopteran suborder (Ensifera) tympanal hearing occurs in several groups, including the Tettigoniidae, where sound signalling is found in all species that possess tympanal hearing. In the Blattodea (Cockroaches) however, sound signalling is more prevalent than hearing, it has been argued that signalling may have evolved initially as a means of startling predators.

3. Ecology of insect communication

The plethora of insect communication sounds shape and often give unique character to terrestrial soundscapes across the globe (Fig. 1). Acoustic signalling is vital for numerous extant insects and there is evidence for communication via airborne sound in fossils of crickets, katydids and cicadas that suggest insects have been doing so for millions of years (Senter, 2008), and even the song of a Jurassic cricket was reconstructed

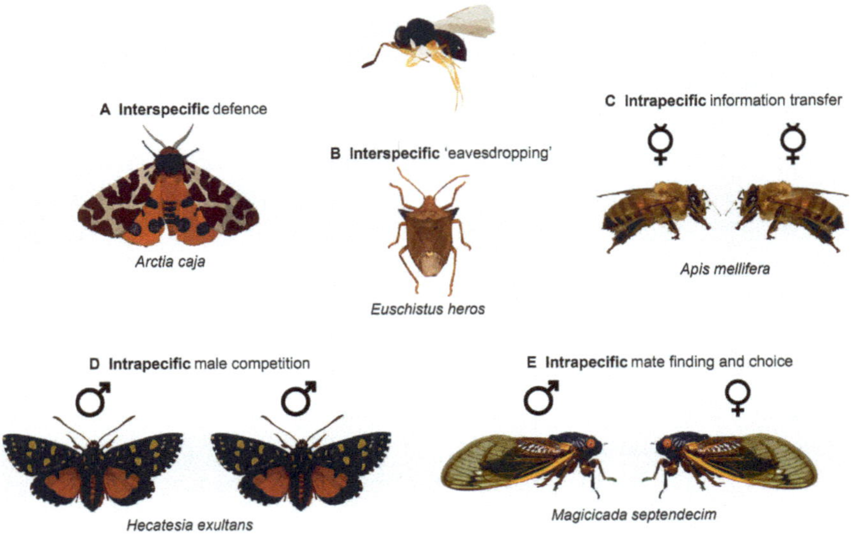

Fig. 1 Summary of ecological acoustic functions. (A) Defence, adult *Arctia caja* moths emit ultrasonic clicks during flight to warn bats that they are unpalatable. (B) Eaves dropping, the parasitoid wasp *Telenomus podisi* will localise their host species *Euschistus heros* by listening for their mating calls. (C) Information transfer, honeybees (*Apis mellifera*) transfer information on resource location through a 'waggle dance', which includes an acoustic component. (D) Competition, males of the moth *Hecatesia exultans* produce clicks with their wings to defend a territory from which they can try to attract females. (E) Mate finding and choice, males of the cicada species *Magicicada septendecim* produce calling songs to attract females, which prefer certain song characteristics that depict a stronger potential mate. Image credit CC. A, Ben Sale. B, Eirefred. Paulo Roberto Valle da Silva. C, pixabay. D, Biodiversity Heritage Library. E, John Cooley. Images recoloured in Adobe Illustrator.

from its fossilised remains (Gu et al., 2012). A number of extant insect groups have evolved airborne acoustic communication, including certain species of grasshoppers (Song et al., 2015; Von Helversen and Von Helversen, 1994) and moths (Nakano et al., 2008; Nakano et al., 2009). In all these examples, the primary function of the acoustic signal is to facilitate mating, with the male typically producing the courtship song and the female responding.

3.1 Long range acoustic signalling and pair formation

Using tympanal ears, crickets, katydids, cicadas, grasshoppers and moths are capable of hearing sound in the far field (Gerhardt and Huber, 2002). In several species, the calling individual will move through a habitat while signalling, and the receiver may either remain stationary or move towards the signal upon hearing it. This strategy is advantageous if the signalling individual is a male and there is a low density of females in its vicinity as the movement will increase the likelihood of pair formation. The males of bladder grasshoppers are able to communicate over vast distances (~450 m), but will often spend time calling from one location, before moving up to 500 m to start calling again (Van Staaden et al., 2003). Utilising this strategy the male is increasing their chances of locating females that might otherwise be sparsely distributed or occurring at low densities (Römer et al., 2014). Additionally, calling this way may release the male from the attention of predators that may pick up on their acoustic signal and zero in on their location (Zuk and Kolluru, 1998).

Another strategy often used in insect pairing is duetting, where both males and females of the species produce sounds. It is almost always the male that produces the initial signal, to which the female replies. There are examples of duetting in katydids, grasshoppers and cicadas (Bailey, 2003; Marshall and Cooley, 2001). Insect movement during duetting is diverse, with the females in some systems not moving at all while there are examples where both partners will converge on one location through the perception of acoustic signals (Bailey, 2003).

The evolutionary origins of duetting in insects is not very well understood, yet some proposals for its emergence have been suggested. One theory stems from the idea that mate calling was ancestrally male driven, but pressures from eavesdropping predators and parasitoids drove the males to produce shorter, less frequent sounds. This would have the effect of making it harder for females to detect males, leading to a scenario whereby females started to produce calls of their own, allowing the males to locate the females (Bailey, 2003). Alternatively, the evolution may have been driven by costs associated with searching by the females, be that an energetic cost or the increased risk of predation (Raghuram et al., 2015). By duetting the risks are shared between the two sexes rather than just one, a scenario discovered in the tree cricket, *Oecanthus henryi*, where both sexes are equally at risk from one of its predators, the lynx spider (Torsekar et al., 2019).

Another energetic cost to factor in when considering the evolutionary origins of pair forming is the nutritional nuptial gift often provided to the female of the species by the male (McCartney et al., 2012). In many species of katydid the male will provide a nuptial gift of spermatophylax to the female (Gwynne, 2008). The size of this nuptial gift is predicted to affect the likelihood of the female of the species moving towards the male.

Females of species in which the males invest more into the gift, which are then larger and more laden with weight, are predicted to be more likely to be the more mobile of the two in pair formation. There is some evidence to support this theory, with several species of the katydid genus *Poecilimon* providing evidence that correlates nuptial gift size with the movement ecology of the females (McCartney et al., 2012). There are even some cases where the gift has become so large, that the males undergo large time periods where they are unable to mate. In these scenarios, the females become the competitive sex, with males only calling for short durations, leaving the onus on the females to locate and find them (Mccartney et al., 2008).

Research into pair forming in the bush cricket, *Ephippiger diurnus*, found no evidence between male song calling traits, such as call rate, syllable number and peak frequency and the size of the nuptial gift (Jarrige et al., 2013). This indicates that females are unable to use calling song as a predictor of male nuptial gift quality prior to pair formation. Interestingly this same study did provide convincing evidence of male mate preference, with the size and quality of nuptial gifts offered by males varying with desirable female traits such as age and size. Why then, might females be listening for and choosing males based on their song? It may be that certain characteristics of song are heritable traits that would influence the attractiveness of male offspring. Another reason may be that a male's calling song does not necessarily advertise their true fitness. Signalling can be energetically costly (Prestwich, 1994; Reinhold et al., 1998), so much so that it is predicted to affect the quality of other sexually selected traits, such as nuptial gift size. One such study into *E. diurnus* found an inverse relationship between desirable call characteristics and immune response, suggesting that males exhibit different energy allocation strategies when it comes to reproduction (Barbosa et al., 2016). Some males may invest more heavily in immunity rather than reproduction, while others might invest more in song than in nuptial gifts. This may explain the discrepancy between male song traits and female preference. Female preference should exert directional selection on male call parameters, but it seems acoustic signals are being constrained by trade-offs with other traits in certain populations.

3.2 Signalling increases predation risk

The ecology of acoustic signalling not only concerns the ability of conspecifics to hear one another, but the increased predation pressure that these calls may have on the signaller. The calling song of male katydids has been shown to attract bats, both in the field and in laboratory experiments, with calls that are longer in duration and higher in duty cycle, being more quickly located than shorter calls (Belwood and Morris, 1987). Similarly, at least four species of leaf-nosed bat have been found to hunt by listening out for the mating calls of Orthopterans. Playback experiments using the mating calls of 12 katydid species found that different species of bats prefer different acoustic features of the katydid calls, suggesting sensory niche partitioning among these bats, resulting in low prey overlap and reduced competition (Falk et al., 2015). It has been speculated that this may have driven the evolution of shorter calls seen in some katydids, which may help reduce predation pressure by bats. However, studies on bat diet analysis and behaviour suggest that females may be at higher predation risk due to the movements they make towards calling males (Geipel et al., 2020; Raghuram et al., 2015).

3.3 Parasitoids

There are numerous parasitoid flies which are known to eavesdrop on the mating calls of many insect species in order to locate suitable hosts (Lakes-Harlan and Lehmann, 2015; Miles et al., 1995; Robert et al., 1992). Many of these parasitoids have evolved to specially target one host species and as such have adaptations to achieve this. For example, the parasitoid fly, *Emblemasoma auditrix*, will deposit her eggs into their cicada host species *Okanagana rimosa* by cutting a slit in the tymbal of males with her ovipositor (Fig. 2). The female flies hearing is tuned to the calling song of males of this species and their larvae will not develop if deposited in other related cicadas (Lakes-Harlan and de Vries, 2014). The effect of these parasitoids can be clearly seen in Hawaii (Zuk et al., 2006), where populations of the cricket *Teleogryllus oceanicus* have been infested by a parasitoid fly, *Ormia ochracea*. A 'flatwing' phenotype has rapidly spread among the population in only 12–20 generations. These flatwing males have a mutation that affects the stridulatory structures on their wings, essentially rendering them silent. Silent males have been shown to have much lower rates of infestation compared to non-silent males, essentially wiping out the singing-male phenotype from the population. The same occurrence has since been documented on two other islands (Pascoal et al., 2014). The island settings may well be a catalyst for this unusually rapid evolution though, as the lack of alternate hosts mean the crickets in these areas are under high risk of parasitism. Indeed, there are many habitats in which crickets and parasitoids coexist without the male crickets losing their song, although some modification of the song can still be apparent, such as a shorter time spent calling (Lakes-Harlan and Lehmann, 2015).

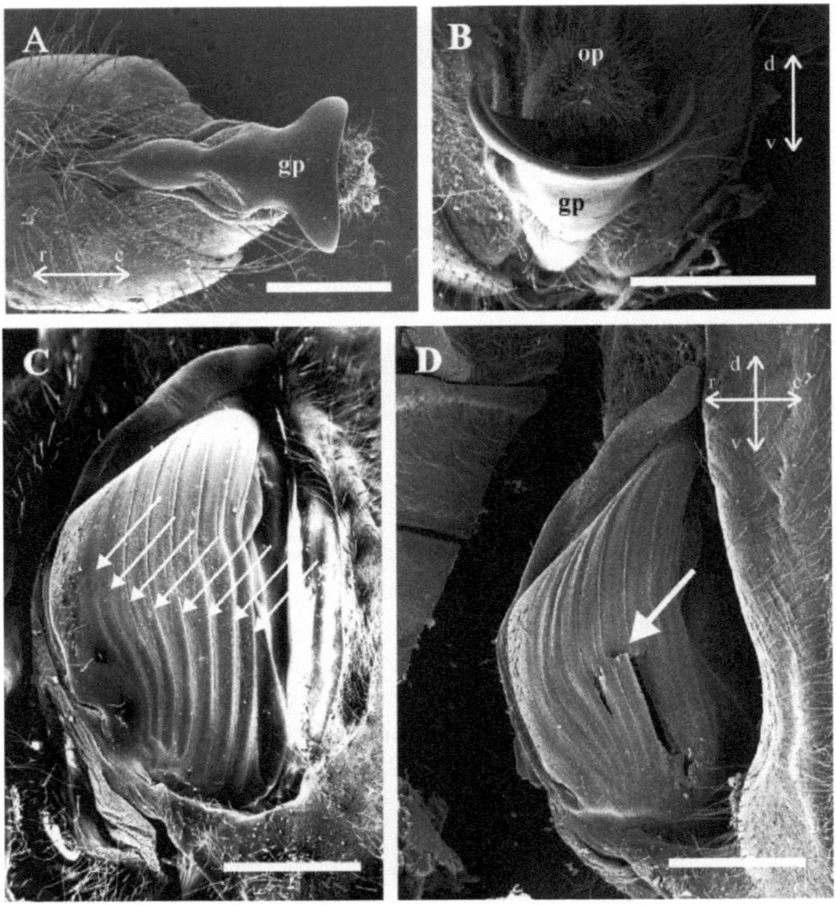

Fig. 2 Scanning electron microscopy images of structures involved in parasitism of *Okanagana rimosa* by *Emblemasoma auditrix*. (A and B) The genital plate (gp) of *E. auditrix* displaying a curved shape which opens dorsally towards the ovipositor (op). (C) Tymbal of an intact male of *O. rimosa* showing the distinct ribs (arrows). (D) Tymbal of a male with parasitoid infection in which the tymbal is perforated. Scale bars = 1 mm (Schniederkötter and Lakes-Harlan, 2004).

Other species that signal acoustically are also the target of parasitoid flies, such as various katydid and cicada species. In katydids, the levels of parasitism have been shown to correlate with the length of the calling song. Two species of katydids coexist with the ormiine fly, *Therobia leonidei*, their calls are similar, but one species calls with 5–9 pulses per call, whereas one only calls with one pulse per call. The species that calls with more pulses has been found to incur significantly higher levels of parasitism. This suggests that the length of the song directly influences the chances of being targeted, although it is not clear whether the advent of the shorter call is a direct response to the threat of fly infestation (Lehmann and Heller, 1998). Laboratory experiments on phonotaxis in *O. ochracea* has demonstrated that flies were able to locate the source of shorter calls just as readily as longer trills. The flies did preferentially move towards calls that were higher in amplitude and length however, suggesting there may be some advantages to modifying call parameters in species that are targeted by these parasitoids (Müller and Robert, 2002).

3.4 Sexual selection

There is some evidence of sexual selection potentially driving the evolution of acoustic characteristics in some insects. Work involving field crickets revealed a difference in mating success of males in which their calls could be differentiated based on acoustic characteristics that may attract females. Surveys conducted on field crickets demonstrated that paired males were often older and larger than unpaired males (Zuk and Simmons, 1997). In this instance sexual selection on acoustic signals could be inferred due to the inherently different calls that would be produced by larger males. A more comprehensive study explored this concept further in which the behaviour and reproductive success of an entire population of field crickets was tracked. In this study genetic analysis enabled for the fitness of males and females to be quantified over two generations of offspring. Support for sexual selection theory was given by the finding that males were found to have an increased variance in lifetime reproductive success, compared to females (Rodríguez-Muñoz et al., 2010).

Assessing the role of acoustic parameters on mate choice is simpler within a laboratory setting, where acoustic signals can be easily controlled and manipulated. Several studies have sought to try and unpick what acoustic features may be influencing mate choice, including crickets (Klappert and Reinhold, 2003; Nandi and Balakrishnan, 2013; Wagner, 2011) and katydids (Snedden and Greenfield, 1998). These studies have revealed that some call features generally dominate others in terms of female mate preference, such as call intensity being a better predictor of mate's choice over call rate. While useful for assessing some aspects of mate choice based on call parameters, it is difficult to faithfully replicate the conditions of the wild habitats in the lab, where features such as signaller densities, structure and intensity can all be highly variable. Some studies have tried to go further within the lab to try and simulate natural acoustic signals. One such study involving female field crickets attempted to characterise the acoustics soundscape experienced by females and create a computer model representing a virtual field cricket. These virtual field crickets were then presented with virtual male calling songs that varied in call intensity, and their responses documented. Follow-up behavioural experiments in the field confirmed that their computational model was able to predict the outcomes of pair formation at a population level, potentially paving the way for more experiments of this design in the future, with more complex models that incorporate more call features (Mhatre and Balakrishnan, 2008).

It is however important to note that this demonstration of sexual selection in acoustic signalling is not necessarily direct evidence that the selection is acting on the acoustic signal. There are several other factors that need to be considered, all of which can convey the fitness of one individual to another, beyond just the acoustic signal. Other communicative modalities include visual, tactile, and chemosensory cues, all of which may be acted on by sexual selection. For example, the cuticular hydrocarbons in *Drosophila melanogaster* and field crickets have been shown to be under sexual selection, with both species also producing courtship songs (Hine et al., 2004; Simmons et al., 2013; Thomas and Simmons, 2009).

4. Male competition in acoustic signalling

Acoustic signalling by males could be used competitively to either defend resources that may attract a female or to increase the likelihood of a female encounter by outcompeting their rivals acoustically (Gerhardt and Huber, 2002). Resources that may be attractive for females are desirable foraging or egg-laying sites, while males may compete with each other for prime calling locations that increase the chance of a female hearing them (Greenfield, 1997).

4.1 Defending resources

Males of many species will try to avoid physical confrontation, if at all possible. Males of the grasshopper *Ligurotettix planum* will defend their territory by partaking in acoustic duels, with one eventually ceding the resource, suggesting that males of this species are able to assess the fitness of each other based on acoustic signals alone (Greenfield, 1997). Males of the katydid species *Nesonotus reticulatus* produce a courtship song via stridulation to attract females, but will produce a concurrent tremulatory signal in response to nearby males, presumably to defend a territory (Stumpner et al., 2013). Other insects have been shown to resort to physical aggression if acoustic signalling alone does not settle the conflict, which has been seen in the grig species *Cyphoderris monstrosa* (Mason, 1996). Within this species a male's song was found to be an honest signal of its physical prowess, with the winner of the confrontation often being the male that had been calling with greater sustain. Interestingly, in many species males have evolved to have two distinct acoustic signals, one used for mate attraction and one used for aggression (Alexander, 1961). The reasons for this distinction are not clear but may be related to certain call features being a better predictor of physical fitness while others may better convey intimidation among conspecific males.

4.2 Attracting females

The calling songs of males are diverse both temporally and spectrally. Globally, calling songs of insects can range from a few hundred hertz to hundreds of kilohertz. Extreme examples of this can be seen in some species of katydid, which call at some of the highest frequencies measured among Orthoptera (115–150 kHz; Fig. 3; Sarria-S et al., 2014). Where many males are aggregated in a localised space, calling males will often modify their call rates or alternate their calling patterns such that their calls do not continuously overlap (Gerhardt and Huber, 2002; Greenfield, 2015). Females have

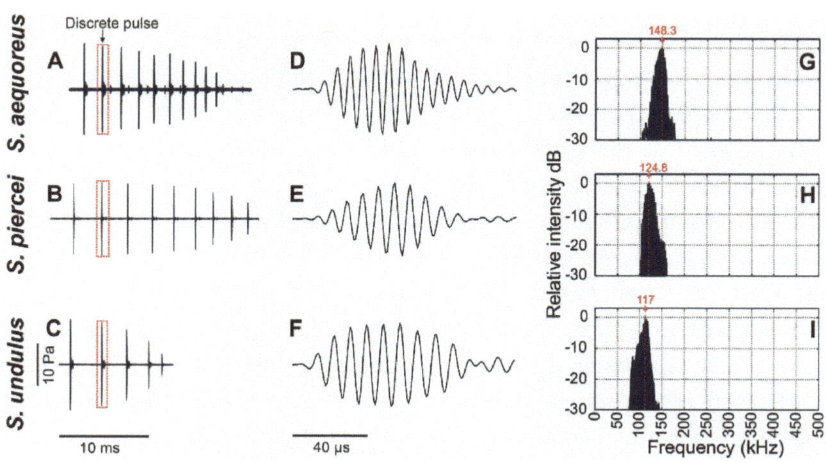

Fig. 3 Example temporal song structure of *Supersonus* spp. (A–C) Pulse train created during wing stroke. (D–F) Waveform of pulses highlighted in red. (G–I) Power spectra of pulses (Sarria-S et al., 2014).

been shown to base their preferences on males based on call rate and duration, but it is generally intensity that is the dominant factor when it comes to mate selection, as such it is more often the loudest male that attract the most females. The simplicity of some female auditory systems exacerbates this effect, with some female cricket and katydid ears only representing the loudest call detected, with lower intensity calls having little influence on their behaviour (Römer and Krusch, 2000).

5. Sound production in insects

Insects are small yet often inhabit ecosystems necessitating communication ranges that span the relatively large distances individuals may be from one another. Sound production in insects has led to a host of elegant solutions that allow insects to produce sounds loud enough to traverse large distances despite their small size, facilitating mating and protective acoustic measures (Fig. 4; Claridge, 2005).

5.1 Vibratory signalling

Historically, research into insect acoustics have focussed on far field signalling, in both a sound production and hearing context. It is only in the last few decades that near field sounds, including solid-borne vibrations, have received increased research efforts (Cocroft, 2011; Cocroft et al., 2014). The prevalence in vibratory signalling in insects continues to expand, with many insects previously considered as non-acoustic being revealed as using substrate-borne vibratory signals, which includes caterpillars (Scott et al., 2010), sawfly larvae (Fletcher, 2007) and beetle pupae (Kojima et al., 2012).

Unlike airborne sounds, vibratory signals are inherently local in their reach, with most vibratory signals only being effective within 1 m of the signalling insect (Van Staaden and Römer, 1997). The propagation of substrate borne signals is more diverse than their airborne counterparts. In air, longitudinal waves are transmitted through a medium, which is relatively uniform in its composition. The transmission of substrate borne vibrations on the other hand can be affected by the type, geometry, and composition of substrate through which they are propagating (Velilla et al., 2020). Vibratory signals used by insects include longitudinal waves, but they additionally utilise bending, torsional and transverse waves. Substrate borne waves can be categorised by their transmission mode, shape, and energy distribution, all of which have important implications for the propagation and reception of these signals by insects (Elias and Mason, 2014).

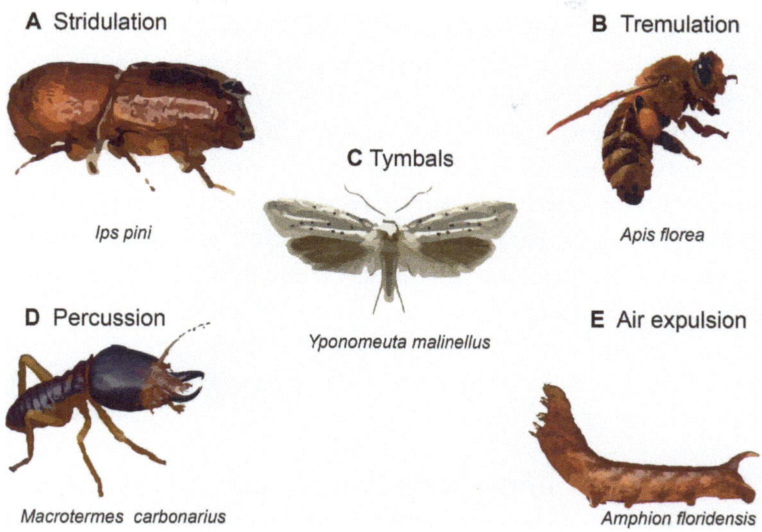

Fig. 4 Mechanisms of sound production in insects. (A) Stridulation. The *Ips pini* beetle produces chirps using a file and scraper mechanism located between the head and pronotum. (B) Tremulation. A piping signal produced by *Apis florea* bees which are produced through vibrations of the flight muscles. (C) Tymbals. The micromoth *Yponomeuta malinellus* produces a train of clicks via a pair of tymbals located on the hind wings which buckle during flight. (D) Percussion. The *Macrotermes carbonarius* termite produces sound by tapping the substrate with its head. (E) Air expulsion. Sound is produced by an *Amphion floridensis* caterpillar by airflow from the alimentary canal. Image credit CC. A, Pennsylvania Department of Conservation and Natural Resources - Forestry Archive. B, pixabay. C, Sarefo. D, Orionmystery. E, Greg Dwyer. Images recoloured in Adobe Illustrator.

Insects can produce both intentional (active) and unintentional (passive) vibratory signals. Insects may produce passive vibrations due to behaviours such as locomotion, feeding and digging (Guedes et al., 2012). Active signals are those which have evolved to convey messages and are much more conspicuous and stereotyped in comparison to passive signals. The mechanisms used by insects to produce vibratory signals can be categorised into four types: stridulation, tremulation, drumming and tymbal buckling. Stridulation involves vibrations being generated due to the friction of different body parts moving over one another or directly contacting the substrate. The structures used to generate vibrations in this way vary among insects but include mandibles and file and scraper mechanisms. The caterpillars of the moth *Drepana arcuata* possess structures near the rear of the body known as 'oars', which they use to scrape against leaves, producing vibratory signals for communication with conspecifics (Ishay et al., 1974;

Yack et al., 2001). Tremulation describes the process by which body movements are used to initiate substrate-borne signals and can include, trembling, shaking and swaying (Hill, 2014). Katydids are known to have a vibratory component included in their courtship displays, which they produce though the movement of their heads (Morris, 1980). Drumming involves an insect striking the substrate with a non-specialised body part to produce vibrations, parasitoid wasps are known to drum on leaves with their antennae to locate potential hosts (Otten et al., 2001). Tymbals are modified areas of insect cuticle, backed by air with muscle attachments on their inner surface. Many species of Hemiptera possess tymbals and use them during courtship songs to produce plant borne vibratory signals (Wessel et al., 2014).

5.2 Air and water borne sound production mechanisms

Sound signals propagate more readily through air and water meaning acoustic signals travelling through these mediums reach further than substrate-borne signals of comparable amplitude. Still, many variables can affect the propagation of sound through air and water, such as attenuation through the medium and diffraction around obstacles. Insects must therefore develop strategies and mechanisms for not only creating sound but also enhancing the sound amplitude to increase the likelihood of it reaching its intended target.

5.2.1 Percussion

Some insects produce percussive sounds by striking two parts of their body together. Males of all three species of whistling moth (genus *Hecatesia*) produce ultrasonic acoustic signals as part of a courtship display, both to attract females and to establish territories. The forewings of the males of these moths have two prominent cuticular knobs on the leading edge of each wing which produce sounds when the opposing forewings are struck together (Bailey, 1978). Two species (*Hecatesia thyridon* and *H. fenestrata*) produce these sounds during display flights, creating broadband sounds, some of which are audible to humans. One species (*H. exultans*) however, produces a pure tone signal at 30 kHz and does so while perched on low vegetation (Alcock and Bailey, 1995).

5.2.2 Stridulation

Stridulation produces sound by frictional mechanisms, involving the movement of two specialised body parts against each other in a systematically coupled way (Claridge, 2005). Stridulation is the mechanism of sound production in certain species of Orthoptera and plays a crucial role in reproductive behaviour (Haskell, 1958). Stridulatory sounds are often produced by the animals running a scraper across a complex of ridges (also known as plectrum and file). In some katydids the sounds are produced by the action of the wings rubbing together, with a row of teeth on the cubital vein of one wing rubbing against a plectrum on the costal margin of the other (Bailey, 1991). Grasshoppers, such as *Chorthippus biguttulus,* produce sound by rubbing a row of pegs on the side of their hind femur against a modified vein on the forewing. Each peg strike produces a click creating audible sounds at 5–6 kHz (Elsner, 1974, 1983). Stridulatory signalling is not limited to terrestrial insects; aquatic insects also produce a plethora of sounds, demonstrated with underwater acoustic surveys which are becoming increasingly vital in ecological research on freshwater ecosystems (Greenhalgh et al., 2020). Numerous aquatic insects have been shown to produce courtship songs to attract mates such as the water boatman *Micronecta scholtzi,* which produces sound by rubbing a scraper on the right paramere against a ridge on the left lobe of the eighth abdominal segment (Jansson, 1989). The sound is extremely loud given the size of the insect (up to 100 dB SPL at 1 m: Fig. 5; Sueur et al., 2011), with the insect thought to utilise a bubble of air around it as it dives to enhance song amplitude underwater (Reid et al., 2018).

5.2.3 Click mechanisms

Tymbals are specialised areas of insect cuticle that have undergone localised thinning (Bennet-Clark, 1997), they often have a muscle attached to one side that initiates buckling, although some tymbals are known to be actuated by aeroelastic deformations of wings during flight (O'Reilly et al., 2019, 2021). Cicadas have one of the best studied tymbals among insects. They possess a large tymbal plate that runs parallel to a series of vertical ribs surrounded by a springy pad of resilin. The tymbal is buckled due to muscular contraction resulting in a click, which is among the loudest sounds measured in insects (158 dB SPL) at their calling song frequency of 4.3 kHz (Young and Bennet-Clark, 1995). Actuated tymbals are also found among Lepidoptera, which use them to produce sound for courtship and defence (Dall'Asta, 1988; Fullard and Heller, 1990).

5.2.4 Air expulsion

There are several examples of insects producing sounds by forcibly squeezing air out of their bodies, often though tracheal spiracles. This kind of sound production is well documented in the cockroach *Gromphadorhina portentosa.* The cockroach produces sound from a pair of modified spiracles, which are

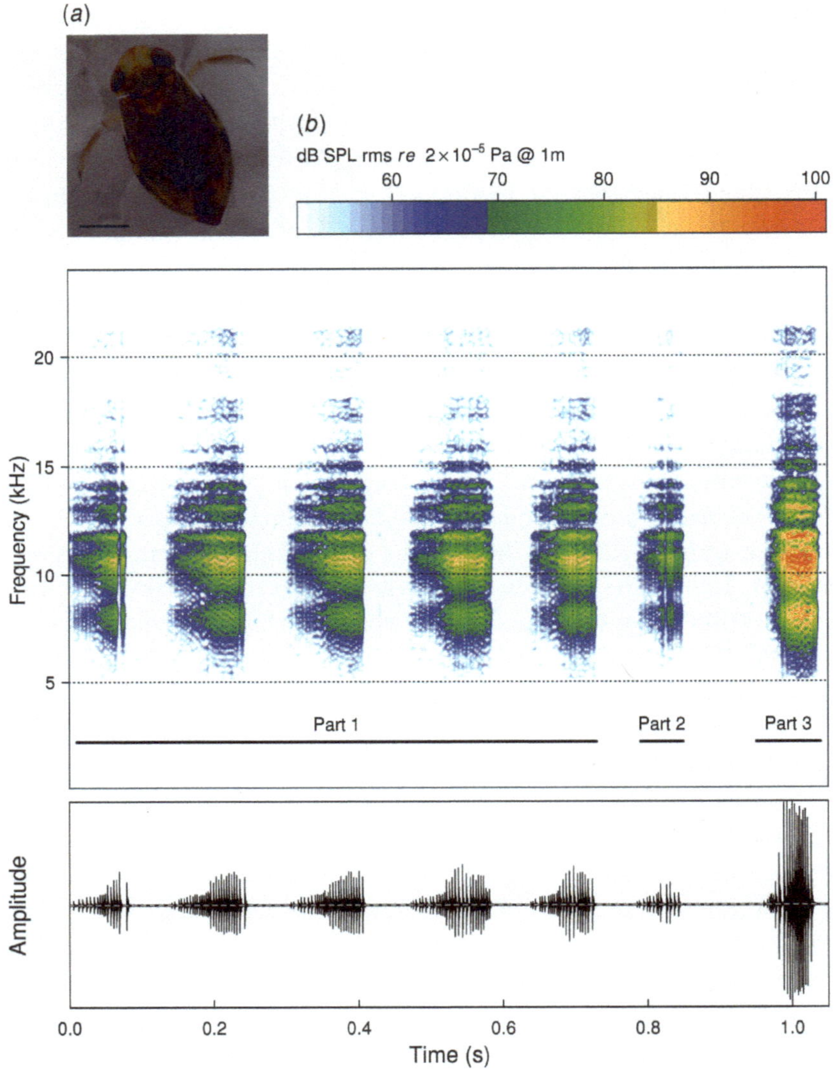

Fig. 5 Calling song of *M. scholtzi.* A Dorsal view of an adult (scale bar = 0.5 mm). B Calling song consisting of three main parts differing in their temporal and amplitude parameters while having a similar frequency structure (Sueur et al., 2011).

connected to its tracheal tubes by a large horn shaped structure. Sound is produced during aggressive encounters, courtship and during copulation. These sounds are relatively broadband but have most of their energy at around 8.5 kHz, the resonant frequency of the horn (Nelson, 1979; Nelson and Fraser, 1980). The Death's-head hawkmoth produces sound by expelling air out of their pharynx as a defence mechanism. The sound production is characterised by having an inflation phase and a deflation phase which collectively last around 200 ms, both of which produce their own unique acoustic signal. The inflation phase consist of around 50 pulses of decreasing frequency with a peak in energy at around 8 kHz, while the deflation phase generally has a less clear acoustic pattern and is produced with lower amplitudes (Brehm et al., 2015). The caterpillars of some moths have also been found to produce sound be contracting their bodies and expelling air from enlarged spiracles on certain body segments. When disturbed, these caterpillars will produce whistling trains at frequencies ranging from 9 to 22 kHz (Bura et al., 2011; Rosi-Denadai et al., 2018), which can also be followed by regurgitation (Brown et al., 2007) or the production of other defensive chemicals (Bura et al., 2009). It is thought that sound produced in this manner is a defence mechanism against avian predators, which startles attacking birds, affording the caterpillar the chance to survive.

6. Hearing in insects

Acoustic perception refers to the detection of sound-induced mechanical waves that propagate through a medium, which are subsequently converted into electrical signals by auditory organs. Acoustic perception among insects is vital to their ecology, with their auditory systems all needing to accomplish the same basic tasks, which influence and inform behaviour. Information gleaned from externally perceived sounds can give the hearer clues about things to avoid, such as predators, or things to be approached, including conspecifics and the locality of a resource. Insect hearing therefore serves the function of detecting, classifying, and reacting to acoustic cues, localising them and filtering out the relevant ones from a backdrop of noise. Hearing in insects is diverse, with some 30 orders and over 900,000 species evolving auditory functionality (Grimaldi and Engel, 2005). The earliest form of hearing in insects was likely through the perception of substrate-borne vibrations (Cocroft, 2011), airborne hearing subsequently evolved independently in at least nine insect orders: Orthoptera, Mantodea, Blattodea, Hemiptera, Hymenoptera, Coleoptera, Neuroptera, Lepidoptera, and Diptera (Fullard and Yack, 1993; Hoy and Robert, 1996; Nadrowski et al., 2011; Yager, 1999).

6.1 Mechanisms of hearing

Among insects, the tools required for auditory perception are almost exclusively derived from chordotonal organs, with a few examples of insects capable of detecting sound using filiform hairs (Eberl, 1999; Field and Matheson, 1998; Taylor and Yack, 2019; Yack, 2004). Chordotonal organs may arise along any segment of an insect's body. They are stretch receptors, which have the primary function of proprioception in insects, detecting changes in the relative movement and elastic deformation of body segments, enabling coordination during locomotion (Field and Matheson, 1998). The evolution of chordotonal organs to auditory perception organs typically occurs on one segment of an insect's body, and often in pairs (Hoy and Robert, 1996). Due to the widespread distribution of the location of chordotonal organs among insects, there is great diversity in the location of the ears across insect orders. Several species within Orthoptera utilise auditory organs in their forelegs and first abdominal segment (Stumpner and Von Helversen, 2001), while mosquitoes, midges and honeybees hear using the Johnston's organ, located at the base of the antenna (Nadrowski et al., 2011). Other insects utilise hearing organs across different segments of their body, such as in the mantids of the *Creobroter* genus, which hears different frequencies with different spatially separated chordotonal organs (Yager, 1996). There are several examples of insects independently converging on the same hearing apparatus. For example, parasitic flies within the Emblemasomatini and Ormiini taxa both independently evolved hearing organs within their neck to hear and localise their larval hosts, singing crickets (Lakes-Harlan and Lehmann, 2015).

To achieve auditory functionality, auditory organs must vibrate in response to acoustic stimuli and be innervated such that these vibrations can be converted to electrical signals in the nervous system that eventually elicit or modify behaviour. In insects, sound receiving structures can be classified into two broad groups, antennal and tympanal ears. Antenna act as particle velocity receivers, detecting sound induced changes to the speed of air around them (Nadrowski et al., 2011). Tympanal ears on the other hand respond to changes in sound pressure (Hoy and Robert, 1996).

6.2 Enhancing the sound

Owing to their small size, insects face many challenges in terms of acoustic signalling and reception. Lower frequency sounds are sometimes desirable because they travel greater distances, yet the longer wavelengths associated with these low frequencies are problematic as they are often many times longer than the size of the insect that is trying to hear them. Furthermore, the acoustic energy that is available to the animal is proportional to its surface area, consequently, most insects struggle to detect low amplitude sounds, with auditory thresholds being typically in the range of 30–50 dB SPL (Pollack, 2017).

6.2.1 Passive improvement of sensitivity

Some insect ears are resonantly tuned to the dominant frequency of the songs of their conspecifics. The tympana of some crickets have a resonance at their preferred calling frequency of 5 kHz, increasing sensitivity at this frequency with acoustic energy being more readily transferred to the ear of the cricket in this ecologically important range (Paton et al., 1977).

The calling songs of katydids are often broadband in composition, spanning frequencies from a few kHz to within the ultrasound range. This limits a katydids' options in terms of signal enhancement, as resonant amplification only works at the resonance frequency. Acoustic energy can impinge on a katydid ear via two routes, directly from the atmosphere, and indirectly through a modified route within their respiratory system. This specialised route is a modification of the trachea through which the katydid breathes, with one end open to the atmosphere and the other end leading to the surface of the katydid's internal tympanum. The trachea is trumpet-like in shape, tapering from an opening of 1–2 mm at the surface of the insect's exoskeleton to around 0.1 mm at the tympanum, facilitating focusing of the acoustic energy entering the trachea. The gain brought about by this unique morphology has been estimated at around 10–20 dB over the entire frequency range utilised by katydids, representing a significant increase in perceived amplitude at the tympanum (Hoffmann and Jatho, 1995).

6.2.2 Active improvement of sensitivity

The mechanics of the resonating cricket ear and the trumpet-like trachea of the katydid are both examples of passive systems that collect, focus and amplify the available acoustic energy thereby improving the ear's sensitivity. A fundamentally different approach to increase sensitivity has been identified in insects such as fruit flies, mosquitoes and tree crickets: they possess active systems of signal enhancement (Mhatre, 2015). In active systems sound waves elicit cellular motility in the hearing structure. The generated mechanical energy is synchronised to and thereby amplifies the available acoustic energy, thus increasing sensitivity. The revealing feature of active systems is the presence of occasional spontaneous activation in the motile elements in the absence of an acoustic stimulus (spontaneous acoustic emissions), a feature long known in vertebrate hearing systems well known for active amplification (Hudspeth, 2008).

The mechanosensory cellular basis of insect ears are comprised of groups of structures called scolopidia, each of which contains one or two auditory nerve cells (Yack, 2004). When the ears are stimulated by an acoustic signal, these nerve cells undergo passive deformation, which opens mechanosensory channels and allows for the flow of ions, which initiate action potentials. Active sound enhancement works by using metabolic energy to exert a force on the ear in addition to the force that is applied acoustically, effectively amplifying the sound signal as heard by the receiver. In insects, it is the combined force of the movement of hundreds of scolopidia that creates this additional vibration of the ears. Like the resonance of cricket ears, there is scope for tuning by active force production in these structures. The mechanics of this tuning can be seen in the spectra of the spontaneous otoacoustic oscillations of the ears, with the spectra often matching the dominant frequency of the songs of the species concerned (Riabinina et al., 2011).

6.3 Antennal ears

The antennae of mosquitoes and *Drosophila* act as acoustic oscillators, they possess a Johnston's organ at the base of the antennae that has evolved acoustic sensing capability. The organ contains numerous mechanosensory scolopidia made up of innervated cilia. These antennae act as near field sound receivers and typically operate at low frequencies (<1 kHz) over short distances.

6.3.1 Mosquitoes

The antennae of male mosquitoes are coated in long, feather like fibrils that are attached to the head just beneath the eyes by a pedicel, which contains the Johnston's organ. The Johnston's organ in male mosquitoes contains around 16,000 mechanosensory neurons, an extremely high number for such an organ. The antennae react to changes in particle velocity by rocking back and forth at the pedicel, with its first resonance matching the flight tone of the female mosquito, making the organ adept at its primary function, mate detection (Roth, 1948). In mosquitoes mating occurs mid-air and is mediated by the exchange of acoustic signals over short distances (Belton, 1994). They mate in swarms, which are highly asymmetrical in their composition, with hundreds of males listening for the flight tones of only a few females (Manoukis et al., 2009). When stimulated with a female-like acoustic stimulus, the antennae of the male mosquito will exhibit nonlinear behaviour above a certain sound level threshold. The antennae have been found to exist in a bi-stable state, whereby the antennae can oscillate at two different amplitudes, depending on how the sound level changes over the course of stimulation (Jackson and Robert, 2006).

6.3.2 Drosophila

Acoustic communication in *Drosophila* is associated with courtship behaviour and is a major determining factor in courtship success (Coen et al., 2014). Males produce the courtship song by vibrating an outstretched wing. The song is made up of two components. In *D. melanogaster* this song consists of a sine component comprising of a pure tone of 160 Hz which is typically 0.2–1 s in duration, and a shorter (5–10 ms) pulse component produced at higher frequencies (200 Hz). The two components are produced in alternation with one another with regular inter-pulse-intervals which are species specific, mediating species recognition during courtship (Morley et al., 2018). The female *Drosophila* detects these songs using her antennal ears (Fig. 6). The Johnston's organ in *Drosophila* has only a few hundred scolopidia, corresponding to around 500 mechanosensory neurons (Boo and Richards, 1975). The *Drosophila* antennae forms a rotating pendulum and can oscillate with only one degree of freedom. The antennae can be approximated as a moderately damped harmonic oscillator (Göpfert et al., 2005; Göpfert and Robert, 2003). The mechanics of *Drosophila* antennae are nonlinear, with the tuning of female antennae being dependent on

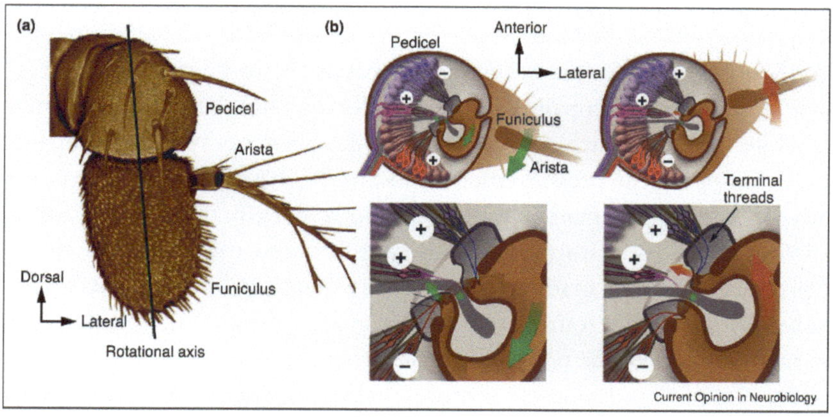

Fig. 6 *Drosophila* Johnston's organ (JO) and direction sensitivities of chordotonal stretch receptor neurons (JONs). (A) Frontal view of the *Drosophila* antenna. When stimulated, the arista and the funiculus vibrate about their longitudinal axis, activating JONs in the pedicel of the antenna. (B) Cross-sections through the pedicel-funiculus joint (top: overviews, bottom: close up), depicting the funicular connection sites of JONs. Deflecting the antenna posteriorly stretch-activates JONs that connect to the posterior side of the funiculus but inactivates JONs connecting to its anterior side. For posterior deflections of the antenna, the signs of activation are inversed (Albert and Göpfert, 2015).

the amplitude of the sound impinging on it (Albert and Kozlov, 2016; Albert et al., 2007). In low amplitude conditions the frequency tuning of the female *Drosophila* matches the content of the male courtship song. As the amplitude increases however, the antennae undergo a spectral tuning shift, moving the resonance of the antennae by several Hz, detuning it from the male's song frequency. The nonlinearity of this amplitude dependent spectral tuning enables the amplification of quiet, low frequency sounds. These key mechanical properties and adaptations are dominated by transducer modules that are housed within the pedicel of the antennae (Karak et al., 2015). These adaptations enable the antennae of *Drosophila* to extract the pulses of conspecifics in an efficient and frequency-dependent way.

6.4 Tympanal ears

Tympanal ears detect changes in sound pressure using a tympanum, an air backed section of insect cuticle that has typically undergone some thinning such that it vibrates when stimulated by an airborne acoustic signal. The air-backed tympanum allows auditory organs to sense the vibrations of the tympanum directly, omitting further impedance matching apparatus (Michelsen and Larsen, 2007). To be precise, such tympanal ears detect differences in pressure acting on the two sides of the tympanum. Depending on whether the sound reaches one or both sides they are either considered pressure or pressure gradient receivers. In pressure receivers the outer surface of the tympanum is open to the medium the sound is travelling in, while the other surface faces a closed air-filled cavity. The tympanum moves following the pressure oscillations on the exposed side of the tympanum. In a pressure gradient receiver, both sides of the tympanum are exposed to the medium of sound propagation and the associated changes in pressure. Pressure gradient tympana are associated with some form of trachea by which the sound enters and is guided through the body towards the inner surface of the tympanum. As the sound reaches both the exterior and interior surfaces on different routes they experience subtle but important differences in time delay (Veitch et al., 2021).

6.4.1 Tree cricket

Research into the hearing capabilities of the tree cricket, *Oecanthus henryi,* has revealed that the mechanics of these forewing borne ears share many similarities of the active hearing previously described in mosquitoes and fruit flies, such as active amplification, adaptive tuning and strong inducible self-oscillation of the tympanal ear (Fig. 7; Mhatre and Robert, 2013). The mechanics allow for high auditory sensitivity and tuning to conspecific

Fig. 7 The Auditory System of *Oecanthus henryi*. (A) The position of the auditory organ is marked on the foreleg of a female tree cricket. It consists of two membranes, the anterior (ATM) and posterior (PTM) tympanal membranes (ATM depicted in inset). (B) The ATM oscillates spontaneously even in the absence of sound stimulation. A frequency distribution of ATM displacements over 25 s exhibits a unimodal pattern. The distribution broadens upon administration of dimethyl sulfoxide (DMSO). There are no periodic oscillations upon death (Mhatre and Robert, 2013).

songs, with the animals also able to adapt their tuning to compensate for variations in conspecific signal characteristics brought about by changing environmental conditions (Mhatre et al., 2016).

7. Hearing and sound production as a defence

The evolution of hearing as a defence in insects is widespread, particularly among those insects which fly at night, where the emergence of insectivorous bats that hunt using echolocation has put them under enormous predation pressure. Around 70% of the ~1400 species of bat are known to be insectivorous, targeting and localising their prey either by echolocation, or by listening to insect borne sounds (Jones and Rydell, 2003). At least 14 of the currently known insect auditory systems are sensitive to the ultrasonic calls of echolocating bats and are thought to have evolved as a direct consequence of the pressure of bat predation (Greenfield, 2016). Other insect orders appear to have evolved hearing apparatus for the primary purpose of intraspecific communication and have then subsequently evolved predator detection as a secondary function.

7.1 Ultrasound sensitivity and bat predation

7.1.1 Strategies for defence

Defensive strategies of hearing insects to avoid bat predation are diverse and range from completely avoiding being detected to attempting to avoid capture through a behavioural response once they are detected.

7.1.2 Detection avoidance

Sounds used by bats for echolocation have to travel twice the distance for the bat to hear the insect's echo (to the insect and back) than for the insect to hear the approaching bat's call. Couple this with the fact that the high frequencies in bat calls attenuate quickly in air and that insects are generally small bodied producing faint echoes gives the insect an intrinsic advantage over the attacking bat in that it can hear the bat approaching over much greater distances than the bat is able to detect the insect (Goerlitz et al., 2010; Nakano and Mason, 2018; Surlykke and Kalko, 2008). Some bat species have broken through this mechanism though by using whispering or stealth echolocation calls which allows them to hear the moth's echo before the moth hears them (Goerlitz et al., 2010).

Ultrasound sensitive moths have been shown to be able to respond accordingly depending on the proximity of the attacking bat. Moths stimulated with low amplitude ultrasound, representing a distant but approaching bat, will steer themselves away from the sound source (Roeder, 1962). However, those same moths stimulated with high amplitude ultrasound, indicating a much closer bat, will fly erratically and often dive for vegetation, as a last-ditch attempt to avoid capture. This type of negative phonotaxis, turning and flying away from a source of sound, has also been demonstrated in crickets (Moiseff et al., 1978) and locusts (Robert, 1989). Ultrasound stimulation elicits a different response in green lacewings, which have been shown to simply cease flight and passively drop out the sky, thus removing them from the approaching bats path (Miller and Olesen, 1979). Some insects do not possess bilateral hearing, such that they have no directional sensitivity to incoming ultrasound. This is the case in the mantids, which respond to ultrasound stimulation in flight by turning in a seemingly random direction with respect to the source of sound and often diving down towards the substrate (Yager and Hoy, 1986; Yager et al., 1990). It is not just during flight that insects are vulnerable though, as many bats are specialist gleaners, plucking insects from substrates and surfaces by listening for the sounds produced by insects, either passively, through locomotion for example, or by courtship calls. Both male wax moths and katydids have been shown to stop calling and fall silent when presented with a bat like stimulus, while female wax moths will not take to the wing and search for the calling males while being stimulated with ultrasound (Faure and Hoy, 2000; Greenfield and Baker, 2003).

Another even simpler defensive strategy used by moths is to reduce the strength of the echoes they reflect. Certain moths that are unable to hear, have been shown to possess scales that are able to absorb the sound energy of a bat's echolocation call, thus giving the moth the ability to disappear into the night sky, seemingly becoming acoustically invisible (Neil et al., 2020a,b; Shen et al., 2018). Such ultrasound absorption may not be exclusive to deaf moths however.

7.1.3 Capture avoidance

Despite all their best efforts, many insects will still be detected by echolocating bats, this does not necessarily mean they are caught, because many insects have another level of defence to call upon. Numerous insects with ultrasound sensitive ears are able to perform evasive manoeuvres upon detecting an approaching bat. These include moths (Roeder, 1962), crickets (Huber and Thorson, 1985; Nolen and Hoy, 1986; Römer and Holderied, 2020), katydids (Schulze and Schul, 2001), lacewings (Miller, 1971), and antlions (Holderied et al., 2018). Insects have been shown to be able to grade their responsiveness to acoustic stimuli based on the temporal structure of the bat's echolocation behaviour. A bat's click rate will change during different phases of a hunt, from a few per second while searching for prey, through to tens and finally >100 per second as the bat localises and closes in on its prey (Fenton et al., 2014). Graded responses to such characteristic indicators of threat are useful as they allow an insect to avoid spending most of its flight time flying maximally erratically rather than foraging or finding a mate. Green lacewings have been found to adjust their avoidance behaviour based on bat call rate. The primary response of a lacewing is to stop flying and passively drop out the sky upon hearing a bat, however, if they detect the high call rates of the final phase of hunting bats echolocation, they will spread their wings, slowing down their fall to descend in a less predictable way (Miller and Olesen, 1979).

7.1.4 Sound production

Many insects have evolved sound production mechanisms of their own, with tiger moths, geometrids, hawkmoths, and tiger beetles all producing ultrasonic clicks in response to bat echolocation calls. In a similar way to the late evasive manoeuvres of the lacewing, most of these insects produce their ultrasonic clicks towards the end of a bat's typical approach, with clicking occurring in response to high repetition rate bat calls (Ratcliffe and Fullard, 2005). There are also examples of deaf insects producing clicks continuously during flight (Krivoruchko et al., 2021; O'Reilly et al., 2019, 2021).

Being deaf, these insects are unable to respond to approaching bats, and instead will continuously produce ultrasonic clicks, typically in synchrony with the wingbeat during flight.

There are many ways in which these ultrasonic clicks might confer some sort of survival advantage to the moth. The simplest explanation is that these clicks startle the bat in some way. Analogous to the visual startling seen in many insect species that are hunted by optical predators, it may be the case that the sudden onset of clicks heard by a hunting bat disrupts their approach sufficiently enough to allow the insect to escape (Bates and Fenton, 1990). Evidence for this occurring in bats is scarce, with some experiments showing that the wing clicks of the butterfly *Aglais io* may startle bats and mice (Møhl and Miller, 1976; Olofsson et al., 2012). There is also evidence that stridulatory clicks of certain species of cockroach startle rats and voles (Guthrie, 1966). If emitted at the correct rate and duty cycle, the clicks could also potentially jam the bat's sonar, making it harder for the bat to localise their prey. Alternatively, the clicks could serve as some sort of aposematic signal, warning the predator that the prey is harmful or toxic in some way. This is the case in several tiger moths, which are known to be unpalatable; bats will quickly learn to associated clicks with inedible moths and will stop targeting them after only a few unpleasant experiences. Different moth species will often converge on the same acoustic warning signal, in a process known as Müllerian mimicry, which can be seen in tiger moths and Tineidae, whose clicks share many of the same spectral and temporal features (Fig. 8; O'Reilly et al., 2019, 2021). Other insects capitalise on this associated learning by mimicking the ultrasonic clicks of unpalatable moths, even if they themselves, are perfectly edible. Known as Batesian mimicry, this allows the insect to piggyback on the acoustic protection afforded by producing sound, offering the same survival benefits as if it were toxic (Barber and Conner, 2007).

Determining exactly which of these strategies is at play involves pitting these nocturnal insects against their bat foes. Experiments involving capture attempts by big brown bats and palatable clicking moths found that startling was an unlikely function of ultrasonic clicks; bats quickly learnt that the clicking moths were edible and happily ate them, indicating that any sort of startle response would be quickly habituated to and lost if this was their function (Bates and Fenton, 1990; Hristov and Conner, 2005). The jamming hypothesis has been demonstrated in one species of tiger moth, *Bertholdia trigona*, which produce ultrasonic clicks of a high enough duty cycle to effectively jam a bat's sonar at the final phase of its attack. Unlike

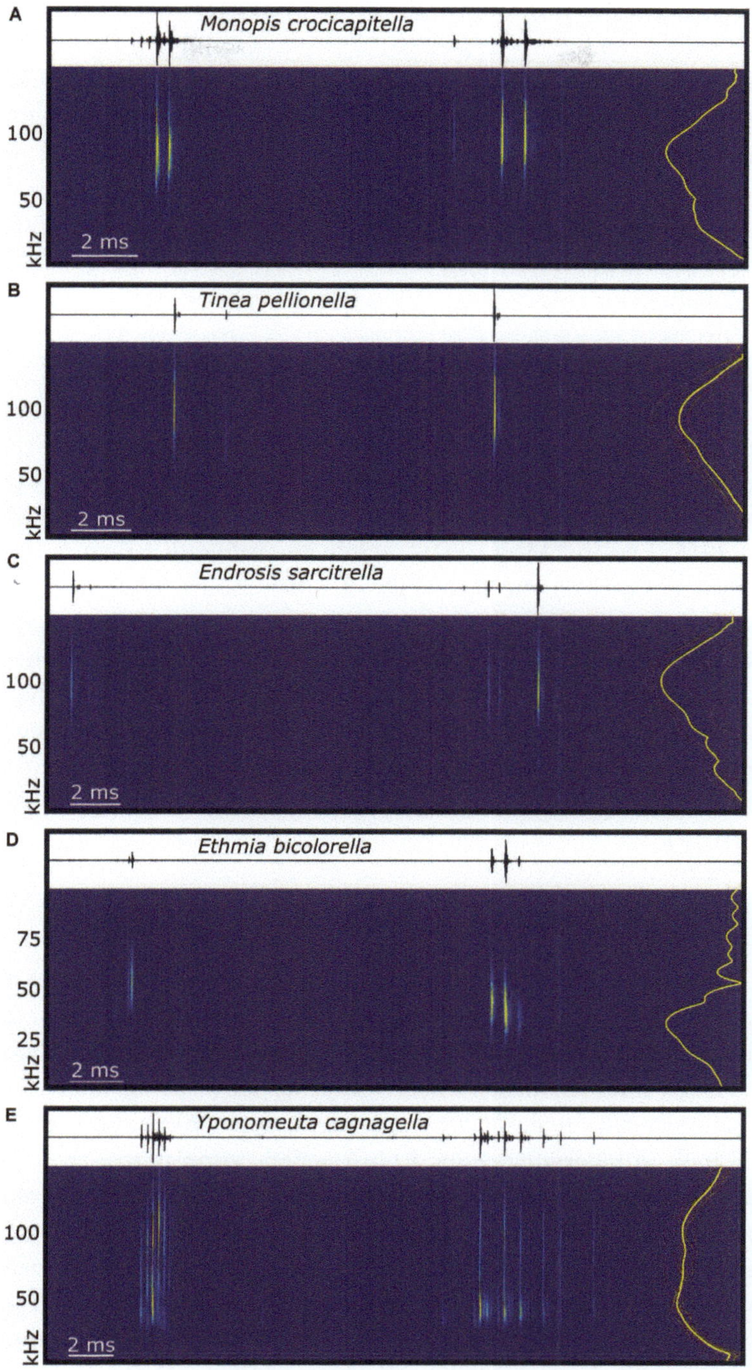

Fig. 8 Spectral and temporal characteristics of microlepidoptera sounds. The waveform and spectrogram of typical examples of in-flight acoustic emissions of five species of microlepidoptera, (A) *Monopis crocicapitella* (Tineidae), (B) *Tinea pellionella* (Tineidae), (C) *Endrosis sarcitrella* (Oecophoridae), (D) *Ethmia bicolorella* (Depressariidae), and (E) *Yponomeuta cagnagella*, with a power spectrum showing the normalised click amplitude for the species on the right (O'Reilly et al., 2021).

a jamming signal, ultrasonic clicks that have evolved to warn of distastefulness tend be of low duty cycle and have relatively few clicks per activation event (Corcoran et al., 2009). This is exactly what was found in two species of deaf moth in the Yponomeutidae family, which produce clicks, with peak frequencies within both the hearing range of bats and the range of frequencies produced by their fellow toxic moths, the arctiines (O'Reilly et al., 2019).

7.2 Other nonauditory defences

The ultrasound absorption found in some moth wings has already been discussed, but there is another example of a structural type of defence, seen in the same moth family, Saturniidae. Many moths within this family possess elongated hindwings, which terminate in a twisted and coiled end (Barber et al., 2015; Lee and Moss, 2016), while an analogous structure was also recently identified on the forewings of other moths (Neil et al., 2021). These structures have been shown to create strong echoes, which may function to divert a bats attack towards them or shift the perceived centre of the echo source, causing the bat to strike towards the periphery of the moth instead of its body. In experiments involving interactions between bats and *Actias luna* moths with intact and ablated tails, the hindwings were found to protect the moths from predation in around 50% of attacks (Rubin et al., 2018).

8. Future directions

Hearing in insects continues to be explored, with both novel species and mechanisms still readily being identified (Taylor and Yack, 2019). Similarly, sound production in insects is still vastly unexplored, with new roles for acoustics in many insects only just now being inferred (Low et al., 2021; Velilla et al., 2020; Vidal-García et al., 2020). Future research should look to incorporate robust phylogenetic comparative analyses into morphological and behavioural research to better understand the evolutionary origins of hearing and sound production and to assess which traits are ancestral or derived. For example, cicada signalling ancestry is relatively well established, yet a precursor to the ear has not been identified. Other groups have more comprehensive ear evolution scenarios such as mantises and grasshoppers (Song et al., 2020). Work should be undertaken to answer questions revolving around the appearance of ears on a given lineage, the selection pressures on those insects and subsequent physiological and behavioural adaptions that diversified the sensory organs further. Beyond evolution there are still many questions to be answered regarding the ecology of insect sounds. In terms of defensive sounds, much work has focused on insects defences against bats, yet there is a growing body of evidence that many other predators may also be the target of these acoustic displays (Low et al., 2021). Key to these studies will be a focus on the ecological function of sounds and identifying the benefits that sound brings to the prey. Coupled with this should be a focus on how these sounds are perceived by the predator, and the mechanisms behind this perception from the position of the predator. Bringing these key elements together will reveal more of the secrets of the fascinating world of insect sound and hearing, exposing more of the brilliant adaptations that have evolved in this enchanting group of animals.

References

Albert, J.T., Göpfert, M.C., 2015. Hearing in *Drosophila*. Curr. Opin. Neurobiol. 34, 79–85.

Albert, J.T., Kozlov, A.S., 2016. Comparative aspects of hearing in vertebrates and insects with antennal ears. Curr. Biol. 26, R1050–R1061.

Albert, J.T., Nadrowski, B., Göpfert, M.C., 2007. Mechanical signatures of transducer gating in the *Drosophila* ear. Curr. Biol. 17, 1000–1006.

Alcock, J., Bailey, W., 1995. Acoustical communication and the mating system of the Australian whistling moth *Hecatesia exultans* (Noctuidae: Agaristinae). J. Zool. 237, 337–352.

Alexander, R.D., 1961. Aggressiveness, territoriality, and sexual behavior in field crickets (Orthoptera: Gryllidae). Behaviour 17, 130–223.

Bailey, W.J., 1978. Resonant wing systems in the Australian whistling moth *Hecatesia* (Agarasidae, Lepidoptera). Nature 272, 444–446.

Bailey, W.J., 1991. Acoustic Behaviour of Insects. An Evolutionary Perspective. Chapman and Hall Ltd.

Bailey, W.J., 2003. Insect duets: underlying mechanisms and their evolution. Physiol. Entomol. 28, 157–174.

Barber, J.R., Conner, W.E., 2007. Acoustic mimicry in a predator–prey interaction. Proc. Natl. Acad. Sci. U. S. A. 104, 9331–9334.

Barber, J.R., Leavell, B.C., Keener, A.L., Breinholt, J.W., Chadwell, B.A., McClure, C.J., Hill, G.M., Kawahara, A.Y., 2015. Moth tails divert bat attack: evolution of acoustic deflection. Proc. Natl. Acad. Sci. U. S. A. 112, 2812–2816.

Barbosa, F., Rebar, D., Greenfield, M.D., 2016. Reproduction and immunity trade-offs constrain mating signals and nuptial gift size in a bushcricket. Behav. Ecol. 27, 109–117.

Bates, D.L., Fenton, M.B., 1990. Aposematism or startle? Predators learn their responses to the defenses of prey. Can. J. Zool. 68, 49–52.

Belton, P., 1994. Attractton of male mosquitoes to sound. J. Am. Mosq. Control Assoc. 10, 297–301.

Belwood, J.J., Morris, G.K., 1987. Bat predation and its influence on calling behavior in neotropical katydids. Science 238, 64–67.

Bennet-Clark, H., 1997. Tymbal mechanics and the control of song frequency in the cicada *Cyclochila australasiae*. J. Exp. Biol. 200, 1681–1694.

Boo, K.S., Richards, A.G., 1975. Fine structure of the scolopidia in the Johnston's organ of male *Aedes aegypti* (L.) (Diptera: Culicidae). Int. J. Insect Morphol. Embryol. 4, 549–566.

Brehm, G., Fischer, M., Gorb, S., Kleinteich, T., Kühn, B., Neubert, D., Pohl, H., Wipfler, B., Wurdinger, S., 2015. The unique sound production of the Death's-head hawkmoth (*Acherontia atropos* (Linnaeus, 1758)) revisited. Sci. Nat. 102, 1–13.

Brown, S.G., Boettner, G.H., Yack, J.E., 2007. Clicking caterpillars: acoustic aposematism in *Antheraea polyphemus* and other Bombycoidea. J. Exp. Biol. 210, 993–1005.

Bura, V.L., Fleming, A.J., Yack, J.E., 2009. What's the buzz? Ultrasonic and sonic warning signals in caterpillars of the great peacock moth (*Saturnia pyri*). Naturwissenschaften 96, 713–718.

Bura, V.L., Rohwer, V.G., Martin, P.R., Yack, J.E., 2011. Whistling in caterpillars (*Amorpha juglandis*, Bombycoidea): sound-producing mechanism and function. J. Exp. Biol. 214, 30–37.

Cator, L.J., Arthur, B.J., Harrington, L.C., Hoy, R.R., 2009. Harmonic convergence in the love songs of the dengue vector mosquito. Science 323, 1077–1079.

Claridge, M.F., 2005. Insect Sounds and Communication: Physiology, Behaviour, Ecology, and Evolution. CRC Press.

Cocroft, R.B., 2011. The Public World of Insect Vibrational Communication. (Wiley Online Library).

Cocroft, R.B., Gogala, M., Hill, P.S., Wessel, A., 2014. Fostering research progress in a rapidly growing field. In: Cocroft, R., Gogala, M., Hill, P., Wessel, A. (Eds.), Studying Vibrational Communication. Animal Signals and Communication, vol. 3. Springer, Berlin, Heidelberg, pp. 3–12. https://doi.org/10.1007/978-3-662-43607-3_1.

Coen, P., Clemens, J., Weinstein, A.J., Pacheco, D.A., Deng, Y., Murthy, M., 2014. Dynamic sensory cues shape song structure in *Drosophila*. Nature 507, 233–237.

Corcoran, A.J., Barber, J.R., Conner, W.E., 2009. Tiger moth jams bat sonar. Science 325, 325–327.

Dall'asta, U., 1988. The tymbal organs of the Lymantriidae (Lepidoptera). Nota Lepid 11, 169–176.

Eberl, D.F., 1999. Feeling the vibes: chordotonal mechanisms in insect hearing. Curr. Opin. Neurobiol. 9, 389–393.

Elias, D.O., Mason, A.C., 2014. The role of wave and substrate heterogeneity in vibratory communication: practical issues in studying the effect of vibratory environments in communication. In: Cocroft, R., Gogala, M., Hill, P., Wessel, A. (Eds.), Studying Vibrational Communication. Animal Signals and Communication, vol. 3. Springer, Berlin, Heidelberg, pp. 215–247. https://doi.org/10.1007/978-3-662-43607-3_12.

Elsner, N., 1974. Neuroethology of sound production in gomphocerine grasshoppers (Orthoptera: Acrididae). J. Comp. Physiol. 88, 67–102.

Elsner, N., 1983. A neuroethological approach to the phylogeny of leg stridulation in gomphocerine grasshoppers. In: Huber, F., Markl, H. (Eds.), Neuroethology and Behavioral Physiology. Springer, Berlin, Heidelberg, pp. 54–68. https://doi.org/10.1007/978-3-642-69271-0_5.

Endler, J.A., Basolo, A.L., 1998. Sensory ecology, receiver biases and sexual selection. Trends Ecol. Evol. 13, 415–420.

Falk, J.J., Ter Hofstede, H.M., Jones, P.L., Dixon, M.M., Faure, P.A., Kalko, E.K., Page, R.A., 2015. Sensory-based niche partitioning in a multiple predator–multiple prey community. Proc. R. Soc. Lond. B Biol. Sci. 282, 20150520.

Faure, P., Hoy, R., 2000. The sounds of silence: cessation of singing and song pausing are ultrasound-induced acoustic startle behaviors in the katydid *Neoconocephalus ensiger* (Orthoptera; Tettigoniidae). J. Comp. Physiol. A 186, 129–142.

Fenton, B.M., Jensen, F.H., Kalko, E.K., Tyack, P.L., 2014. Sonar signals of bats and toothed whales. In: Surlykke, A., Natchtigall, P.E., Fay, R.R., Popper, A.N. (Eds.), Biosonar. Springer-Verlag, New York, pp. 11–60.

Field, L.H., Matheson, T., 1998. Chordotonal organs of insects. Adv. Insect Physiol. 27, 1–228.

Fletcher, L.E., 2007. Vibrational signals in a gregarious sawfly larva (*Perga affinis*): group coordination or competitive signaling? Behav. Ecol. Sociobiol. 61, 1809–1821.

Fullard, J.H., 1994. Auditory changes in noctuid moths endemic to a bat-free habitat. J. Evol. Biol. 7, 435–445.

Fullard, J., Heller, B., 1990. Functional organization of the arctiid moth tymbal (Insecta, Lepidoptera). J. Morphol. 204, 57–65.

Fullard, J.H., Yack, J.E., 1993. The evolutionary biology of insect hearing. Trends Ecol. Evol. 8, 248–252.

Geipel, I., Kernan, C.E., Litterer, A.S., Carter, G.G., Page, R.A., Ter Hofstede, H.M., 2020. Predation risks of signalling and searching: bats prefer moving katydids. Biol. Lett. 16, 20190837.

Gerhardt, H.C., Huber, F., 2002. Acoustic Communication in Insects and Anurans: Common Problems and Diverse Solutions. University of Chicago Press.

Goerlitz, H.R., Ter Hofstede, H.M., Zeale, M.R., Jones, G., Holderied, M.W., 2010. An aerial-hawking bat uses stealth echolocation to counter moth hearing. Curr. Biol. 20, 1568–1572.

Göpfert, M., Robert, D., 2003. Motion generation by *Drosophila* mechanosensory neurons. Proc. Natl. Acad. Sci. U. S. A. 100, 5514–5519.

Göpfert, M., Humphris, A., Albert, J., Robert, D., Hendrich, O., 2005. Power gain exhibited by motile mechanosensory neurons in *Drosophila* ears. Proc. Natl. Acad. Sci. U. S. A. 102, 325–330.

Greenfield, M.D., 1997. Sexual selection and the evolution of advertisement signals. In: Owings, D.H., Beecher, M.D., Thompson, N.S. (Eds.), Communication. Perspectives in Ethology, vol. 12. Springer, Boston, MA, pp. 145–177. https://doi.org/10.1007/978-1-4899-1745-4_6.

Greenfield, M.D., 2015. Signal interactions and interference in insect choruses: singing and listening in the social environment. J. Comp. Physiol. A 201, 143–154.

Greenfield, M.D., 2016. Evolution of acoustic communication in insects. In: Pollack, G., Mason, A., Popper, A., Fay, R. (Eds.), Insect Hearing. Springer Handbook of Auditory Research, vol. 55. Springer, Cham, pp. 17–47. https://doi.org/10.1007/978-3-319-28890-1_2.

Greenfield, M.D., Baker, M., 2003. Bat avoidance in non-aerial insects: the silence response of signaling males in an acoustic moth. Ethology 109, 427–442.

Greenhalgh, J.A., Genner, M.J., Jones, G., Desjonquères, C., 2020. The role of freshwater bioacoustics in ecological research. Wiley Interdiscip. Rev. Water 7, e1416.

Grimaldi, D., Engel, M.S., 2005. Evolution of the Insects. Cambridge University Press.

Gu, J.-J., Montealegre-Z, F., Robert, D., Engel, M.S., Qiao, G.-X., Ren, D., 2012. Wing stridulation in a Jurassic katydid (Insecta, Orthoptera) produced low-pitched musical calls to attract females. Proc. Natl. Acad. Sci. U. S. A. 109, 3868–3873.

Guedes, R., Matheson, S., Frei, B., Smith, M., Yack, J., 2012. Vibration detection and discrimination in the masked birch caterpillar (*Drepana arcuata*). J. Comp. Physiol. A 198, 325–335.

Guthrie, D., 1966. Sound production and reception in a cockroach. J. Exp. Biol. 45, 321–328.

Gwynne, D.T., 2008. Sexual conflict over nuptial gifts in insects. Annu. Rev. Entomol. 53, 83–101.

Haskell, P., 1958. Stridulation and associated behaviour in certain Orthoptera. 2. Stridulation of females and their behaviour with males. Anim. Behav. 6, 27–42.
Heller, K.-G., Von Helversen, D., 1986. Acoustic communication in phaneropterid bushcrickets: species-specific delay of female stridulatory response and matching male sensory time window. Behav. Ecol. Sociobiol. 18, 189–198.
Hill, P.S., 2014. Stretching the paradigm or building a new? Development of a cohesive language for vibrational communication. In: Cocroft, R., Gogala, M., Hill, P., Wessel, A. (Eds.), Studying Vibrational Communication. Animal Signals and Communication, vol. 3. Springer, Berlin, Heidelberg, pp. 13–30. https://doi.org/10.1007/978-3-662-43607-3_2.
Hine, E., Chenoweth, S.F., Blows, M.W., 2004. Multivariate quantitative genetics and the lek paradox: genetic variance in male sexually selected traits of *Drosophila serrata* under field conditions. Evolution 58, 2754–2762.
Hoffmann, E., Jatho, M., 1995. The acoustic trachea of tettigoniids as an exponential horn: theoretical calculations and bioacoustical measurements. J. Acoust. Soc. Am. 98, 1845–1851.
Holderied, M.W., Thomas, L.A., Korine, C., 2018. Ultrasound avoidance by flying antlions (Myrmeleontidae). J. Exp. Biol. 221. jeb189308.
Hoy, R.R., 1992. The evolution of hearing in insects as an adaptation to predation from bats. In: Webster, D.B., Popper, A.N., Fay, R.R. (Eds.), The Evolutionary Biology of Hearing. Springer, New York, NY, pp. 115–129. https://doi.org/10.1007/978-1-4612-2784-7_8.
Hoy, R.R., Robert, D., 1996. Tympanal hearing in insects. Annu. Rev. Entomol. 41, 433–450.
Hristov, N.I., Conner, W.E., 2005. Sound strategy: acoustic aposematism in the bat–tiger moth arms race. Naturwissenschaften 92, 164–169.
Huber, F., Thorson, J., 1985. Cricket auditory communication. Sci. Am. 253, 60–73.
Hudspeth, A., 2008. Making an effort to listen: mechanical amplification in the ear. Neuron 59, 530–545.
Hunt, J., Richard, F.-J., 2013. Intracolony vibroacoustic communication in social insects. Insectes Soc. 60, 403–417.
Ishay, J., Motro, A., Gitter, S., Brown, M.B., 1974. Rhythms in acoustical communication by the oriental hornet, *Vespa orientalis*. Anim. Behav. 22, 741–744.
Jackson, J.C., Robert, D., 2006. Nonlinear auditory mechanism enhances female sounds for male mosquitoes. Proc. Natl. Acad. Sci. U. S. A. 103, 16734–16739.
Jansson, A., 1989. Stridulation of Micronectinae (Heteroptera, Corixidae). Ann. Entomol. Fenn. 55, 161–175.
Jarrige, A., Greenfield, M.D., Goubault, M., 2013. Male song as a predictor of the nuptial gift in bushcrickets: on the confounding influence of male choice. Anim. Behav. 85, 1427–1434.
Jones, G., Rydell, J., 2003. Attack and defense: interactions between echolocating bats and their insect prey. Bat Ecol., 301–345.
Karak, S., Jacobs, J.S., Kittelmann, M., Spalthoff, C., Katana, R., Sivan-Loukianova, E., Schon, M.A., Kernan, M.J., Eberl, D.F., Göpfert, M.C., 2015. Diverse roles of axonemal dyneins in *Drosophila* auditory neuron function and mechanical amplification in hearing. Sci. Rep. 5, 1–12.
Kawahara, A.Y., Plotkin, D., Espeland, M., Meusemann, K., Toussaint, E.F., Donath, A., Gimnich, F., Frandsen, P.B., Zwick, A., Dos Reis, M., 2019. Phylogenomics reveals the evolutionary timing and pattern of butterflies and moths. Proc. Natl. Acad. Sci. U. S. A. 116, 22657–22663.
Klappert, K., Reinhold, K., 2003. Acoustic preference functions and sexual selection on the male calling song in the grasshopper *Chorthippus biguttulus*. Anim. Behav. 65, 225–233.

Kojima, W., Ishikawa, Y., Takanashi, T., 2012. Deceptive vibratory communication: pupae of a beetle exploit the freeze response of larvae to protect themselves. Biol. Lett. 8, 717–720.
Krivoruchko, K., Goldshtein, A., Boonman, A., Eitan, O., Ben-Simon, J., Thong, V.D., Yovel, Y., 2021. Fireflies produce ultrasonic clicks during flight as a potential aposematic anti-bat signal. Iscience 24, 102194.
Lakes-Harlan, R., De Vries, T., 2014. Experimental infection of a periodical cicada (*Magicicada cassinii*) with a parasitoid (*Emblemasoma auditrix*) of a proto-periodical cicada (*Okanagana rimosa*). BMC Ecol. 14, 1–6.
Lakes-Harlan, R., Lehmann, G.U., 2015. Parasitoid flies exploiting acoustic communication of insects—comparative aspects of independent functional adaptations. J. Comp. Physiol. A 201, 123–132.
Lande, R., 1981. Models of speciation by sexual selection on polygenic traits. Proc. Natl. Acad. Sci. U. S. A. 78, 3721–3725.
Lee, W.-J., Moss, C.F., 2016. Can the elongated hindwing tails of fluttering moths serve as false sonar targets to divert bat attacks? J. Acoust. Soc. Am. 139, 2579–2588.
Lehmann, G.U.C., Heller, K.-G., 1998. Bushcricket song structure and predation by the acoustically orienting parasitoid fly *Therobia leonidei* (Diptera: Tachinidae: Ormiini). Behav. Ecol. Sociobiol. 43, 239–245.
Low, M.L., Naranjo, M., Yack, J.E., 2021. Survival sounds in insects: diversity, function, and evolution. Front. Ecol. Evol. 9, 91.
Manoukis, N.C., Diabate, A., Abdoulaye, A., Diallo, M., Dao, A., Yaro, A.S., Ribeiro, J.M., Lehmann, T., 2009. Structure and dynamics of male swarms of *Anopheles gambiae*. J. Med. Entomol. 46, 227–235.
Marshall, D., Cooley, J., 2001. Sexual signaling in periodical cicadas, *Magicicada spp.* (Hemiptera: Cicadidae). Behaviour 138, 827–855.
Mason, A.C., 1996. Territoriality and the function of song in the primitive acoustic insect *Cyphoderris monstrosa* (Orthoptera: Haglidae). Anim. Behav. 51, 211–214.
McCartney, J., Potter, M.A., Robertson, A.W., Telscher, K., Lehmann, G., Lehmann, A., Von-Helversen, D., Reinhold, K., Achmann, R., Heller, K.-G., 2008. Understanding nuptial gift size in bush-crickets: an analysis of the genus *Poecilimon* (Tettigoniidae: Orthoptera). J. Orthopt. Res. 17, 231–242.
Mccartney, J., Kokko, H., Heller, K.-G., Gwynne, D.T., 2012. The evolution of sex differences in mate searching when females benefit: new theory and a comparative test. Proc. R. Soc. B Biol. Sci. 279, 1225–1232.
Mhatre, N., 2015. Active amplification in insect ears: mechanics, models and molecules. J. Comp. Physiol. A 201, 19–37.
Mhatre, N., Balakrishnan, R., 2008. Predicting acoustic orientation in complex real-world environments. J. Exp. Biol. 211, 2779–2785.
Mhatre, N., Robert, D., 2013. A tympanal insect ear exploits a critical oscillator for active amplification and tuning. Curr. Biol. 23, 1952–1957.
Mhatre, N., Pollack, G., Mason, A., 2016. Stay tuned: active amplification tunes tree cricket ears to track temperature-dependent song frequency. Biol. Lett. 12, 20160016.
Michelsen, A., Larsen, O.N., 2007. Pressure difference receiving ears. Bioinspir. Biomim. 3, 011001.
Miles, R., Robert, D., Hoy, R., 1995. Mechanically coupled ears for directional hearing in the parasitoid fly *Ormia ochracea*. J. Acoust. Soc. Am. 98, 3059–3070.
Miller, L.A., 1971. Physiological responses of green lacewings (Chrysopa, Neuroptera) to ultrasound. J. Insect Physiol. 17, 491–506.
Miller, L.A., Olesen, J., 1979. Avoidance behavior in green lacewings. J. Comp. Physiol. 131, 113–120.

Møhl, B., Miller, L.A., 1976. Ultrasonic clicks produced by the peacock butterfly: a possible bat-repellent mechanism. J. Exp. Biol. 64, 639–644.
Moiseff, A., Pollack, G.S., Hoy, R.R., 1978. Steering responses of flying crickets to sound and ultrasound: mate attraction and predator avoidance. Proc. Natl. Acad. Sci. U. S. A. 75, 4052–4056.
Morley, E.L., Jonsson, T., Robert, D., 2018. Auditory sensitivity, spatial dynamics, and amplitude of courtship song in *Drosophila melanogaster*. J. Acoust. Soc. Am. 144, 734–739.
Morris, G.K., 1980. Calling display and mating behaviour of *Copiphora rhinoceros* Pictet (Orthoptera: Tettigoniidae). Anim. Behav. 28, 42–51.
Müller, P., Robert, D., 2002. Death comes suddenly to the unprepared: singing crickets, call fragmentation, and parasitoid flies. Behav. Ecol. 13, 598–606.
Nadrowski, B., Effertz, T., Senthilan, P.R., Göpfert, M.C., 2011. Antennal hearing in insects–new findings, new questions. Hear. Res. 273, 7–13.
Nakano, R., Mason, A.C., 2018. Early erratic flight response of the lucerne moth to the quiet echolocation calls of distant bats. PLos One 13, e0202679.
Nakano, R., Skals, N., Takanashi, T., Surlykke, A., Koike, T., Yoshida, K., Maruyama, H., Tatsuki, S., Ishikawa, Y., 2008. Moths produce extremely quiet ultrasonic courtship songs by rubbing specialized scales. Proc. Natl. Acad. Sci. U. S. A. 105, 11812–11817.
Nakano, R., Ishikawa, Y., Tatsuki, S., Skals, N., Surlykke, A., Takanashi, T., 2009. Private ultrasonic whispering in moths. Commun. Integr. Biol. 2, 123–126.
Nandi, D., Balakrishnan, R., 2013. Call intensity is a repeatable and dominant acoustic feature determining male call attractiveness in a field cricket. Anim. Behav. 86, 1003–1012.
Neil, T.R., Shen, Z., Robert, D., Drinkwater, B.W., Holderied, M.W., 2020a. Moth wings are acoustic metamaterials. Proc. Natl. Acad. Sci. U. S. A. 202014531.
Neil, T.R., Shen, Z., Robert, D., Drinkwater, B.W., Holderied, M.W., 2020b. Thoracic scales of moths as a stealth coating against bat biosonar. J. R. Soc. Interface 17, 20190692.
Neil, T.R., Kennedy, E.E., Harris, B.J., Holderied, M.W., 2021. Wingtip folds and ripples on saturniid moths create decoy echoes against bat biosonar. Curr. Biol. 31, 1–7.
Nelson, M.C., 1979. Sound production in the cockroach, *Gromphadorhina portentosa*: the sound-producing apparatus. J. Comp. Physiol. 132, 27–38.
Nelson, M.C., Fraser, J., 1980. Sound production in the cockroach, *Gromphadorhina portentosa*: evidence for communication by hissing. Behav. Ecol. Sociobiol., 305–314.
Nolen, T.G., Hoy, R.R., 1986. Phonotaxis in flying crickets. J. Comp. Physiol. A 159, 423–439.
Olofsson, M., Jakobsson, S., Wiklund, C., 2012. Auditory defence in the peacock butterfly (*Inachis io*) against mice (*Apodemus flavicollis* and *A. sylvaticus*). Behav. Ecol. Sociobiol. 66, 209–215.
O'Reilly, L.J., Agassiz, D.J., Neil, T.R., Holderied, M.W., 2019. Deaf moths employ acoustic Müllerian mimicry against bats using wingbeat-powered tymbals. Sci. Rep. 9, 1–9.
O'Reilly, L.J., Harris, B.J., Agassiz, D.J.L., Holderied, M.W., 2021. Convergent evolution of wingbeat-powered anti-bat ultrasound in the microlepidoptera. Front. Ecol. Evol. 9, 252.
Otten, H., Wäckers, F., Battini, M., Dorn, S., 2001. Efficiency of vibrational sounding in the parasitoid *Pimpla turionellae* is affected by female size. Anim. Behav. 61, 671–677.
Pascoal, S., Cezard, T., Eik-Nes, A., Gharbi, K., Majewska, J., Payne, E., Ritchie, M.G., Zuk, M., Bailey, N.W., 2014. Rapid convergent evolution in wild crickets. Curr. Biol. 24, 1369–1374.
Paton, J.A., Capranica, R.R., Dragsten, P.R., Webb, W.W., 1977. Physical basis for auditory frequency analysis in field crickets (Gryllidae). J. Comp. Physiol. 119, 221–240.
Pollack, G.S., 2017. Insect bioacoustics. Acoust. Today 13, 26–34.
Prestwich, K., 1994. The energetics of acoustic signaling in anurans and insects. Am. Zool. 34, 625–643.

Raghuram, H., Deb, R., Nandi, D., Balakrishnan, R., 2015. Silent katydid females are at higher risk of bat predation than acoustically signalling katydid males. Proc. R. Soc. B Biol. Sci. 282, 20142319.

Ratcliffe, J.M., Fullard, J.H., 2005. The adaptive function of tiger moth clicks against echolocating bats: an experimental and synthetic approach. J. Exp. Biol. 208, 4689–4698.

Regier, J.C., Mitter, C., Zwick, A., Bazinet, A.L., Cummings, M.P., Kawahara, A.Y., Sohn, J.-C., Zwickl, D.J., Cho, S., Davis, D.R., Baixeras, J., Brown, J., Parr, C., Weller, S., Lees, D.C., Mitter, K.T., 2013. A large-scale, higher-level, molecular phylogenetic study of the insect order Lepidoptera (Moths and Butterflies). PLos One 8, e58568.

Reid, A., Hardie, D.J., Mackie, D., Jackson, J.C., Windmill, J.F., 2018. Extreme call amplitude from near-field acoustic wave coupling in the stridulating water insect *Micronecta scholtzi* (Micronectinae). J. R. Soc. Interface 15, 20170768.

Reinhold, K., Greenfield, M.D., Jang, Y., Broce, A., 1998. Energetic cost of sexual attractiveness: ultrasonic advertisement in wax moths. Anim. Behav. 55, 905–913.

Riabinina, O., Dai, M., Duke, T., Albert, J.T., 2011. Active process mediates species-specific tuning of *Drosophila* ears. Curr. Biol. 21, 658–664.

Robert, D., 1989. The auditory behaviour of flying locusts. J. Exp. Biol. 147, 279–301.

Robert, D., Amoroso, J., Hoy, R.R., 1992. The evolutionary convergence of hearing in a parasitoid fly and its cricket host. Science 258, 1135–1137.

Rodríguez-Muñoz, R., Bretman, A., Slate, J., Walling, C.A., Tregenza, T., 2010. Natural and sexual selection in a wild insect population. Science 328, 1269–1272.

Roeder, K.D., 1962. The behaviour of free flying moths in the presence of artificial ultrasonic pulses. Anim. Behav. 10, 300–304.

Römer, H., Holderied, M., 2020. Decision making in the face of a deadly predator: high-amplitude behavioural thresholds can be adaptive for rainforest crickets under high background noise levels. Phil. Trans. R. Soc. B 375, 20190471.

Römer, H., Krusch, M., 2000. A gain-control mechanism for processing of chorus sounds in the afferent auditory pathway of the bushcricket *Tettigonia viridissima* (Orthoptera; Tettigoniidae). J. Comp. Physiol. A 186, 181–191.

Römer, H., Smith, A.R., Van Staaden, M., 2014. Hearing and sensory ecology of acoustic communication in bladder grasshoppers. In: Hedwig, B. (Ed.), Insect Hearing and Acoustic Communication. Animal Signals and Communication, vol. 1. Springer, Berlin, Heidelberg, pp. 27–43. https://doi.org/10.1007/978-3-642-40462-7_3.

Rosi-Denadai, C.A., Scallion, M.L., Merrett, C.G., Yack, J.E., 2018. Vocalization in caterpillars: a novel sound-producing mechanism for insects. J. Exp. Biol. 221, jeb169466.

Roth, L.M., 1948. A study of mosquito behavior. An experimental laboratory study of the sexual behavior of *Aedes aegypti* (Linnaeus). Am. Midl. Nat. 40, 265–352.

Rubin, J.J., Hamilton, C.A., McClure, C.J., Chadwell, B.A., Kawahara, A.Y., Barber, J.R., 2018. The evolution of anti-bat sensory illusions in moths. Sci. Adv. 4, eaar7428.

Sarria-S, F.A., Morris, G.K., Windmill, J.F.C., Jackson, J., Montealegre-Z, F., 2014. Shrinking wings for ultrasonic pitch production: hyperintense ultra-short-wavelength calls in a new genus of neotropical katydids (Orthoptera: Tettigoniidae). PLos One 9, e98708.

Schniederkötter, K., Lakes-Harlan, R., 2004. Infection behavior of a parasitoid fly, *Emblemasoma auditrix*, and its host cicada *Okanagana rimosa*. J. Insect Sci. 4, 36.

Schubnel, T., Legendre, F., Roques, P., Garrouste, R., Cornette, R., Perreau, M., Perreau, N., Desutter-Grandcolas, L., Nel, A., 2021. Sound vs. Light: wing-based communication in Carboniferous insects. Commun. Biol. 4, 1–11.

Schulze, W., Schul, J., 2001. Ultrasound avoidance behaviour in the bushcricket *Tettigonia viridissima* (Orthoptera: Tettigoniidae). J. Exp. Biol. 204, 733–740.

Scott, J.L., Kawahara, A.Y., Skevington, J.H., Yen, S.-H., Sami, A., Smith, M.L., Yack, J.E., 2010. The evolutionary origins of ritualized acoustic signals in caterpillars. Nat. Commun. 1, 1–9.

Senter, P., 2008. Voices of the past: a review of Paleozoic and Mesozoic animal sounds. Hist. Biol. 20, 255–287.

Shen, Z., Neil, T.R., Robert, D., Drinkwater, B.W., Holderied, M.W., 2018. Biomechanics of a moth scale at ultrasonic frequencies. Proc. Natl. Acad. Sci. U. S. A. 201810025.

Simmons, L.W., Thomas, M.L., Simmons, F.W., Zuk, M., 2013. Female preferences for acoustic and olfactory signals during courtship: male crickets send multiple messages. Behav. Ecol. 24, 1099–1107.

Snedden, W.A., Greenfield, M.D., 1998. Females prefer leading males: relative call timing and sexual selection in katydid choruses. Anim. Behav. 56, 1091–1098.

Song, H., Amédégnato, C., Cigliano, M.M., Desutter-Grandcolas, L., Heads, S.W., Huang, Y., Otte, D., Whiting, M.F., 2015. 300 million years of diversification: elucidating the patterns of orthopteran evolution based on comprehensive taxon and gene sampling. Cladistics 31, 621–651.

Song, H., Béthoux, O., Shin, S., Donath, A., Letsch, H., Liu, S., McKenna, D.D., Meng, G., Misof, B., Podsiadlowski, L., 2020. Phylogenomic analysis sheds light on the evolutionary pathways towards acoustic communication in Orthoptera. Nat. Commun. 11, 1–16.

Stumpner, A., Von Helversen, D., 2001. Evolution and function of auditory systems in insects. Naturwissenschaften 88, 159–170.

Stumpner, A., Dann, A., Schink, M., Gubert, S., Hugel, S., Ziesmann, J., 2013. True katydids (Pseudophyllinae) from Guadeloupe: acoustic signals and functional considerations of song production. J. Insect Sci. 13.

Sueur, J., Mackie, D., Windmill, J.F., 2011. So small, so loud: extremely high sound pressure level from a pygmy aquatic insect (Corixidae, Micronectinae). PLos One 6, e21089.

Surlykke, A., Kalko, E.K.V., 2008. Echolocating bats cry out loud to detect their prey. PLos One 3, e2036.

Taylor, C.J., Yack, J.E., 2019. Hearing in caterpillars of the monarch butterfly (*Danaus plexippus*). J. Exp. Biol. 222. jeb211862.

Ter Hofstede, H.M., Goerlitz, H.R., Ratcliffe, J.M., Holderied, M.W., Surlykke, A., 2013. The simple ears of noctuoid moths are tuned to the calls of their sympatric bat community. J. Exp. Biol. 216, 3954–3962.

Thomas, M.L., Simmons, L.W., 2009. Sexual selection on cuticular hydrocarbons in the Australian field cricket, *Teleogryllus oceanicus*. BMC Evol. Biol. 9, 162.

Torsekar, V.R., Isvaran, K., Balakrishnan, R., 2019. Is the predation risk of mate-searching different between the sexes? Evol. Ecol. 33, 329–343.

Van Staaden, M., Römer, H., 1997. Sexual signalling in bladder grasshoppers: tactical design for maximizing calling range. J. Exp. Biol. 200, 2597–2608.

Van Staaden, M.J., Rieser, M., Ott, S.R., Pabst, M.A., Römer, H., 2003. Serial hearing organs in the atympanate grasshopper *Bullacris membracioides* (Orthoptera, Pneumoridae). J. Comp. Neurol. 465, 579–592.

Veitch, D., Celiker, E., Aldridge, S., Pulver, C., Soulsbury, C.D., Jonsson, T., Woodrow, C., Montealegre-Z, F., 2021. A narrow ear canal reduces sound velocity to create additional acoustic inputs in a microscale insect ear. Proc. Natl. Acad. Sci. U. S. A. 118, e2017281118.

Velilla, E., Polajnar, J., Virant-Doberlet, M., Commandeur, D., Simon, R., Cornelissen, J.H., Ellers, J., Halfwerk, W., 2020. Variation in plant leaf traits affects transmission and detectability of herbivore vibrational cues. Ecol. Evol. 10, 12277–12289.

Vidal-García, M., O'Hanlon, J.C., Svenson, G.J., Umbers, K.D., 2020. The evolution of startle displays: a case study in praying mantises. Proc. R. Soc. B 287, 20201016.

Von Helversen, O., Von Helversen, D., 1994. Forces driving coevolution of song and song recognition in grasshoppers. In: Schildberger, K., Elsner, N. (Eds.), Neural Basis of Behavioural Adaptations. Jena, Gustav Fischer Verlag, pp. 253–284.

Wagner Jr., W.E., 2011. Direct benefits and the evolution of female mating preferences: conceptual problems, potential solutions, and a field cricket. Adv. Study Behav. 43, 273–319.

Wessel, A., Mühlethaler, R., Hartung, V., Kuštor, V., Gogala, M., 2014. The tymbal: evolution of a complex vibration-producing organ in the Tymbalia (Hemiptera excl. Sternorrhyncha). In: Cocroft, R., Gogala, M., Hill, P., Wessel, A. (Eds.), Studying Vibrational Communication. Animal Signals and Communication, vol. 3. Springer, Berlin, Heidelberg, pp. 395–444. https://doi.org/10.1007/978-3-662-43607-3_20.

Yack, J.E., 2004. The structure and function of auditory chordotonal organs in insects. Microsc. Res.Tech. 63, 315–337.

Yack, J., Smith, M., Weatherhead, P., 2001. Caterpillar talk: acoustically mediated territoriality in larval Lepidoptera. Proc. Natl. Acad. Sci. U. S. A. 98, 11371–11375.

Yager, D., 1996. Serially homologous ears perform frequency range fractionation in the praying mantis, *Creobroter* (Mantodea, Hymenopodidae). J. Comp. Physiol. A 178, 463–475.

Yager, D.D., 1999. Structure, development, and evolution of insect auditory systems. Microsc. Res. Tech. 47, 380–400.

Yager, D.D., Hoy, R.R., 1986. The cyclopean ear: a new sense for the praying mantis. Science 231, 727–729.

Yager, D., May, M., Fenton, M., 1990. Ultrasound-triggered, flight-gated evasive maneuvers in the praying mantis *Parasphendale agrionina*. I. Free flight. J. Exp. Biol. 152, 17–39.

Young, D., Bennet-Clark, H., 1995. The role of the tymbal in cicada sound production. J. Exp. Biol. 198, 1001–1020.

Zuk, M., Kolluru, G.R., 1998. Exploitation of sexual signals by predators and parasitoids. Q. Rev. Biol. 73, 415–438.

Zuk, M., Simmons, L.W., Choe, J., 1997. Reproductive strategies of the crickets (Orthoptera: Gryllidae). In: Crespi, B. (Ed.), The Evolution of Mating Systems in Insects and Arachnids. Cambridge University Press, Cambridge, pp. 89–109.

Zuk, M., Rotenberry, J.T., Tinghitella, R.M., 2006. Silent night: adaptive disappearance of a sexual signal in a parasitized population of field crickets. Biol. Lett. 2, 521–524.

CHAPTER THREE

Sound production in *Drosophila melanogaster*: Behaviour and neurobiology

Bijayalaxmi Swain and Anne C. von Philipsborn
Danish Research Institute of Translational Neuroscience (DANDRITE), Aarhus University, Aarhus C, Denmark

Contents

Abstract

Drosophila melanogaster produce different communication sounds during interactions with conspecifics by vibrating their wings. Males sing a pre-copulatory courtship song with pulse and sine elements. Females sing a structurally different song during copulation. Both sexes also produce a third type of sound pulses during agonistic interactions, i.e. during sexual rejection and inter-male aggression. It is well established that male courtship song increases female receptivity to mating. Female copulation song depends on male inseminate composition and might influence the transfer of the latter. The behavioural effect and function of agonistic sounds are not yet fully understood. Male courtship song production relies on sex specific neurons. As a precisely quantifiable and very specific element of courtship with defined motor parameters, it is an ideal model system for behavioural circuit neuroscience and has provided

Advances in Insect Physiology, Volume 61
ISSN 0065-2806
https://doi.org/10.1016/bs.aiip.2021.08.001

basic insight to how the nervous system generates, modulates and patterns adaptive behaviour. Central brain networks integrate multimodal sensory information and flexibly control the initiation and maintenance of singing. The circuits generating the species-specific song pattern and shaping motor control of the wing are mainly located in the ventral nerve cord. First studies have started to address how other *Drosophila* species with divergent sound producing behaviour differ on the level of neuronal circuits from *D. melanogaster*. In the future, advances in connectomics are expected to greatly facilitate neuronal circuit analysis in general and allow for a comprehensive synthesis of described circuit motifs for different aspects of sound production and its coordination in behavioural context.

1. Introduction

Insect sounds have fascinated humans for a long time. The loud and persistent songs of cicadas and crickets appear in ancient folk tales and mythology and were imagined to represent a carefree attitude towards life, or the lament of transformed lovers (Clausen, 1954; Myers, 1929). Sound production by Diptera is less conspicuous and many people might not be aware of it at all. However, everybody who was kept awake by a mosquito buzzing close to her ear in a summer night can appreciate that vibrating wings of smaller flies produce tones well in the audible range of humans. Apart from being a side product of flight, sounds generated by wing vibrations serve as acoustic signals in conspecific communication.

Here, we review sound production during social interactions in *Drosophila melanogaster*, the most widely used insect model organism in biological and biomedical research. We first discuss the different types of sound producing behaviours that have been observed, the structure of the acoustic signals and their biological function. In the second part of the article, we will outline the neurobiology of sound production. We will describe what is known about the neuronal circuits controlling sound producing behaviour: sensory and central processing circuits involved in triggering, gating or modulating sound production, as well as motor output circuits shaping the signal pattern. Lastly, we will elaborate how sound production behaviour in *D. melanogaster* serves as an excellent model to study basic questions in neuroscience. Connectome-assisted circuit mapping, improved genetic tools and imaging techniques will allow addressing open questions in the field from a new angle.

Our focus is on sound generation, i.e. the side of the signal sender. The topic of hearing, acoustic pattern detection and neuronal responses to auditory signals in *D. melanogaster* has been covered in several reviews (Albert and Göpfert, 2015; Baker et al., 2019; Ishikawa and Kamikouchi, 2016; Ishimoto and Kamikouchi, 2021; Kamikouchi, 2013) and will be not discussed in detail here. We will only review what is known about the behavioural effect of acoustic signals in order to delineate which aspects of it are thought to be crucial for their function and how their function might benefit the signal sender.

Sound production in *D. melanogaster* has also been extensively studied in regards to its genetic control. Here, we will only mention genes that have been of considerable relevance for exploring and understanding the neurobiology of sound production. Readers interested in a comprehensive overview of the genetics underlying sound production, its intra- and interspecific variability and role in speciation are directed to articles with specific focus on this subject (Anholt et al., 2020; Ding et al., 2016; Fedotov et al., 2014; Gleason, 2005; Markow and O'Grady, 2005; Turner et al., 2013).

2. Ethology of sound production

Due to its small size, *D. melanogaster* produces sounds that are only audible to humans after amplification. All acoustic signals described so far for this species are generated by rapid wing movements, i.e. vibrations or flicks, while the animal stands or walks on the substrate. They are near-field signals directed at conspecifics in close vicinity (Bennet-Clark, 1971; Morley et al., 2018). Sound production mainly occurs during reproduction, but also plays a role in agonistic interactions (Fig. 1). Males sing a pre-copulatory courtship song and females sing during copulation. Both sexes can produce sexual rejection sounds when courted and males emit aggression sounds during fights over females or food. Male courtship song and female copulation song are clearly distinct in their sound pattern and also distinct from agonistic sounds. Sexual rejection sounds and male-male aggression sounds appear very similar in structure (Fig. 1). Male courtship song has been studied much more than the other sound producing behaviours and will therefore be discussed in more detail.

2.1 Male courtship song

Male *D. melanogaster* court readily—as can be observed around degrading fruit around human dwellings in summer or in the laboratory. Courtship is immediately recognized by the characteristic unilateral wing extension and vibration of the male. When this behavioural element was first described

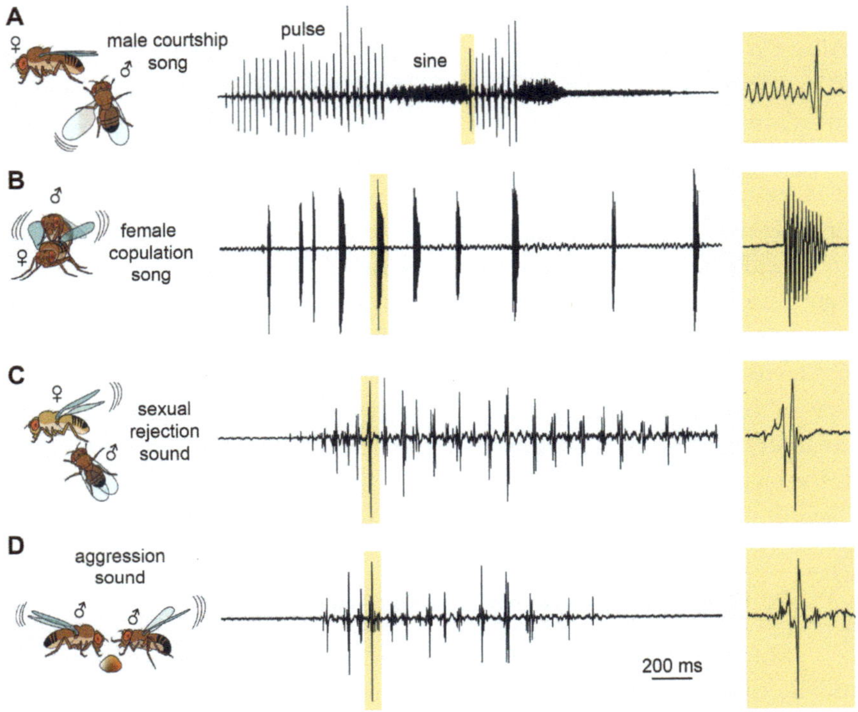

Fig. 1 Sound production behaviours and sample sound oscillograms. (A) Male courtship song: A male sings by vibrating a unilaterally extended wing towards a female. The song contains pulse trains and sine song components. (B) Female copulation song: A female sings during copulation by vibrating both wings. Song pulses are polycyclic and spaced at wider intervals than in male song. (C) Sexual rejection sound: An immature virgin (or a male fly) responds to male courtship with bilateral wing flicks, producing a train of irregular pulses. (D) Aggression sound: Two males fighting over a food resource (or a female) perform bilateral wing flicks, producing a train of irregular pulses. Oscillograms are 2.5 s long, with a selected pulse (yellow boxes) shown at higher temporal resolution. Sounds were recorded in the Philipsborn laboratory.

and found to be of key importance for female receptivity to mating (Bastock and Manning, 1955; Spieth, 1952; Sturtevant, 1921), it was not clear if the wing movements were a primarily visual display or dispersed pheromones. Only recordings with a sensitive microphone close to the flies revealed a structured sound produced by the small amplitude wing movements (Shorey, 1962). Playback of this courtship song could rescue the delay in copulation success of wing-amputated males, establishing that females are indeed evaluating the acoustic component of wing vibration during courtship (Bennet-Clark and Ewing, 1967, 1969; von Schilcher, 1976b).

Work over many decades has provided a good understanding of the acoustic song parameters, the context in which the male is likely to sing and the function of male song in stimulating female receptivity.

Additional to air-borne sound, *D. melanogaster* males also generate substrate-transmitted vibrations during courtship (Fabre et al., 2012; Mazzoni et al., 2013), which will not be discussed in detail here.

2.1.1 Acoustic properties of courtship song and corresponding wing movements

Male courtship song has two main elements: pulse song and sine song (Fig. 1A). The most commonly quantified acoustic parameters of male courtship song are depicted in Fig. 2.

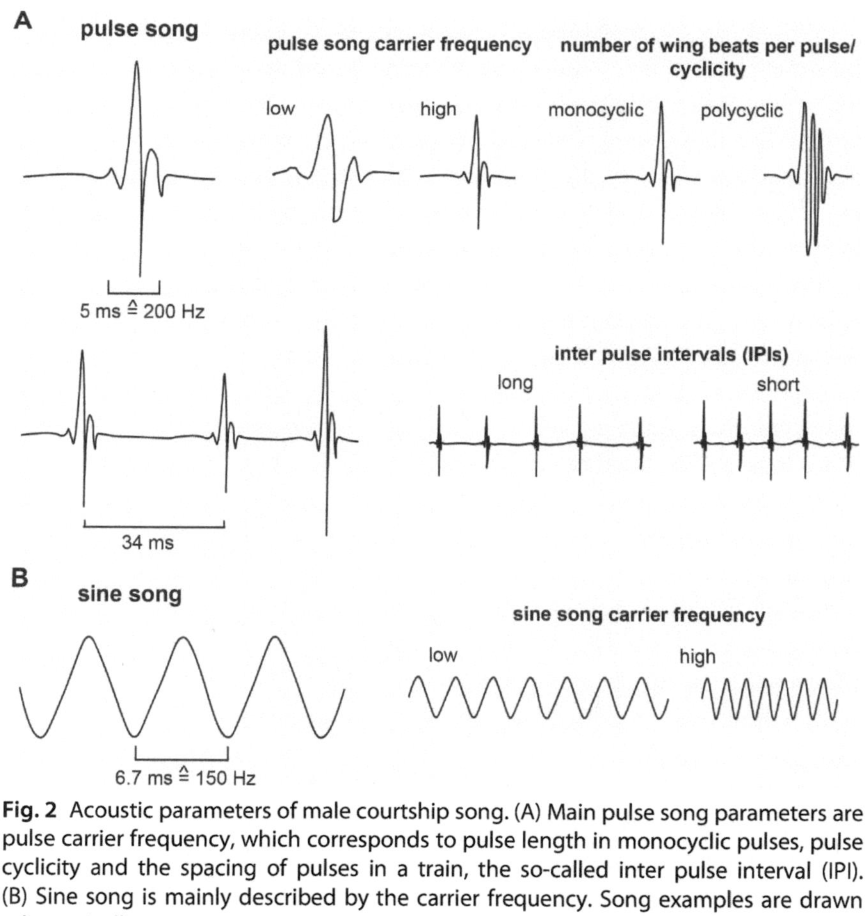

Fig. 2 Acoustic parameters of male courtship song. (A) Main pulse song parameters are pulse carrier frequency, which corresponds to pulse length in monocyclic pulses, pulse cyclicity and the spacing of pulses in a train, the so-called inter pulse interval (IPI). (B) Sine song is mainly described by the carrier frequency. Song examples are drawn schematically.

During courtship song, one wing is extended to up to 90°. Both wings vibrate, but the extended wing does so at much larger amplitude (Bennet-Clark and Ewing, 1968). There is no indication that males have an intrinsic preference for the left or right wing (Perelle and Ehrman, 1983). Typically, the wing closer to the female is extended, and males can quickly switch between wings when circling the female (Bastock and Manning, 1955; Morley et al., 2018).

Each pulse arises from one to two up- and downward strokes of an extended wing. The speed of these strokes determines the carrier frequency of an individual pulse, which is approx. 200 Hz, corresponding to the duration of 5 ms of one stroke cycle (Fig. 2A, Bennet-Clark and Ewing, 1968; O'Sullivan et al., 2018). Around 20% of pulses have a higher carrier frequency of up to 400 Hz (Clemens et al., 2018).

The internal carrier frequency of single, spaced courtship song pulses is similar to the frequency of continuous wing beat during flight. Compared to flight, however, wing beats during song are of smaller amplitude (around a quarter of a flight wing beat) and the plane of the wing undergoes less rotation but stays closer to the horizontal. Since the sound level increases with the square of the vibration amplitude, courtship song is not as loud as the tone produced during flight (Bennet-Clark and Ewing, 1968).

Pulses come in trains of various lengths. Most trains are short (around 4 pulses), but median train length can be up to 10–20 in most courtship conditions. In a train, pulses are spaced from each other at a characteristic distance, the so-called inter pulse interval (IPI) (Fig. 2A). The IPI has received much attention, since it is species-specific. IPIs in *D. melanogaster* are approx. 34 ms long and in the three closest relatives *D. simulans* 49 ms, *D. mauritiana* 46 ms and *D. sechellia* 86 ms (Cowling and Burnet, 1981; Ewing and Bennet-Clark, 1968; Tomaru and Yamada, 2011). IPI length shortens with increasing temperature by about 1 ms/°C (Shorey, 1962). At 25 °C, approx. 90% of all IPIs in *D. melanogaster* range from 30 to 40 ms (Arthur et al., 2013; O'Sullivan et al., 2018). Within a pulse train, the IPIs tend to increase towards the end of the train (Arthur et al., 2013; Ewing, 1988).

IPI durations were previously proposed to cycle of over a longer period of approx. 1 min (Kyriacou and Hall, 1980), but this phenomenon has a long history of controversy and has not been unanimously confirmed (Arthur et al., 2013; Crossley, 1988; Ewing, 1988; Kyriacou et al., 2017).

Pulse trains alternate with episodes of sine song. Sine song bouts arise from continuous sinusoidal vibration with a carrier frequency of 140–160 Hz (Fig. 2B, Arthur et al., 2013; O'Sullivan et al., 2018; von Schilcher, 1976a).

The duration of sine song episodes has a similar long-tailed distribution as pulse train length, with a mean of 300–400 ms (Arthur et al., 2013).

2.1.2 Behavioural context of courtship song generation

Natural male song always occurs when the male is in close pursuit of a female or orienting towards her. When courtship is initiated, the male taps the female abdomen at least once with his foreleg before he starts singing (Bastock and Manning, 1955; Spieth, 1952). He thus comes in contact with her female specific cuticular pheromones (Kohatsu et al., 2011). As discussed in Section 3.1.1, integration of several sensory modalities guides the initiation and maintenance of male courtship singing.

In most studies of courtship behaviour, singing is evaluated by measuring the time the male spends extending one of his wings. This serves as a good proxy, but does not allow for the exact quantification of pulse trains and sine episodes, since these cannot be detected in videos with conventional frame rates of around 30 fps. By combining audio and video recordings, it has been shown that the sequencing of the pulse and sine is flexible and closely correlates with female behaviour, namely her locomotion. When a female moves fast, male pulse song dominates, when she is slower and the distance to the male is small, there is a higher probability of sine song. It has been thus suggested that males adjust their song choice and the proportion of the song modes in response to the female (Coen et al., 2014; Trott et al., 2012). Since a receptive female is known to decrease her movement in response to song (see next section), this interplay is very dynamic and shows no absolute uni-directional causality (Trott et al., 2012). In 30 s long windows of courtship interactions, the proportion of pulse song among total song production was found to vary between 40% and 80%, depending on how much the female moved. Of note, even when the female was decapitated and completely motionless (and despite this still a target of vigorous courtship), pulse song did not completely disappear, but still took up around 30% of the total song. Interestingly, immature virgins, which are not receptive and run away when courted by males, elicited song with a proportion of around 50% pulse. The specific pheromone profile of these unreceptive virgins might also modulate male song choice (Trott et al., 2012). Song amplitude also varies to some extent over the course of pulse and sine bouts. Pulse amplitude is higher when the male is further away from the female, suggesting that he can perform amplitude modulation with distance, a behavioural phenomenon also observed in humans and songbirds (Coen et al., 2016).

Many aspects of *D. melanogaster* courtship are innate and male flies do not have to learn or practice their courtship song before they perform for the first time to a female. Immature males, i.e. males younger than 3 days after eclosion, court and sing less than mature males. Pulse song IPIs from immature males have a broader distribution and are shifted towards longer durations (Moulin et al., 2001). In the mature male, pulse song IPI and sine song carrier frequency are not dependent on social experience during early life or afterwards (Marie-Orleach et al., 2019). Social exposure or successful mating experience can however increase the length of singing bouts (Marie-Orleach et al., 2019; Saleem et al., 2014). In contrast, the experience of unsuccessful courtship towards an unreceptive female suppresses subsequent male singing and courtship behaviour in general (Griffith and Ejima, 2009).

In summary, males vary their exact song composition depending on cues from the female and their previous experience. Song produced during an individual courtship interaction, even when all experimental conditions are kept constant, always has a unique pattern regarding the sequencing and duration of pulse trains and sine episodes (Fig. 3).

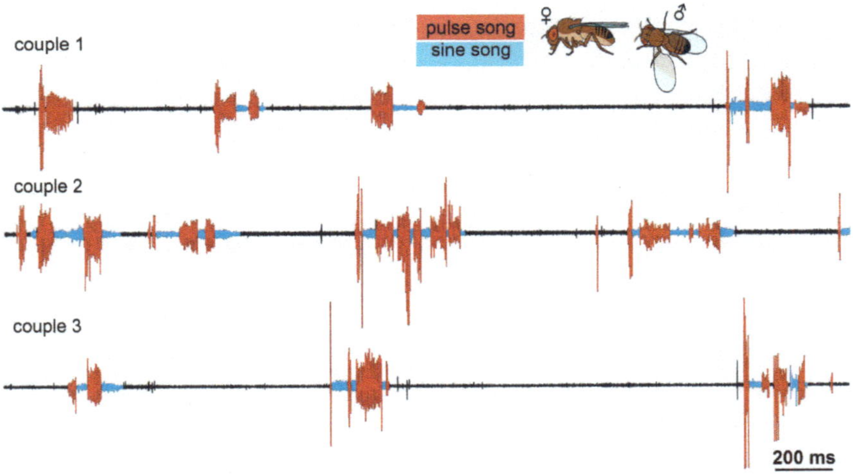

Fig. 3 Variability in male courtship song. Example song traces from three couples where a mature male courts a receptive virgin in a small chamber of 1 cm diameter. Pulse song is marked in red, sine song in blue. The two song modes alternate in a non-stereotyped way and show amplitude modulations, making each song individual even when experimental conditions are kept constant. Sounds were recorded in the Philipsborn laboratory.

2.1.3 Behavioural effect of courtship song

Male courtship song has a strong effect on the copulation success of males when they are paired with virgin females in smaller observation chambers. It is not required for copulation, but greatly accelerates mating. When males are muted by wing amputations (Bennet-Clark and Ewing, 1969; Bussell et al., 2014; Rybak et al., 2002; von Schilcher, 1976b) or silencing of song producing motor neurons (O'Sullivan et al., 2018), the probability of copulation in a given time severely decreases. Likewise, less copulations are observed when females are deaf due to amputation of their external hearing organ, the sound-transducing arista on the antenna (Manning, 1967a), carry mutations impairing or abolishing hearing (Burnet et al., 1971; Markow, 1987; Yorozu et al., 2009) or have their auditory neurons silenced (Vaughan et al., 2014).

Which parameters of courtship song are important for the effect of courtship song on female receptivity? Playback of pulse song facilitates te mating of mute males, but only when it has the *D. melanogaster* specific IPI, as opposed to shorter or longer IPIs (Bennet-Clark and Ewing, 1969; von Schilcher, 1976b; Yamada et al., 2018). When females are exposed to conspecific song for several days before being subjected to this assay, their selectivity for the conspecific IPI is more strongly tuned than in naïve females (Li et al., 2018). How parameters other than IPI affect the receptivity enhancing effect of song on females is less clear. In one early playback experiment, pulse carrier frequencies of 167, 334 and 667 Hz did not result in different copulation rates (Bennet-Clark and Ewing, 1969). Some classical "behavioural mutants", with telling names such as *cacophony* or *dissonance* (*nonA*), produce pulse song that is polycyclic and the males have decreased copulation success. A causal relationship between the two phenotypes has not been established and seems doubtful, given additional defects of the mutants (Kulkarni et al., 1988; von Schilcher, 1976c). Males which lack sine song due to the silencing of the motor neurons innervating the wing control muscles hg1 (Shirangi et al., 2013) or hg2 (O'Sullivan et al., 2018) do not achieve copulation. However, the pulse song of these males also shows longer IPIs and polycyclic pulse structure (O'Sullivan et al., 2018). Since playback of pulse song without sine episodes is sufficient for female receptivity stimulation, sine song might not be required for male copulation success. From partial amputation experiments, there is indication that male wing area, which is expected to correlate with song amplitude, is positively correlated with male mating speed (Ewing, 1964). Playback of song that is amplitude modulated at a pattern found in natural song was found to be more effective in stimulating mating with wingless males than song with different or no amplitude modulations (Brüggemeier et al., 2018).

In summary, only pulse song IPI and song amplitude have been convincingly shown to determine the effect of male courtship song on copulation rate.

An alternative way to explore the behavioural effect of male courtship song is the monitoring of fly locomotion. Slowing down or pausing is required, but not sufficient for the female acceptance response, which also includes the opening of the vaginal plates (sometimes called: partial ovipositor extrusion) (Aranha et al., 2017; Bussell et al., 2014; Tompkins et al., 1982). Vaginal plate opening occurs in response to song and is necessary for the male to achieve copulation (Mezzera et al., 2020; Wang et al., 2021). When courtship song is played back to receptive virgins, they slow down their locomotion. In some studies, this effect is only seen in the presence of males. Males, on the other hand, robustly increase their locomotion when exposed to song (Bussell et al., 2014; Crossley et al., 1995; Ishikawa et al., 2019; Kowalski et al., 2004; Vaughan et al., 2014; von Schilcher, 1976a). Recently, the tuning of this song induced locomotion response to different courtship song parameters was comprehensively and systematically tested with single flies of both sexes. Both females and males match the strength of their response surprisingly well to the naturally occurring distributions of IPI, pulse carrier frequency and pulse cyclicity/length. The response to sine song, as previously reported, was not very pronounced or frequency tuned (Deutsch et al., 2019).

Notably, playback of courtship song can also lower the defensive response (leg kicking) in virgin females to mechanical stimulation (Liu et al., 2020).

Male flies do not only speed up during playback of courtship song, but also start to approach each other with elements of courtship behaviour. Several males often closely and persistently follow each other, oriented head to abdomen, a behaviour that has been termed "chaining". This effect is most pronounced when their wings are amputated. The lack of wings may facilitate this male-male courtship chaining, since males cannot produce rejection wing flicks (Eberl et al., 1997; Inagaki et al., 2010; Kamikouchi et al., 2009; von Schilcher, 1976a; Yoon et al., 2013; Zhou et al., 2015). Indeed around 40% of single, winged males respond to playback of courtship song with brief, mostly bilateral wing extensions. This response (only documented in video recordings without audio) has been interpreted as courtship singing (Deutsch et al., 2019), but might as well represent the emission of sexual rejection sounds (see Section 2.3).

2.1.4 Function of courtship song in mate choice

Pre-copulatory displays such as *D. melanogaster* male courtship song are proposed to function in mate choice. They may serve in species and sex recognition and/or signal sender fitness and act in sexual selection (Greenfield, 2002; Smith and Harper, 2003).

In some *Drosophila* species, male courtship song indeed plays a crucial role in reproductive isolation, e.g. in the *D. buzzatii* complex (Iglesias and Hasson, 2017) or between *D. montana* and *D. lummei* (Liimatainen and Hoikkala, 1998; Saarikettu et al., 2005). In these species pairs, pre-mating sexual isolation can be overturned by manipulating acoustic signalling: females accept heterospecific males with amputated wings for copulation only when conspecific song of the female is played back, but not in the presence of heterospecific songs.

In *D. melanogaster,* the fine-tuning of behavioural effects to acoustic song parameters suggest that females can distinguish between con- and heterospecific males based on auditory cues. The evidence is limited however, that females do so extensively in the wild and that male courtship song prevents interbreeding or prolonged interspecific courtship in *D. melanogaster* across a wide range of species. *D. simulans*, *D. yakuba* and *D. virilis* males avoid *D. melanogaster* females via chemosensation and hardly even start singing to them. Likewise, male *D. melanogaster* rarely court *D. simulans*, *D. yakuba* and *D. virilis* females in a sustained way, mainly because they detect inhibitory cuticular hydrocarbons on the females as soon as they tap and taste them with the foreleg (Billeter and Levine, 2013; Fan et al., 2013; Sato and Yamamoto, 2020). Heterospecific courtship and interbreeding between *D. melanogaster* and *D. sechellia*, an island species from the Seychelles, is more easily observed. Females from both species copulate more readily with males of their own species when the conspecific song is present. *D. melanogaster* females copulate at higher rate with singing *D. sechellia* males than with wingless ones. In contrast, *D. sechellia* females show a higher copulation rate with mute, wingless *D. melanogaster* males than with intact singers. This suggests that *D. sechellia* courtship song is stimulating receptivity in *D. melanogaster* females, at least as compared to no song. *D. melanogaster* courtship song, on the other hand, might prevent *D. sechellia* females from mating with *D. melanogaster* males (Tomaru and Oguma, 2000).

If male courtship song does not contribute much to widespread species recognition and sexual isolation in *D. melanogaster*, why is male song still so specifically patterned and why do females detect and respond to the precision of the pattern? *Drosophila* courtship displays with complex, precisely timed and coordinated motor elements have been hypothesized to function as "honest signals" in sexual selection. Honest signals cannot be faked, since they are energetically costly and directly display male condition, such as strength, agility or sensory acuity. Females with preference for such signals can obtain high quality mates (Smith, 1956; Smith and Harper, 2003). For a signal to be honest, some of its characteristics have to be correlated with male quality. Females have to choose a mate based on them and derive a fitness benefit from doing so. There is some evidence that *D. melanogaster* courtship song depends on male condition. For example, large males sing more and louder (Partridge et al., 1987). Song patterning defects can be indicative for defects in the neuromuscular apparatus for flight, such as mutations in muscle proteins (Chakravorty et al., 2014). Silencing of single classes of the motor neurons innervating wing control muscles also severely alters song patterning. Malfunction of several of these neurons results in an out of tune song while some basic flight ability is retained, indicating that song patterning could be a sensitive readout of neuromuscular coordination (O'Sullivan et al., 2018). Likewise, impairment in mechanosensation results in defect song patterning (Tauber and Eberl, 2001).

In summary, phenotypes of mutants and neuronal manipulations suggest that courtship song conveys information about nervous system and muscle function of the singer. Do females use this information to preferentially mate with males producing "correctly" patterned song? As elaborated in the previous section, the locomotion response of females to courtship song playback is strongly tuned to the naturally occurring range of song parameters (Deutsch et al., 2019). To which extent this locomotion response transfers to mating preferences is not yet fully explored. Several playback studies have shown that IPI and song amplitude clearly impact mating rate (Bennet-Clark and Ewing, 1969; Brüggemeier et al., 2018; Li et al., 2018; von Schilcher, 1976b; Yamada et al., 2018). Of note, effects seen in no-choice tests of isolated pairs of flies may not translate to larger groups of flies aggregating on food patches, where male-male scramble competition and reproductive interference prevails (see Section 2.3.3; Sturtevant, 1915; Partridge et al., 1987; Tauber and Eberl, 2002). Under these conditions, mate choice may also largely be exerted by previously mated females (Kohlmeier et al., 2021; von Philipsborn, 2020) and arise from complex re-mating and sperm competition patterns (Laturney et al., 2018; Manier et al., 2010)—aspects of sexual selection that so far have not been taken into consideration in studies of courtship song function.

2.2 Female copulation song

D. melanogaster females emit structured sound pulses during copulation by short bilateral wing vibration (Kerwin et al., 2020). These female copulation songs are also observed in other species such as *D. simulans*, *D. mauritiana*, *D. sechellia* (Kerwin et al., 2020) and *D. orena* (Cobb et al., 1989). Since female song in *D. melanogaster* has been overlooked for many decades, many aspects of it are not yet investigated. Female song occurs throughout the 15–20 min of copulation. Female song pulses are polycyclic (on average five cycles per pulse) and have a higher carrier frequency than male song (approx. 350 Hz). They appear isolated or in short trains with a median IPI of approx. 100 ms (Fig. 1B, Kerwin et al., 2020). Female copulation song depends on the composition of inseminate received from the male partner. Spermless males elicit more song pulses. In contrast, males depleted of inseminate by multiple recent matings elicit very little copulation song (Fig. 4). Female copulation song is also severely decreased in matings with male *iab-6cocu* mutants (Kerwin et al., 2020). These mutants have impaired

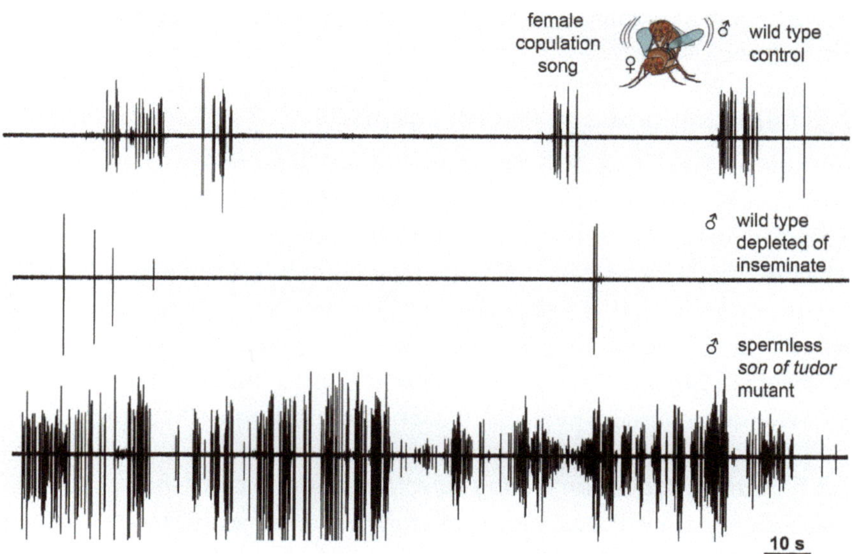

Fig. 4 Female copulation song depends on male inseminate. Example song traces from females copulating with a control male without previous mating history, a male that has depleted his seminal fluid store by multiple recent matings and mutant male (*son of tudor*) devoid of sperm. The oscillograms show a randomly chosen 3 min time interval during copulation. Sounds were recorded in the Philipsborn laboratory.

function of a particular secretory cell type in the male accessory gland and produce altered seminal fluid (Gligorov et al., 2013; Maeda et al., 2018; Sitnik et al., 2016).

The current hypotheses about the behavioural effect and biological function of female copulation song are still rather speculative (Kerwin and von Philipsborn, 2020; Sirot and Wolfner, 2020). Playback of female song to copulating pairs with mute females has a subtle effect on the post-mating behaviour of the female. When her partner was exposed to the female song during copulation, she is more hesitant to mate with a subsequent male. This effect could indicate that the male, having heard the female song during copulation, increased his allocation of receptivity suppressing seminal fluid molecules (Kerwin and von Philipsborn, 2020; Kerwin et al., 2020). Female copulation song might thus serve to fine-tune inseminate transfer by a behavioural-molecular copulatory dialogue (Kerwin and von Philipsborn, 2020; Sirot and Wolfner, 2020). A female might benefit from influencing amount and/or composition of seminal fluid she receives. Seminal fluid molecules have both beneficial and detrimental effects on females and can be a source of sexual conflict. Balancing and titrating their transfer by sending an acoustic feedback signal to the copulation partner could be an adaptive female strategy in response to this conflict (Kerwin and von Philipsborn, 2020).

2.3 Rejection and aggression sounds

There is evidence that *D. melanogaster* also uses acoustic signals during agonistic encounters, namely when rejecting unwanted courtship or when fighting over resources. From the few studies with audio recordings of sexual rejection (Pailette et al., 1991) and aggression (Jonsson et al., 2011; Versteven et al., 2017) behaviour, it seems possible that the two sounds are very similar. So far, they have not been thoroughly analyzed for potential differences. Neither has it been tested if they might elicit the same behavioural response when played back in the same social context.

2.3.1 Sound production during rejection of male courtship

Males do not always court receptive targets, but also mated females, immature virgins or immature males. Mated females and immature (i.e. within the first day after eclosion) virgins both have very low receptivity, but differ in the way they respond to male courtship. Immature virgins frequently flick their wings, whereas mated females extrude their ovipositor when males sing towards them and lick their genitalia (Connolly and Cook, 1973; Ewing and

Bennet-Clark, 1968). Only one study has paid closer attention to the sounds produced by these bilateral wing flicks and proposed that they are a rejection signal (Pailette et al., 1991). Rejection signals are observed in immature virgins or males in response to male courtship song (Fig. 1C). They consist of pulses with carrier frequency similar to male courtship song, spaced at longer more variable IPIs with a mean of around 80 ms. Individual pulses can have multiple cycles. There is anecdotal mention that males stop courting after rejection signals of their target, but this potential behavioural effect was not quantified or further tested (Pailette et al., 1991). If sexual rejection sounds are indeed effective in deterring male courtship, immature virgins could benefit from producing them until their ovaries are mature and their physiological state poised for reproduction.

2.3.2 Sound production during inter-male aggression

D. melanogaster males show aggressive behaviour when competing for small and defendable resources, such as food or a mate. During these interactions, they flick their wings, which can be accompanied by sound generation. Notably, not all wing movements during fights produce sounds: the so-called wing-threat, a bilateral raising of wings to an approx. 45° angle, is for example silent (Jonsson et al., 2011). Aggression sounds arise from the vibration of both wings and consist of pulse trains with longer and more variable IPIs than in courtship song (Fig. 1D). The individual aggression pulses have a carrier frequency similar to courtship pulses (Jonsson et al., 2011; Versteven et al., 2017). Ethogram analysis shows that aggression sounds are likely to occur after a male taps the abdomen of his opponent with the foreleg, a behaviour that is connected with sampling of the sex-specific pheromone makeup. After sound emission, the likelihood for high-level aggressive interactions is slightly decreased (Jonsson et al., 2011).

Both male and female flies show same-sex aggression over limited resources such as food, but the orchestration of their agonistic interactions is different. There is evidence that wing flicking only occurs during inter-male aggression (Nilsen et al., 2004). Aggression sounds are thus probably only produced by males.

Males that are deafened by manipulation of their hearing organ, by inactivation of neurons required for hearing or by knockdown of hearing relevant genes display lower aggression. Playback of aggression sound, on the other hand, was found to increase the number of aggressive encounters such as kicking, chasing, wing threats, boxing and head butts between male flies (Versteven et al., 2017). Somewhat contrary to this, a different study reports that playback of aggression sounds induces male-male chaining in wing amputated males, in a similar, but more transient way than does the playback of courtship song (see above, Yoon et al., 2013). It is possible that the difference is due to the loss of functional wings. Loss of wings can alter behavioural preferences and may influence neuromodulatory state (Gorostiza et al., 2016). In summary, these studies suggest that sound production, although not being necessarily required for inter-male aggression, plays some modulatory role in the behaviour.

Acoustic displays during aggression have been hypothesized to resolve conflicts and prevent escalation, when they reliably signal body size, strength or other traits that determine the resource holding potential of an opponent (Smith and Harper, 2003). In insects, this idea has been most extensively explored in crickets, which sing specific aggression or rivalry songs. When a male cricket stridulates in an aggressive encounter, its opponent may retreat or answer with its own song. After this acoustic interchange, either male may retreat or the pair starts to grapple (Alexander, 1961). The acoustic parameters of the aggression song of an individual male were found to contain information about his size and ability to win a fight (Brown et al., 2006). From these observations, it indeed seems plausible that aggression songs help males to decide if they should engage in physical combat with an opponent or retreat. Vigorous aggression songs intimidate an opponent and inhibit his aggression. In line with this, deafened crickets become more aggressive and defeat opponents that have formerly dominated them (Phillips and Konishi, 1973).

There is no obvious explanation of why the manipulation of aggression songs has opposite effects in crickets and *D. melanogaster*. Maybe aggression songs in *D. melanogaster* do not serve as a reliable signal of fighting ability of an individual opponent. When played back to a group of flies, they may mainly convey the information that there is a contested resource (which is worth fighting for) close by and therefore bias flies towards more aggression.

The establishment and maintenance of dominance and social hierarchy has been extensively researched in *D. melanogaster* (Duistermars et al., 2018; Simon and Heberlein, 2020; Trannoy and Kravitz, 2017) but a potential role of sound production has not been analyzed. Moreover, the classification and quantification of aggressive actions varies between different research groups (Asahina, 2017), making it difficult to conclude if wing movements reported in the literature correspond to the "wing flicks with sound production" that were confirmed by audio recordings by Jonsson et al. (2011) and Versteven et al. (2017).

2.3.3 Sound production during sexual competition

As mentioned above, it is currently not clear if sexual rejection signals and aggression sounds are distinct in regards to their acoustic parameters and the wing kinematics generating them. In a natural setting, two or more males frequently compete for a female. All involved parties may emit overlapping sounds: the males courtship songs towards the female and aggression sounds while fighting each other, and the female (if an immature virgin) rejection signals. Moreover, one male may emit rejection signals in response to the courtship song of the other male. Sounds from a male could jam the courtship song of his rival and prevent the female from properly detecting it. One study addressed some aspects of such a situation, analyzing sounds produced when one winged male competed with a wingless male for a mature virgin. Males produced fewer and shorter courtship pulse song bouts when in competition, indicating that sound production is modulated by the presence of rivals (Tauber and Eberl, 2002). The phenomenon of conspecific signal jamming during sexual competition is well known from birds (Tobias and Seddon, 2009) or other groups of insects such as *Orthoptera* (Shuker and Burdfield-Steel, 2017), but has not been studied in more detail in *D. melanogaster*.

Alternatively, males may profit from courtship song of rivals in a competitive situation. When a wing-amputated male competes with a winged rival for one female, his chance of obtaining copulation is greatly increased as compared to a situation where he is alone (Sturtevant, 1915; Watanabe et al., 2018). It is unclear if in a natural setting, *D. melanogaster* males may sometimes employ "sneaker" behaviour in acoustic signalling, as it is known from other species. In field crickets, satellite males use the alternative reproductive strategy of exploiting the song of calling males. They intercept approaching phonotacting females, while staying mute and protected from parasitoids themselves (Rotenberry et al., 2015). In the *D. nasuta* subgroup satellite-like behaviour, with a courting male dancing in front of the female and a competing rival meanwhile mounting without display, has been described (Spieth, 1974).

3. Neurobiology of sound production

Different acoustic signals allow *D. melanogaster* to communicate in various ways with conspecifics. How does the nervous system control and modulate the context-dependent production of sounds and their specific patterning? We will mainly discuss the neurobiology of male courtship song production, since little is known for the other sound producing behaviours.

In *D. melanogaster*, versatile genetic tools allow for reproducible, temporally and spatially controlled manipulation of genetically defined neuronal subsets in intact, behaving animals. Most recent studies make use of expression systems such as GAL4/UAS, large collections of driver lines and intersectional genetics strategies to sparsely target identifiable neurons and explore their behavioural phenotype upon activation and silencing, their connectivity and response pattern (Guo et al., 2019; Pfeiffer et al., 2010; Simpson and Looger, 2018).

3.1 Neuronal control of male courtship song

Only males produce courtship song. An obvious assumption is that this is due to difference in the male vs the female nervous system. What parts of the nervous system have to be male for an animal to sing? The first studies of the neuronal control of male *D. melanogaster* courtship song addressed this question with sex mosaics (gynandromorphs). Sex mosaics are created by random loss of an X chromosome and subsequent cell autonomous masculinization in neurons of the developing nervous system. By analyzing many of such animals, Benzer, Hall and coworkers concluded that a small part of the brain, rather than any external sensory or effector organ, has to be male for a fly to sing (Hall, 1979; Hotta and Benzer, 1976). Sex change in areas of the ventral nerve cord is required for correct song patterning (von Schlicher and Hall, 1979). The findings of these ground-breaking early works have since been confirmed, refined and expanded with more advanced genetic tools.

The neuronal circuits for courtship are especially well studied, not least because two major sex determination genes, *fruitless* and *doublesex*, have provided an early entry point to identifying neurons whose activity is necessary and sufficient for male courtship. Expression of the male specific transcription factor FruitlessM is required for male sexual behaviour, and was early on argued to delineate a neuronal circuit spanning from brain to ventral nerve cord and coordinating the different steps of mating (Baker et al., 2001). FruitlessM is found in around 2000 neurons of the adult male nervous system (Manoli et al., 2005; Stockinger et al., 2005). Many of these also express the male specific isoform of the transcription factor Doublesex, present in around 650 male neurons (Rideout et al., 2010). Inhibition of synaptic transmission in all FruitlessM expressing neurons or in all DoublesexM expressing neurons causes severe reduction of courtship song production (along with other elements of male courtship) (Manoli et al., 2005; Rideout et al., 2010; Stockinger et al., 2005). As discussed below, different cell types underlie these phenotypes and FruitlessM and/or DoublesexM expression is found in almost all neurons that have been identified so far to be relevant for male courtship song production, from sensory neurons and central brain interneurons to ventral nerve cord pre-motor neurons.

3.1.1 Sensory input pathways for song initiation, modulation and wing choice

From behavioural studies it is evident that male courtship song production is triggered and maintained by a combination of several sensory stimuli, each of them contributing, but not absolutely essential for initiation of the behaviour (Krstic et al., 2009; Markow, 1987). In the vast majority of cases, song production only starts after the male touches the female with his foreleg and tastes her cuticular pheromones. Before this contact, he has normally seen and tracked his target. Song production in male flies can be elicited, albeit with low efficiency and reliability, by visual inputs alone, i.e. by a small moving dummy that the male starts to chase when it has the right shape and size (Agrawal et al., 2014). A good candidate neuron for encoding the percept of such a dummy is the visual projection neuron class LC10a that connects the optic lobe with the central brain. LC10a specifically responds to small, moving visual objects and can elicit short wing extensions when activated. When LC10a is activated in males with a recent history of activation of the central brain neuronal class P1, their wing extensions are greatly increased (Ribeiro et al., 2018). Such a pre-activation protocol is thought to introduce a longer lasting courtship state in the male, similar to the effect of a pheromonal stimulation in a natural scenario (Ribeiro et al., 2018). Indeed, adding gustatory stimuli to a visual dummy by perfuming it with female cuticular pheromones strongly enhances the response of the male (Agrawal et al., 2014).

The gustatory song initiation pathway, from foreleg gustatory neurons to courtship song eliciting centre P1 in the brain, has been dissected in some detail (Fig. 5A). On the male foreleg, sensory bristles are innervated by chemosensitive neurons expressing FruitlessM and the cation channel Pickpocket23. In each bristle, there is a cell responding to the female cuticular pheromones 7–11-heptacosadiene and 7–11-nonacosadiene, the so-called F-cell, and an M-cell responding to male 7-tricosene. The F-cells are required for efficient courtship initiation, whereas the M-cells

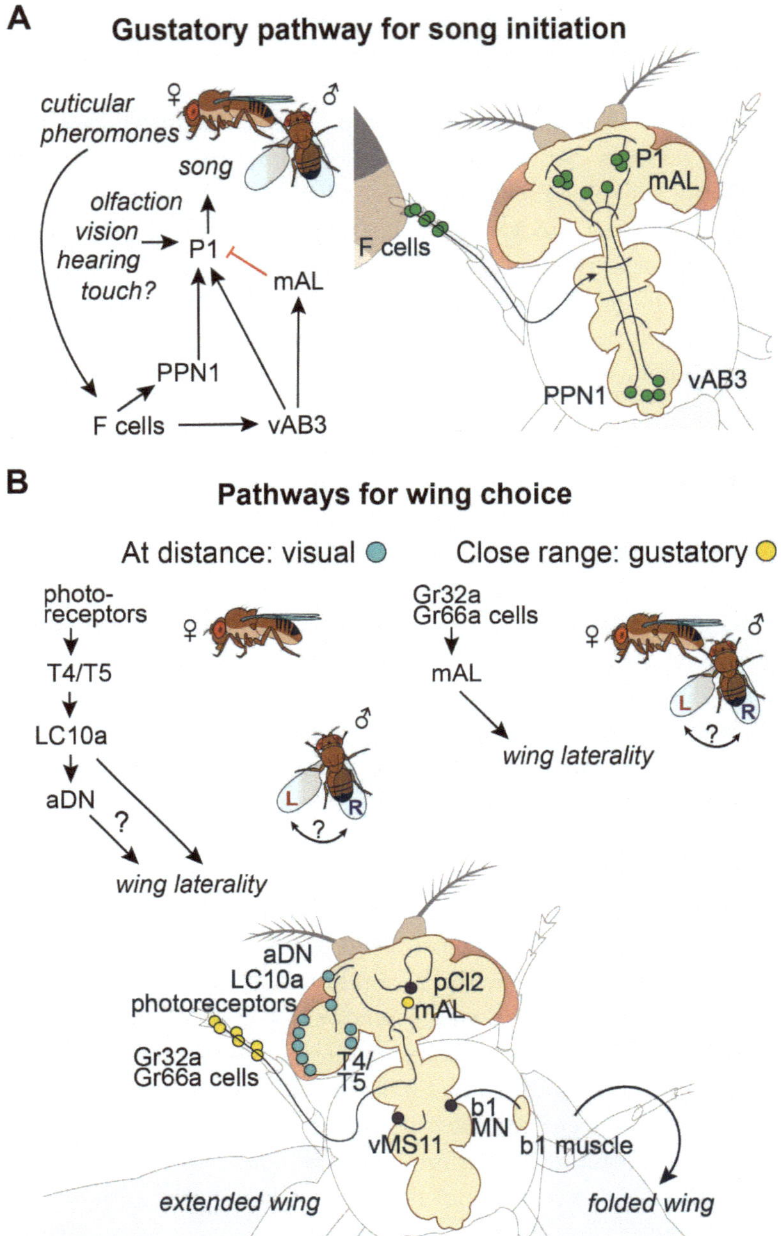

Fig. 5 Sensory control of male song initiation and wing choice. (A) Female cuticular pheromones are sensed by the gustatory F cells in the male foreleg and relayed via ventral nerve cord neurons PPN1 and vAB3 to brain integration neurons P1. The inhibitor neuronal class mAL serves as gain control module. P1 mediates the initiation of singing once the male has touched the female abdomen. It also receives sensory input from other modalities. (B) Two pathways for wing choice have been described that ensure the male sings with the wing closer to the female. At a distance, visual information activates via photoreceptors and components of the elementary motion detector, T4/T5 cells, the visual projection neurons LC10a. Among LC10a's outputs is aDN. At close range, wing choice is also mediated via gustatory information through foreleg gustatory Gr32a/Gr66a cells, putatively connecting to mAL brain neuron. Unilateral activation of pCl2 can determine wing choice, but its potential connection to the sensory pathways is not known. In the ventral nerve cord, vMS11 interneuron and b1 motor neuron (b1 MN) are implicated in wing choice.

are required to suppress courtship towards other males (Kimura et al., 2019; Lu et al., 2012; Thistle et al., 2012; Toda et al., 2012).

The sensory bristles on the forelegs that contain F-cells are also innervated by a mechanosensory neuron (Stocker, 1994), but the role of this modality in pheromone processing is not known. F-cells project to the ventral nerve cord, where they activate two types of interneurons, vAB3 and TTN1 neurons. Both send processes to the brain and activate a prominent cluster of male specific cells called P1. vAB3 also activates an inhibitor neuronal class, mAL, that inhibits P1. This at first sight paradoxical circuit motif of direct activation and indirect inhibition of P1 by vAB3 is thought to provide gain control and normalize input stimuli of a wide range of intensities (Clowney et al., 2015; Kallman et al., 2015). Although male foreleg tapping is behaviourally closely correlated with song initiation (Kohatsu et al., 2011), experimental activation of F-cells or TTN1 in absence of a female does not elicit singing of the male (Kallman et al., 2015; Toda et al., 2012). This is in contrast to activation of P1 cells, which do elicit singing when activated in solitary males (Kohatsu et al., 2011; von Philipsborn et al., 2011). P1 is commonly viewed as an "integration centre" and also receives input from other sensory modalities such as vision, olfaction and hearing (reviewed in: Auer and Benton, 2016; Ellendersen and von Philipsborn, 2017; Rings and Goodwin, 2019). Which of these are necessary in combination with F-cell activity to release the first song bout in a male that has made contact with a female is not known. In general, P1 activity levels and responsiveness to sensory input also depends on the internal state (see next section).

Concomitant with the initiation of singing, the male chooses the amplitude of pulse song, song mode (sine or pulse song) and wing laterality. Modulation of pulse song amplitude has been shown to depend on distance to the female within a range of around 5 mm. Visual cues play a main role for distance estimation. Males show amplitude modulation when the size and azimuthal position of a black square in their visual field changes in a way a female target would during a courtship interaction (Coen et al., 2016). Flies perceive moving and expanding objects by the elementary motion detection neuronal pathway and special looming sensitive neurons in their large optic lobes (Silies et al., 2014), but these pathways seem not to be involved in visually induced song amplitude modulation (Coen et al., 2016).

Female distance and locomotion speed is also correlated with choice of pulse vs sine song mode (Coen et al., 2014; Trott et al., 2012). There is no consensus which sensory input is mainly employed for this song mode modulation: Trott et al. (2012) report the importance of olfactory pathways, while Coen et al. conclude that vision is required (Coen et al., 2014).

The male typically extends the wing closer to the female and vibrates it with much higher amplitude than the folded wing. He uses vision and taste to make this choice (Fig. 5B). A small visual stimulus is sufficient to reliably bias wing extension to one side in a male whose brain neuronal class P1 (or pC1, a related and partially overlapping population) is activated (Kohatsu and Yamamoto, 2015; Pan et al., 2012). P1/pC1 is thought to be upstream of the sensory input informing wing choice, since activating it in only one hemisphere leads to equal left and right wing usage (Kohatsu and Yamamoto, 2015; von Philipsborn et al., 2011). In contrast, unilateral activation of the brain neurons pC2l leads to extension of the contralateral wing only (Kohatsu and Yamamoto, 2015). Differential activity of pC2l in the two brain hemispheres might thus instruct downstream circuits about wing usage. How this information is relayed to the ventral nerve cord and the motor effector circuits is unknown. Unilateral activation of the only known song descending interneuron pIP10 leads to singing with equal left and right wing usage, as it is the case for P1/pC1 (von Philipsborn et al., 2011). pIP10 is therefore not a good candidate for relaying wing choice information to the ventral nerve cord.

In contrast to pulse song amplitude modulation, vision instructed wing choice requires the elementary motion detection pathway, i.e. the optic lobe interneurons called T4 and T5-cells. It also depends on LC10a neurons, the visual projection neurons that convey courtship relevant information from the optic lobe to the brain. When LC10a neurons are silenced, the male extends either wing during singing without regard to the female position. Unilateral activation of LC10a elicits ipsilateral wing extensions (Ribeiro et al., 2018). In males, LC10a connects to the sexually dimorphic neuronal class aDN that leads to bilateral wing extensions and misdirected singing when silenced (Nojima et al., 2021). Wing choice defects in LC10a silenced and aDN silenced males seem to differ to some extent, making it likely that the laterality information from LC10a is also relayed to neurons other than aDN (Nojima et al., 2021).

Blind males or males with impairment in the LC10a visual pathway show strong wing usage defects mostly when they are at some distance (more than 4 mm) away from the female (Ribeiro et al., 2018; I. Ribeiro, personal communication). When the male is closer, gustatory information contributes more strongly to wing choice. This is concluded from the observation that silencing of foreleg sensory neurons coexpressing the gustatory receptors Gr32a and Gr66a (these are distinct from the Pickpocket23 expressing F-cells) or amputation of the foreleg increases simultaneous extension of both wings when visual input is lacking (Koganezawa et al., 2010). Anatomical analysis suggests that Gr32a/Gr66a neurons from one leg connect to the contralateral inhibitory brain neurons mAL that are highly asymmetric in their arborization pattern and mainly innervate the ipsilateral brain hemisphere. Silencing of mAL also increases bilateral wing extensions (Koganezawa et al., 2010). mAL expresses FruitlessM and requires this transcription factor for correct neurite outgrowth and development of its anatomy. In the absence of FruitlessM, mAL arborizations lose their potential contact areas with Gr32a/Gr66a neuron arborizations. This may explain why *fruitless* mutant males show high levels of bilateral wing extensions (Koganezawa et al., 2010).

The phenotypes reported for vision and taste deprived males, i.e. randomized unilateral vs bilateral wing extension, appear somewhat different. Defects resulting from manipulation of the visual pathway might be slightly obfuscated by a functional gustatory pathway. It is plausible that the pathways mainly function at different distances, but show some redundancy (Fig. 5B). To date, it is unclear how the two pathways are integrated, if pC2l is part of either of them and how they connect to the motor network for wing extension. In the ventral nerve cord, interneuron vMS11 and wing motor neuron b1 MN have been implicated in executing the wing choice during singing (Fig. 5B, see also Section 3.1.3).

3.1.2 Central neuronal control

When Hotta and Benzer delineated a small male brain area that was sufficient to make an otherwise female gynandromorph sing male courtship song (Hotta and Benzer, 1976), this area almost certainly included the P1 neuronal class. P1 consists of around 20 neurons with cell bodies in the posterior central brain that send their large, ring shaped arbours around the peduncle of the mushroom body (Fig. 6A). P1 only exists in adult males, since it undergoes apoptosis during pupal development in females by the action of female-specific DoublesexF. Presence of male-specific FruitlessM in P1 is required for development of its arborization pattern, the outgrowth of specific neurites and, assumedly establishment of its full connectivity pattern. Masculinization of P1 precursors in the female is sufficient to generate an ectopic P1 in a female. As in the early gynandromorph experiments, such a female displays elements of male courtship behaviour, including short bouts of courtship song production (Kimura et al., 2008). In males, activation of P1 in the absence of females triggers courtship song production (Kohatsu et al., 2011; von Philipsborn et al., 2011). As outlined above,

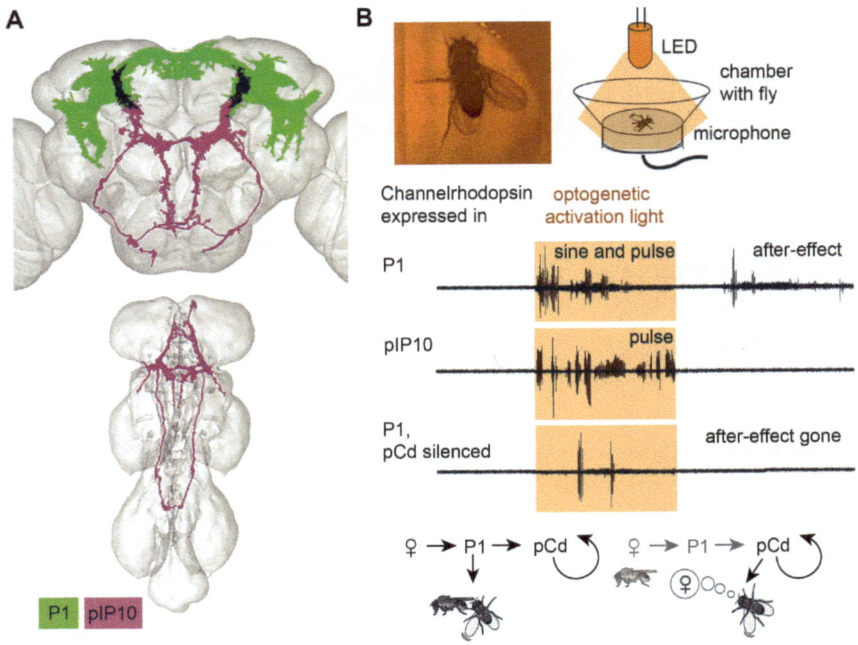

Fig. 6 Central neuronal control of male song. (A) Central brain and ventral nerve cord with reconstructed neuronal classes P1 (green) and the pair of descending pIP10 neurons. Regions where the two neuronal classes overlap with their arbours are shown in black. Neuron reconstructions from Philipsborn laboratory. (B) A single male fly expressing red shifted channelrhodopsin in P1 responds to optogenetic activation light by unilateral wing extension and singing (top left). Schematics of recording chamber with LED illumination for optogenetic activation experiments (top right). Oscillograms of optogenetically evoked song (drawn semi-schematically): Activation of P1 leads to sine and pulse song, with singing outlasting the acute activation period (light). Activation of pIP10 triggers pulse song only, with no after-effect. Activation of P1, with simultaneous silencing of neuronal class pCd, abolishes the after-effect. pCd is thought to enable the neuronal encoding of a persistent state and longer lasting behavioural changes during and after courtship when direct stimuli from the female are intermittent.

P1 receives multimodal sensory input and is thought to integrate information about external stimuli from the female with internal physiological information from the male to compute probability and persistence of courtship behaviours, including singing. The concept of such a circuit node for decision making or action selection that encodes arousal, motivation and persistent internal state has spurred numerous studies and models about P1 (reviewed in: Anderson, 2016; Lee and Wu, 2020; Rings and Goodwin, 2019). It has also become apparent that P1 is a heterogeneous group of cells, close by and similar to the pC1 cell group that is targeted by genetic reagents used in some studies of P1. The functional diversity of P1/pC1 subclasses is not yet fully explored, but is important to consider when interpreting behavioural phenotypes seen upon activation or silencing (Asahina, 2018).

Regarding song production, it is of interest that activation of P1 stimulates the male to sing close to natural courtship song bouts with sine and pulse elements. P1 directly connects to a pair of single descending neurons, pIP10, that relay information from central brain to the wing neuropil in the mesothoracic ganglia of the ventral nerve cord. pIP10 activation leads to pulse song only, lacking the alternation with sine bouts (von Philipsborn et al., 2011; Fig. 6B). As observable with optogenetic activators that can be temporally controlled with high resolution, pIP10 induced pulse song is only seen during acute neuronal stimulation, i.e. exposure to optogenetic activation light. In contrast, P1 stimulation produces a probabilistic onset of song bouts. Moreover, once activated males continue to sing after the experimental activation of P1 has ended (Inagaki et al., 2014; Fig. 6B). This indicates that P1 can support a state of arousal and continued singing when sensory input is intermittent as during courtship. Brain pCd neurons that are indirectly connected to P1, play an important role in the sustained effect of P1 activation. Acute silencing of this group of "P1 follower neurons" in the wake of brief P1 activation abolishes the after-effect of the latter. When pCd neurons are chronically silenced, males still sing towards females, but are less persistent in maintaining this behaviour over longer periods of time (Jung et al., 2020; Fig. 6B).

3.1.3 Neuronal control of motor output and signal patterning

The species-specific pattering of male courtship song is important for its behavioural effect on conspecifics. How does the nervous system generate this pattern and structure motor output?

The motor and premotor centres for control of wing movements are located in the ventral nerve cord (Venkatasubramanian and Mann, 2019). Flies move their wings with 15 control muscles and two sets of large antagonistic power muscles. The control muscles are each innervated by one excitatory motor neuron that effects fast muscle contraction for steering during flight. The power muscles, i.e. the dorsal longitudinal muscles (dlms) and the dorsal ventral muscles (dvms) are stretch-activated: firing of the motor neurons innervating them increases their output, but is not directly coupled to their contraction. As indicated by their name, they are powering the large wing strokes during flight and are required to keep the animal airborne (Dickinson and Tu, 1997; Fig. 7A). Specific targeting of the motor neurons

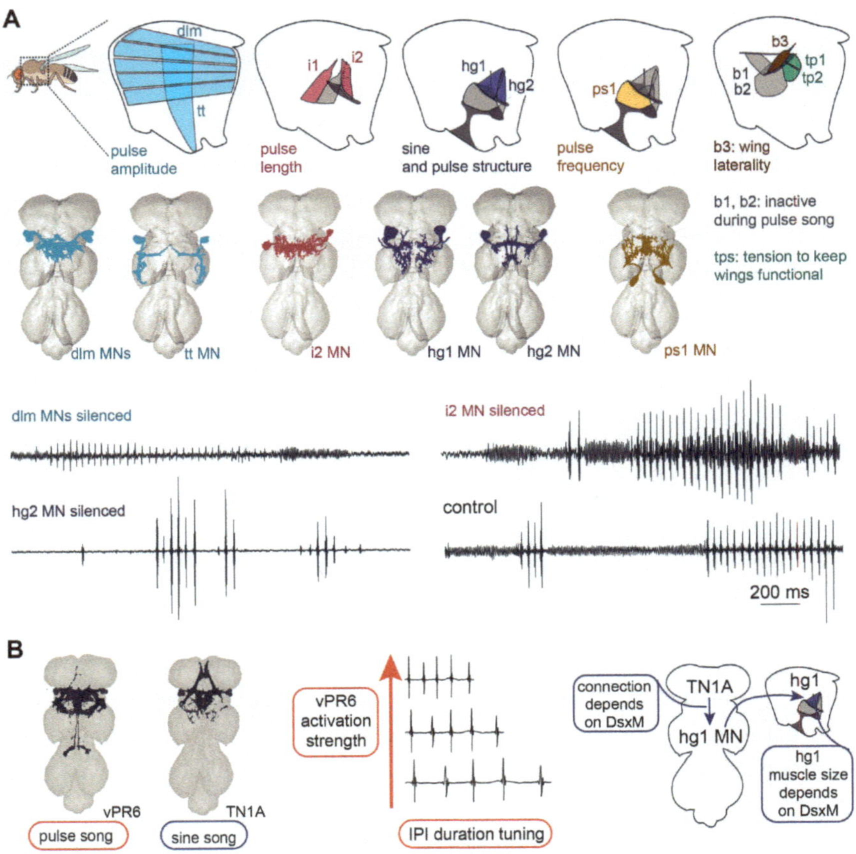

Fig. 7 Neuromuscular and pre-motor control of male song patterning. (A) Wing muscles and their innervating motor neurons (MNs) in the ventral nerve cord that are important for male song generation. Exemplary oscillograms (2 s long) of song from males with silenced dlm MNs (low amplitude), i2 MN (oscillations between pulses, polycyclic pulses) and hg2 MN (no sine song, longer IPIs), with a control wild type song. dlm, dorsal longitudinal muscles; tt, tergotrochanter; i1–2, muscles of the first axillary; hg1–2, muscles of the fourth axillary; ps1, pleurosternal muscle; b1–3, basalar muscles; tp1–2, tergopleural muscles. (B) Ventral nerve cord interneurons vPR6 and TN1A, required for pulse song and sine song, respectively. Experimental activation of vPR6 with increasing activation strength decreases IPI. TN1A forms synaptic connection with hg1 MN in a DoublesexM (DsxM) dependent way. The innervated muscle hg1 is larger in males, a dimorphism that also depends on the action of DsxM. Neuron reconstructions from Philipsborn laboratory. *Panel (A): Drawn with modifications after O'Sullivan, A., Lindsay, T., Prudnikova, A., Erdi, B., Dickinson, M., von Philipsborn, A.C., 2018. Multifunctional wing motor control of song and flight. Curr. Biol. 28, 2705–2717.*

has elucidated the role of the different wing muscles for male courtship song production. The activity of many motor neurons is required for producing correctly patterned song, with some motor neurons differentially affecting sine and pulse song (O'Sullivan et al., 2018; Shirangi et al., 2013). This widespread activity during pulse song is also seen by non-invasive calcium imaging of control muscles. Disabling a single muscle often alters the species-specific parameters of song in a characteristic way, without strongly affecting the amount of overall song production and wing extension patterns (Fig. 7A).

Silencing of the motor neurons innervating the dlms (dlm MNs) renders flies flightless and strongly decreases pulse song amplitude and pulse carrier frequencies. A similar phenotype is seen upon silencing of the tergotrochanter motor neuron (tt MN). The tt muscle initiates the first contraction of the stretch activated power muscle system at the start of flying and might have a similar role to recruit the power muscles during song for increasing pulse amplitude (O'Sullivan et al., 2018). A hallmark of male song, as compared to flight, is single, mostly monocyclic pulses. This led to the hypothesis that a specific control muscle has to stop the stretch activated system at the end of each pulse, preventing additional wing oscillations and polycyclic pulses (Ewing, 1979a,b). Good candidates for such "pulse stopper" muscles are i2 and its synergistic partner i1, muscles of the first axillary. Silencing of the i2 innervating motor neuron, i2 MN, leads to polycyclic song pulses and continuing oscillations between pulses. Acute activation of i2 MN can stop wing beats in flight or decrease their amplitude. The two muscles i1 and i2 are highly active during pulse song and tend to decrease their activity when the fly switches to sine song, a song mode that does not require termination or strong dampening of each wing oscillation (O'Sullivan et al., 2018). Required for sine song are muscles of the fourth axillary hg1 (Shirangi et al., 2013) and hg2 (O'Sullivan et al., 2018). They also contribute to pulse structure and IPI, both of which are defect upon silencing of hg1 MN and hg2 MN (O'Sullivan et al., 2018). Pulse carrier frequency is decreased upon disabling pleurosternal muscle 1 (ps1) (O'Sullivan et al., 2018; Shirangi et al., 2013). The three basalar muscles vary in their involvement in pulse song generation. While b1 and b2 muscles are not active during pulse song, b3 is the only muscle that shows strong unilateral activation during singing, with a stronger activity at the side of the non-extended wing. b3 MN might thus function in pulling the folded wing in and restricting in this way the amplitude of its vibration. It might be the motor output for pathways determining wing laterality during singing based on sensory input from the female (Fig. 5B). So far, exclusive genetic access to b3 MN has been lacking and this proposed function has not yet been tested. The two tergopleural muscles together keep the wing in its natural resting position. Disabling them paralyzes all wing movement to an extent that no song is produced (O'Sullivan et al., 2018; Fig. 7A).

Male and female flies have the same requirements for flight control. It is therefore not surprising that the motor neurons innervating wing control muscles are sexually monomorphic (Rideout et al., 2007; Trimarchi and Schneiderman, 1994). Interestingly, one of the control muscles, hg1, is larger in males. It has not been tested if this sexual dimorphism is relevant for song (Shirangi et al., 2013). Sexual dimorphisms are also found in the premotor neuronal circuits of the ventral nerve cord and these are important for male song. Mutations in either of the sex determination genes *doublesex* or *fruitless* lead to defect song. Mutants for male-specific transcripts of *doublesex* (*dsxM*) do not produce sine song (Villella and Hall, 1996). Mutants for all male-specific transcripts of *fruitless* (*fruM*) show reduced wing extensions and no sound production during these wing extensions (Ryner et al., 1996). FruM has three splicing isoform that are co-expressed in most neurons and are to some extent redundant in their function in song generation. Loss of one of the isoforms, FruMC, causes loss of sine song, while loss of the two other isoforms (FruMA and FruMB) on their own gives rise to changes in IPI and pulse structure (Neville et al., 2014; von Philipsborn et al., 2014). The combined action of all FruM isoforms and DsxM thus shapes the circuits required for song patterning in the ventral nerve cord (Rideout et al., 2007).

It is estimated that around 650 neurons in the ventral nerve cord of an adult male express FruM, and around 300 express DsxM, many of them also FruM positive (Rideout et al., 2010; Stockinger et al., 2005). Activation of all FruM expressing neurons in a decapitated male elicits the production of structured courtship song, underlining the notion that while initiation of singing is controlled by the brain, patterning of song arises from neurons in the ventral nerve cord (Clyne and Miesenböck, 2008). Activation of neuronal class dPR1 and vPR6 elicits pulse song in an intact, solitary male, while activation of TN1A elicits sine song. These neuronal classes each consist of around two to six neurons and express both FruM and DsxM. The cell bodies of dPR1 are located in the prothoracic neuromere. vPR6 and TN1A cell bodies are located close together in the mesothoracic neuromere and arise from the same developmental cell lineage. The three neuronal classes have a similar arborization pattern centered around the wing neuropil in the tectulum region of the mesothoracic neuromere (von Philipsborn et al., 2011; Shirangi et al., 2016; Fig. 7B). vPR6 might be directly involved in shaping the IPI of pulse song, since experimental activation of this neuronal class at increasing strength leads to pulse song with decreasing IPIs (von Philipsborn et al., 2011; Fig. 7B). TN1A offers a good mechanistic explanation of why *dsxM* mutant males do not sing sine song: it makes an excitatory connection to hg1 MN, one of the motor neurons required for sine song (Shirangi et al., 2016). The formation of these connections depends on the presence of DsxM during development. Of note, the above mentioned sexual size dimorphism of the hg1 muscle also depends on DsxM expression (Shirangi et al., 2013). The entire sine circuit motif TN1A > hg1 MN > hg1 is thus specified by DsxM (Fig. 7B). Additional to the three song interneuron classes dPR1, vPR6 and TN1A, the FruM positive interneurons dMS2, vMS11 and vMS12 are also involved in wing movement control and song pattern generation. When activated together, they lead to wing extensions resembling the ones during singing, but without vibration or sound generation. Unilateral activation of vMS11 leads to ipsilateral wing extension. Together with b1 MN, vMS11 might thus be part of the motor output side of the wing choice pathway (Fig. 5B). Synaptic silencing of dMS2, vMS11 and vMS12 together leads to strong structural changes in song, resembling the ones seen upon hg2 MN silencing (von Philipsborn et al., 2011). It remains to be clarified which of these neurons contribute to this phenotype and if any of them connect to hg2 MN.

According to current view, sex specific pre-motor circuits comprise a central pattern generator that generates the species-specific acoustic parameters of male song. This central pattern generator is switched on and off by signals from the brain, where sensory stimuli, context and internal information are combined to execute descending commands (Ellendersen and von Philipsborn, 2017; Rings and Goodwin, 2019). Though overall supported by experimental evidence, there are some open questions remaining about this simple model: Is there really a "central pattern generator", as it has been defined in sensu stricto? And if so, is it only located in the ventral nerve cord? Central pattern generators are usually conceptualized as networks of neurons that can generate rhythmic motor patterns (such as seen during locomotion, chewing, breathing…) in the absence of constant sensory feedback. The rhythm might arise from intrinsic rhythmic firing of pacemaker neurons or from reciprocally connected circuit motifs (reviewed in Mulloney and Smarandache, 2010; Selverston, 2010). Regarding male courtship song, we do not conclusively know to what extent sensory feedback maintains pulse parameters such as pulse length/cyclicity or IPI duration. There is good evidence that the male does not have to hear his own song to produce it properly, as removal or impairment of the antennal ear (Johnston's organ) does not affect song patterning (Burnet et al., 1977; Ewing, 1979b). Hearing and mechanosensation, both achieved by chordotonal organs such as the Johnston's organ, rely on similar molecular mechanisms. Mutants with compromised chordotonal organs (such as *atonal* and *beethoven*) show unusual wing postures during singing and song patterning defects (Tauber and Eberl, 2001). This supports a model proposed by Ewing (1979b), according to which proprioceptive feedback from the wing determines pulse length and cyclicity. The neurons and circuit motifs constituting such potential sensory motor control loop remain to be identified. What about the notion that song patterning arises exclusively from ventral nerve cord circuits? To date, it is not entirely clear to what extent brain circuit motifs are involved in tuning song structure. Bilateral wing usage during singing goes along with some changes in song acoustic parameters, i.e. mildly polycyclic pulses (Koganezawa et al., 2010). Since wing usage is under the control of brain neurons, manipulating these can affect patterning. Recently, a song patterning mutant, *croaker*, has been mapped to the gene *camta* (encoding Calmodulin-binding transcription factor), which is required in brain, but not ventral nerve cord neurons for correctly structured song (Sato et al., 2019). In summary, the role of sensory feedback and brain circuit nodes should be further investigated for a complete understanding of song patterning.

3.1.4 Species specificity and evolution of neuronal circuits

Manipulation of wing motor neurons or pre-motor interneurons in the ventral nerve cord leads to changes in male courtship song patterning, and often the resulting "defect" songs resemble songs of other *Drosophila* species in some parameters. For example, *D. simulans*, *D. mauritiana* and *D. sechellia* all have longer IPIs than *D. melanogaster*. Species from the *virilis* group, such as *D. montana*, sing polycyclic pulses (Hoikkala and Mazzi, 2009; Tomaru and Yamada, 2011). Faced with the fascinating multitude of acoustic patterns of different species, one cannot but wonder if this diversity could be mechanistically explained by variation in existence, connectivity pattern or cellular physiology of a few pattern generator neurons. Do new song types need new neurons? How conserved is the male *fruitless* circuit between species? Development of genetic tools in species other than *D. melanogaster* (Auer et al., 2020; Sato et al., 2020; Stern et al., 2017) has made it possible to start addressing such questions. A very interesting study employed *D. subobscura*, whose males show a distinct courtship pattern with nuptial feeding by regurgitation and do not generate any sound producing wing vibrations. *D. melanogaster* and *D. subobscura* are separated by around 25 million years of evolution. At the level of the brain, the neurons expressing FruM in the two species look overall very similar. Activation of *D. subobscura* FruM expressing neurons in the whole nervous system leads to *D. subobscura* specific courtship elements, i.e. regurgitation of a nuptial gift and silent bilateral wing extension (Tanaka et al., 2017). If *D. subobscura* thus has neurons clearly homologous to *D. melanogaster* pIP10, dPR1, vPR6 and TNA1, they might have a function different from courtship song generation. In the future, it will be very interesting to investigate if such homologous neurons exist and how they function in courtship.

For *D. yakuba* that diverged from *D. melanogaster* around 15 million years ago, specific genetic manipulation of descending song neuron pIP10 has been achieved. *D. yakuba* sings two types of pulse song, louder and higher frequency clack song and more silent, lower frequency pulse song. In this species, pIP10 looks very similar to *D. melanogaster* pIP10, with nearly identical arborizations in brain and ventral nerve cord and very similar electrophysiological properties. As in *D. melanogaster*, the neuron expresses FruM and is cholinergic. Activation of pIP10 in *D. yakuba* elicits clack song at lower activation strength and both clack and pulse song at higher activation strength (Ding et al., 2019). These findings indicate that the species-specific parameters of pulse song depend on circuit motifs downstream of pIP10.

For both *D. subobscura* and *D. yakuba*, it seems likely circuit motifs in the ventral nerve cord will provide an explanation for why the species does not produce song or produces song different from *D. melanogaster*.

3.2 Neuronal control of female copulation song

Females almost exclusively sing during copulation, with very few song pulses occurring after copulations or during copulation attempts (Kerwin et al., 2020). Since females do not sing in copulations with males that are depleted of inseminate or have altered inseminate composition, it appears likely that the sensory input triggering or gating female singing is chemosensory. It has been hypothesized that seminal fluid molecules could act in the fashion of pheromones, and are sensed by receptor neuron in the female reproductive tract (Kerwin and von Philipsborn, 2020), but sensory input pathways important for female singing have not yet been identified.

Ventral nerve cord neurons expressing *dsx* are necessary for song production. These neurons might be either sensory neurons or interneurons. At the motor output side, female song requires the activity of the motor neurons innervating the indirect dorsal longitudinal muscles (dlm MNs), part of the indirect musculature powering flight. There is indication that neuronal circuits in the ventral nerve cord are sufficient for generating or at least maintaining already initiated copulation song: when females are beheaded during copulation, they continue to produce copulation song (Kerwin et al., 2020). So far, it is not known if female copulation singing is modulated by female physiological condition or internal state and which circuits might mediate such effects.

Another largely unanswered question is to which extent the neuronal substrate for male courtship song and female copulation song might be shared. As mentioned above, many neurons implicated in male song generation are only found in males. On the level of motor neurons, dlm MNs are important for both song types, but play different roles. Whereas inactivation of dlm MNs in the male only decreases courtship song amplitude, it completely abolishes copulation song in females (Kerwin et al., 2020; O'Sullivan et al., 2018).

3.3 Neuronal control of agonistic sound production

Very little is known about the neuronal control of agonistic sound production. Since the sounds produced during sexual rejection look very similar to aggression sounds emitted during fights between two males, one may speculate that they rely on the same neuronal substrate, at least in regards to motor output circuits.

3.3.1 Neuronal control of sexual rejection sounds

Sexual rejection sounds are produced by immature virgins or males, but not mature virgins or mated females. It is not known if specific aspects of male courtship elicit immature rejection and which sensory modalities are involved.

The neuronal control of female receptivity has been studied in great detail, but mostly in mature virgins and mated females, and with a focus on how the nervous system mediates the transition between these two stages (the so-called post-mating switch) (reviewed in: Aranha and Vasconcelos, 2018; Ellendersen and von Philipsborn, 2017; Ishimoto and Kamikouchi, 2021; Laturney and Billeter, 2014; von Philipsborn, 2020).

Since mated females and immature virgins respond to male courtship in a different way, with ovipositor extrusion and rejection wing flicks, respectively, there is the notion that the low receptivity in these two types of females depends on different neuronal mechanisms. During the first 48 h after eclosion, female ovaries mature and pheromone production develops under the influence of juvenile hormone and ecdysone (Bilen et al., 2013; Billeter and Wolfner, 2018; Manning, 1967b). Juvenile hormone is thus a good candidate for configuring the neuronal circuits for receptivity in immature virgins (Riddiford, 2012), and might also modulate neurons involved in rejection sound production. Interestingly, mature virgins with reduced ecdysteroid signalling were reported to produce sound pulses resembling male song towards courting males (Ganter et al., 2012). Ganter et al. (2012) interpreted this phenotype as masculinization and an indication that ecdysteroids are involved in development of sex specific circuits of the nervous system. Alternatively, the sound production in these mutants could hypothetically be due to preservation of wild type immature virgin circuits for rejection sounds in the absence of ecdysteroid signalling.

3.3.2 Neuronal control of inter-male aggression sounds

The neuronal control of inter-male aggressive behaviour in *D. melanogaster* has been very well studied (reviewed in: Asahina, 2017; Ellendersen and von Philipsborn, 2017; Hoopfer, 2016), but little attention has been paid to wing flicking and aggression sounds in particular. A recent study investigated in detail one element of aggressive displays, the so-called wing threat, during which a male extends and elevates his wings for some seconds while running towards and following a target fly. This element is controlled by a small group of around three brain neurons (Duistermars et al., 2018). If sounds are produced during wing threats has not been investigated in this work. From the earlier work of Jonsson et al. (2011), it appears that wing threats and silent and distinct from sound producing wing flicks.

Aggression behaviour is sexually dimorphic. There is indication that sound producing wing flicks are male-specific behavioural elements (Nilsen et al., 2004). Studies focusing on the differences between the sexes and their neuronal control have mostly focused on male-characteristic attack lunges and female-characteristic headbutts (Chiu et al., 2021; Palavicino-Maggio et al., 2019). For these behavioural elements, sex specific neuron types have been identified downstream of sexually monomorphic neurons that drive a general aggressive approach (Chiu et al., 2021).

The neuronal control of aggression and courtship is interlinked. Intermale aggression is enhanced by the same sensory cues (female pheromones and food odours) as courtship. Fighting and courting may happen in short succession in the presence of both sexes. The nervous system is thus challenged to execute the two behaviours in an appropriate pattern towards the right target without intermixing them to an extent at which communication signals and motor acts become ineffective. In the central brain, in part overlapping subsets of a larger and functionally heterogenous cell group P1/pC1 can promote both aggression and courtship when activated (Hoopfer, 2016; Ishii et al., 2020; Koganezawa et al., 2016; Wohl et al., 2020). So far, it is not known if courtship and aggression circuits are also sharing bifunctional neuronal classes in the ventral nerve cord, e.g. for pre-motor control of aggression sound pulses and courtship song.

In general, functional diversity of groups of neurons has sometimes been either overlooked or overstated when different genetic tools used for targeting them lacked resolution or were wrongly assumed to be interchangeable (Asahina, 2018).

4. Sound production in *Drosophila melanogaster* as model system for behavioural circuit neuroscience

D. melanogaster song production is an attractive model system for behavioural circuit neuroscience for several reasons.

First, song behaviour is easily and precisely quantifiable by recording song. For this, inexpensive multiplexed recording systems, as well as software scripts for automated pulse and sine detection have been developed (Arthur et al., 2013). Sound oscillograms track wing latitudinal velocity and position in a precise fashion (Bennet-Clark and Ewing, 1968), so that wing movements and any small alterations can be accurately quantified at millisecond resolution from audio recordings. Male courtship singing can be also quantified by evaluating wing extensions from video recordings only. This is possible, since courtship song involves very characteristic, longer lasting unilateral wing extensions that are practically never seen outside a courtship context.

Second, song production is sex-specific. This means that focus on sex-specific neurons and neuronal subpopulations expressing sex determination genes such as *dsx* and *fru* provides a good entry point into circuit analysis. The other sex can serve as an instructive control or comparison for the main sex studied.

Third, song production is a multifaceted and versatile model system. Many different questions of behavioural circuit neuroscience can be addressed: sensory processing and multimodal sensory integration, central control and state-dependency, motor patterning and sensory-motor coordination, as well as the evolution of circuits for species-specific behaviour.

Fourth, song production is crucial for reproduction, an essential and robust part of *D. melanogaster's* behavioural repertoire. Reproductive behaviour provides a rich conceptual framework from behavioural ecology and evolutionary biology for more ultimate interpretations about behaviour and its underlying neurobiology. Even for neuroscientists focused on proximate mechanisms, Theodosius Dobzhansky's "Light of Evolution" (Dobzhansky, 1973) resonates, inspires and guides the formulation of hypotheses.

4.1 Open questions

Many aspects of sound production behaviour and its neuronal control in *D. melanogaster* are still not fully understood. While male courtship song is clearly an important signal, the behavioural significance of female copulation song and agonistic sounds await further investigation. In regard to female song, important questions are: Which seminal fluid molecules, if any, trigger female song and how are they detected by the female nervous system? Can it be directly demonstrated that males hearing female song alter their seminal fluid allocation, and if so, how do they alter it? Does female physiological state affect the amount of copulation song? On the level of neuronal control, it will be interesting to compare action selection, context dependence and motor patterning circuits for female and male song. Since many of the interneurons for male song generation are male specific, neurons for female song might be female specific or part of a unisex interneuron network for wing movements during flight.

Sexual rejection sounds have never been probed directly for an effect on behaviour of conspecifics. Since they look very similar to sounds produced during inter-male aggression, it is not clear if flies show specific reaction to them when there are no additional cues or signals defining the behavioural context. Likewise, it is not known if they constitute reactions to general irritation by a nearby fly and are elicited by activation of the same neuronal circuit.

Male courtship song is well studied and many neuronal classes and small circuit motifs involved in its control have been described. Brain circuits and sensory input pathways are somewhat better understood than motor control circuits of the ventral nerve cord. As already elaborated in Section 3.1.3, we still do not know how the male courtship song pattern is generated at the level of interconnected neurons and their cellular physiology. For example, how is the timing of the IPI computed? What aspects of song patterning are under the control of mechanosensory feedback?

Song production is not linked to a very well defined anatomical substructure in the brain that by its architecture alone supports hypotheses about information flow and processing logic. Such is the case for example for visual processing in the optic lobe (Takemura et al., 2013), for learning in the mushroom body (Li et al., 2020) and for navigation in the central complex (Hulse et al., 2020). The anatomical survey of all *fruitless* expressing neurons in the male nervous system has yielded a somewhat hypothetical "courtship circuit" from sensory input to motor output (Yu et al., 2010).

Once a full male nervous system connectome is available (see next section), it will greatly facilitate connecting all described circuit modules involved in male courtship song production into a larger circuit. We might then better understand how well FruM expression delineates this circuit and to what extent a defined idiosyncratic behavioural element such as male singing relies on dedicated neurons that do not have any other function.

The last topic also relates to the many open questions regarding differences in the neuronal control of sound production between behaviourally differing *Drosophila* species. From the few existing studies it seems homologous neurons are easily found—how easily they can alter their function and drive new motor patterns remains to be seen.

4.2 Outlook: The impact of connectomics and genetic tools

D. melanogaster circuit neuroscience is permeated by Benzer's vision for behavioural genetics and the idea that "the path from the gene to behaviour" leads through the nervous system (Benzer, 1971). It has been fuelled in the past by the discovery of behavioural mutants, the genes affected in them and the analysis where and how these genes act in the nervous system. Importantly, it has also been strongly guided by concepts and mechanistic principles developed from neurophysiology research with larger, genetically less amendable species (Burrows, 1996; Strausfeld, 1976, 2012).

In the future, it will be exciting to see how electron microscopy connectomes of the *D. melanogaster* nervous system will accelerate and transform behavioural circuit neuroscience. Currently, such connectomes with synaptic resolution exist for the female brain (a full brain by transmission electron microscopy and a "hemibrain" by focused ion beam scanning electron microscopy) and the female ventral nerve cord (Phelps et al., 2021; Scheffer et al., 2020; Zheng et al., 2018). Neurons and synapses in the female hemibrain have been traced and reconstructed by machine learning algorithms and the derived connectivity map is freely available. Since many circuit nodes of male courtship song are sex specific, study of this particular behaviour has not yet profited to a full extent from the new "*Drosophila* connectome era". Future male connectomes are expected to change this quickly.

From behavioural circuit neuroscience studies that make use of the connectome, it becomes clear that anatomically very similar neurons can have substantially different connectivity patterns (e.g. Schretter et al., 2020). In order to manipulate such subsets of neuronal classes, rational design of expression systems and intersectional strategies is important (Luan et al., 2020). For further developing the latter, single-cell RNA sequencing that allows for classification of different cell types based on expression profiles (Li, 2020; Mohr et al., 2021), is a promising technology. Integrating data from single-cell RNA sequencing with the connectome is also expected to enhance the usefulness of the latter to make predictions about circuit function (Bates et al., 2019).

The effector toolkit for circuit neuroscience, comprising proteins for visualization, connectivity mapping, monitoring of neuronal activity and manipulation of neuronal activity, is constantly increasing and becoming more refined (Guo et al., 2019; Simpson and Looger, 2018).

Technical developments in circuit neuroscience inspire to ask new questions about old behaviours. Likewise, the newest concepts in vogue encourage sophisticated modelling of neuronal control of behaviour. We believe that understanding behaviour in detail, spending time in observation and thought about ecological and adaptive relevance, is as important as ever for research with long lasting impact.

Acknowledgements

We thank Kenta Asahina for thoughtful input on aggression behaviour, Ines Ribeiro for sharing insight on visual control of courtship, Peter Kerwin and Bardur Ellendersen for providing oscillograms and members of the Philipsborn lab for feedback on the manuscript. We apologize to all colleagues whose work we did not mention out of ignorance or oversight. This work was supported by Lundbeckfonden grant no. DANDRITE-R248-2016-2518.

References

Agrawal, S., Safarik, S., Dickinson, M., 2014. The relative roles of vision and chemosensation in mate recognition of *Drosophila melanogaster*. J. Exp. Biol. 217, 2796–2805.
Albert, J.T., Göpfert, M.C., 2015. Hearing in *Drosophila*. Curr. Opin. Neurobiol. 34, 79–85.
Alexander, R.D., 1961. Aggressiveness, territoriality, and sexual behavior in field crickets (Orthoptera: Gryllidae). Behaviour 17, 130–223.
Anderson, D.J., 2016. Circuit modules linking internal states and social behaviour in flies and mice. Nat. Rev. Neurosci. 17, 692–704.
Anholt, R.R.H., O'Grady, P., Wolfner, M.F., Harbison, S.T., 2020. Evolution of reproductive behavior. Genetics 214, 49–73.
Aranha, M.M., Vasconcelos, M.L., 2018. Deciphering *Drosophila* female innate behaviors. Curr. Opin. Neurobiol. 52, 139–148.
Aranha, M.M., Herrmann, D., Cachitas, H., Neto-Silva, R.M., Dias, S., Vasconcelos, M.L., 2017. Apterous brain neurons control receptivity to male courtship in *Drosophila melanogaster* females. Sci. Rep. 7, 46242.
Arthur, B., Sunayama-Morita, T., Coen, P., Murthy, M., Stern, D., 2013. Multi-channel acoustic recording and automated analysis of *Drosophila* courtship songs. BMC Biol. 11, 11.
Asahina, K., 2017. Neuromodulation and strategic action choice in *Drosophila* aggression. Annu. Rev. Neurosci. 40, 51–75.
Asahina, K., 2018. Sex differences in *Drosophila* behavior: qualitative and quantitative dimorphism. Curr. Opin. Physiol. 6, 35–45.
Auer, T.O., Benton, R., 2016. Sexual circuitry in *Drosophila*. Curr. Opin. Neurobiol. 38, 18–26.
Auer, T.O., Khallaf, M.A., Silbering, A.F., Zappia, G., Ellis, K., Álvarez-Ocaña, R., Arguello, J.R., Hansson, B.S., Jefferis, G.S.X.E., Caron, S.J.C., Knaden, M., Benton, R., 2020. Olfactory receptor and circuit evolution promote host specialization. Nature 579, 402–408.
Baker, B.S., Taylor, B.J., Hall, J.C., 2001. Are complex behaviors specified by dedicated regulatory genes? Reasoning from *Drosophila*. Cell 105, 13–24.
Baker, C.A., Clemens, J., Murthy, M., 2019. Acoustic pattern recognition and courtship songs: insights from insects. Annu. Rev. Neurosci. 42, 129–147.
Bastock, M., Manning, A., 1955. The courtship of *Drosophila melanogaster*. Behaviour 8, 85–110.
Bates, A.S., Janssens, J., Jefferis, G.S., Aerts, S., 2019. Neuronal cell types in the fly: single-cell anatomy meets single-cell genomics. Curr. Opin. Neurobiol. 56, 125–134.
Bennet-Clark, H.C., 1971. Acoustics of insect song. Nature 234, 255–259.
Bennet-Clark, H.C., Ewing, A.W., 1967. Stimuli provided by courtship of male *Drosophila melanogaster*. Nature 215, 669–671.
Bennet-Clark, H.C., Ewing, A.W., 1968. The wing mechanism involved in the courtship of *Drosophila*. J. Exp. Biol. 49, 117–128.
Bennet-Clark, H.C., Ewing, A.W., 1969. Pulse interval as a critical parameter in the courtship song of *Drosophila melanogaster*. Anim. Behav. 17, 755–759.
Benzer, S., 1971. From the gene to behavior. JAMA 218, 1015–1022.
Bilen, J., Atallah, J., Azanchi, R., Levine, J.D., Riddiford, L.M., 2013. Regulation of onset of female mating and sex pheromone production by juvenile hormone in *Drosophila melanogaster*. Proc. Natl. Acad. Sci. U. S. A. 110, 18321–18326.
Billeter, J.-C., Levine, J.D., 2013. Who is he and what is he to you? Recognition in *Drosophila melanogaster*. Curr. Opin. Neurobiol. 23, 17–23.
Billeter, J.-C., Wolfner, M.F., 2018. Chemical cues that guide female reproduction in *Drosophila melanogaster*. J. Chem. Ecol. 44, 750–769.

Brown, W.D., Smith, A.T., Moskalik, B., Gabriel, J., 2006. Aggressive contests in house crickets: size, motivation and the information content of aggressive songs. Anim. Behav. 72, 225–233.

Brüggemeier, B., Porter, M.A., Vigoreaux, J.O., Goodwin, S.F., 2018. Female *Drosophila melanogaster* respond to song-amplitude modulations. Biol. Open 7, bio032003.

Burnet, B., Connolly, K., Dennis, L., 1971. The function and processing of auditory information in the courtship behaviour of *Drosophila melanogaster*. Anim. Behav. 19, 409–415.

Burnet, B., Eastwood, L., Connolly, K., 1977. The courtship song of male *Drosophila* lacking aristae. Anim. Behav. 25, 460–464.

Burrows, M., 1996. The Neurobiology of an Insect Brain. Oxford University Press.

Bussell, J.J., Yapici, N., Zhang, S.X., Dickson, B.J., Vosshall, L.B., 2014. Abdominal-B neurons control *Drosophila* virgin female receptivity. Curr. Biol. 24, 1584–1595.

Chakravorty, S., Vu, H., Foelber, V., Vigoreaux, J.O., 2014. Mutations of the *Drosophila* myosin regulatory light chain affect courtship song and reduce reproductive success. PLoS One 9, e90077.

Chiu, H., Hoopfer, E.D., Coughlan, M.L., Pavlou, H.J., Goodwin, S.F., Anderson, D.J., 2021. A circuit logic for sexually shared and dimorphic aggressive behaviors in *Drosophila*. Cell 184, 507–520.

Clausen, L.W., 1954. Insect Fact and Folklore. Macmillan, New York.

Clemens, J., Coen, P., Roemschied, F.A., Pereira, T.D., Mazumder, D., Aldarondo, D.E., Pacheco, D.A., Murthy, M., 2018. Discovery of a new song mode in *Drosophila* reveals hidden structure in the sensory and neural drivers of behavior. Curr. Biol. 28, 2400–2412.

Clowney, E.J., Iguchi, S., Bussell, J.J., Scheer, E., Ruta, V., 2015. Multimodal chemosensory circuits controlling male courtship in *Drosophila*. Neuron 87, 1036–1049.

Clyne, J.D., Miesenböck, G., 2008. Sex-specific control and tuning of the pattern generator for courtship song in *Drosophila*. Cell 133, 354–363.

Cobb, M., Burnet, B., Blizard, R., Jallon, J.-M., 1989. Courtship in *Drosophila sechellia*: its structure, functional aspects, and relationship to those of other members of the *Drosophila melanogaster* species subgroup. J. Insect Behav. 2, 63–89.

Coen, P., Clemens, J., Weinstein, A.J., Pacheco, D.A., Deng, Y., Murthy, M., 2014. Dynamic sensory cues shape song structure in *Drosophila*. Nature 507, 233–237.

Coen, P., Xie, M., Clemens, J., Murthy, M., 2016. Sensorimotor transformations underlying variability in song intensity during *Drosophila* courtship. Neuron 89, 629–644.

Connolly, K., Cook, R., 1973. Rejection responses by female *Drosophila melanogaster*: their ontogeny, causality and effects upon the behaviour of the courting male. Behaviour 44, 142–166.

Cowling, D.E., Burnet, B., 1981. Courtship songs and genetic control of their acoustic characteristics in sibling species of the *Drosophila melanogaster* subgroup. Anim. Behav. 29, 924–935.

Crossley, S.A., 1988. Failure to confirm rhythms in *Drosophila* courtship song. Anim. Behav. 36, 1098–1109.

Crossley, S.A., Bennet-Clark, H.C., Evert, H.T., 1995. Courtship song components affect male and female *Drosophila* differently. Anim. Behav. 50, 827–839.

Deutsch, D., Clemens, J., Thiberge, S.Y., Guan, G., Murthy, M., 2019. Shared song detector neurons in *Drosophila* male and female brains drive sex-specific behaviors. Curr. Biol. 29, 3200–3215.

Dickinson, M.H., Tu, M.S., 1997. The function of Dipteran flight muscle. Comp. Biochem. Physiol. A 116, 223–238.

Ding, Y., Berrocal, A., Morita, T., Longden, K.D., Stern, D.L., 2016. Natural courtship song variation caused by an intronic retroelement in an ion channel gene. Nature 536, 329–332.

Ding, Y., Lillvis, J.L., Cande, J., Berman, G.J., Arthur, B.J., Long, X., Xu, M., Dickson, B.J., Stern, D.L., 2019. Neural evolution of context-dependent fly song. Curr. Biol. 29, 1089–1099.

Dobzhansky, T., 1973. Nothing in biology makes sense except in the light of evolution. Am. Biol. Teach. 35, 125–129.

Duistermars, B.J., Pfeiffer, B.D., Hoopfer, E.D., Anderson, D.J., 2018. A brain module for scalable control of complex, multi-motor threat displays. Neuron 100, 1474–1490.

Eberl, D.F., Duyk, G.M., Perrimon, N., 1997. A genetic screen for mutations that disrupt an auditory response in *Drosophila melanogaster*. Proc. Natl Acad. Sci. USA 94, 14837–14842.

Ellendersen, B.E., von Philipsborn, A.C., 2017. Neuronal modulation of *D. melanogaster* sexual behaviour. Curr. Opin. Insect Sci. 24, 21–28.

Ewing, A.W., 1964. The influence of wing area on the courtship behaviour of *Drosophila melanogaster*. Anim. Behav. 12, 316–320.

Ewing, A.W., 1979a. The neuromuscular basis of courtship song in *Drosophila*: the role of the direct and axillary wing muscles. J. Comp. Physiol. 130, 87–93.

Ewing, A.W., 1979b. The role of feedback during singing and flight in *Drosophila melanogaster*. Physiol. Entomol. 4, 329–337.

Ewing, A.W., 1988. Cycles in the courtship song of male *Drosophila melanogaster* have not been detected. Anim. Behav. 36, 1091–1097.

Ewing, A.W., Bennet-Clark, H.C., 1968. The courtship songs of *Drosophila*. Behaviour 31, 288–301.

Fabre, C.C., Hedwig, B., Conduit, G., Lawrence, P.A., Goodwin, S.F., Casal, J., 2012. Substrate-borne vibratory communication during courtship in *Drosophila melanogaster*. Curr. Biol. 22, 2180–2185.

Fan, P., Manoli, D.S., Ahmed, O.M., Chen, Y., Agarwal, N., Kwong, S., Cai, A.G., Neitz, J., Renslo, A., Baker, B.S., Shah, N.M., 2013. Genetic and neural mechanisms that inhibit *Drosophila* from mating with other species. Cell 154, 89–102.

Fedotov, S.A., Bragina, J.V., Besedina, N.G., Danilenkova, L.V., Kamysheva, E.A., Panova, A.A., Kamyshev, N.G., 2014. The effect of neurospecific knockdown of candidate genes for locomotor behavior and sound production in *Drosophila melanogaster*. Fly 8, 176–187.

Ganter, G.K., Desilets, J.B., Davis-Knowlton, J.A., Panaitiu, A.E., Sweezy, M., Sungail, J., Tan, L.C., Adams, A.M., Fisher, E.A., O'Brien, J.R., Kincaid, K.M., Heinrich, R., 2012. *Drosophila* female precopulatory behavior is modulated by ecdysteroids. J. Insect Physiol. 58, 413–419.

Gleason, J.M., 2005. Mutations and natural genetic variation in the courtship song of *Drosophila*. Behav. Genet. 35, 265–277.

Gligorov, D., Sitnik, J.L., Maeda, R.K., Wolfner, M.F., Karch, F., 2013. A novel function for the Hox gene *abd-B* in the male accessory gland regulates the long-term female post-mating response in *Drosophila*. PLoS Genet. 9, e1003395.

Gorostiza, E.A., Colomb, J., Brembs, B., 2016. A decision underlies phototaxis in an insect. Open Biol. 6, 160229.

Greenfield, M.D., 2002. Signalers and Receivers: Mechanisms and Evolution of Arthropod Communication. Oxford University Press.

Griffith, L.C., Ejima, A., 2009. Courtship learning in *Drosophila melanogaster*: diverse plasticity of a reproductive behavior. Learn. Mem. 16, 743–750.

Guo, C., Pan, Y., Gong, Z., 2019. Recent advances in the genetic dissection of neural circuits in *Drosophila*. Neurosci. Bull. 35, 1058–1072.

Hall, J.C., 1979. Control of male reproductive behavior by the central nervous system of *Drosophila*: dissection of a courtship pathway by genetic mosaics. Genetics 92, 437–457.

Hoikkala, A., Mazzi, D., 2009. Evolution of complex acoustic signals in *Drosophila* species. In: Kim, Y.-K. (Ed.), Handbook of Behavior Genetics. Springer New York, New York, NY, pp. 187–196.

Hoopfer, E.D., 2016. Neural control of aggression in *Drosophila*. Curr. Opin. Neurobiol. 38, 109–118.

Hotta, Y., Benzer, S., 1976. Courtship in *Drosophila* mosaics: sex-specific foci for sequential action patterns. Proc. Natl. Acad. Sci. U. S. A. 73, 4154–4158.

Hulse, B.K., Haberkern, H., Franconville, R., Turner-Evans, D.B., Takemura, S., Wolff, T., Noorman, M., Dreher, M., Dan, C., Parekh, R., Hermundstad, A.M., Rubin, G.M., Jayaraman, V., 2020. A connectome of the *Drosophila* central complex reveals network motifs suitable for flexible navigation and context-dependent action selection. bioRxiv. 2020.2012.2008.413955.

Iglesias, P.P., Hasson, E., 2017. The role of courtship song in female mate choice in South American cactophilic *Drosophila*. PLoS One 12, e0176119.

Inagaki, H.K., Kamikouchi, A., Ito, K., 2010. Protocol for quantifying sound-sensing ability of *Drosophila melanogaster*. Nat. Protoc. 5, 26–30.

Inagaki, H.K., Jung, Y., Hoopfer, E.D., Wong, A.M., Mishra, N., Lin, J.Y., Tsien, R.Y., Anderson, D.J., 2014. Optogenetic control of *Drosophila* using a red-shifted channelrhodopsin reveals experience-dependent influences on courtship. Nat. Methods 11, 325–332.

Ishii, K., Wohl, M., DeSouza, A., Asahina, K., 2020. Sex-determining genes distinctly regulate courtship capability and target preference via sexually dimorphic neurons. elife 9, e52701.

Ishikawa, Y., Kamikouchi, A., 2016. Auditory system of fruit flies. Hear. Res. 338, 1–8.

Ishikawa, Y., Okamoto, N., Yoneyama, Y., Maeda, N., Kamikouchi, A., 2019. A single male auditory response test to quantify auditory behavioral responses in *Drosophila melanogaster*. J. Neurogenet. 33, 64–74.

Ishimoto, H., Kamikouchi, A., 2021. Molecular and neural mechanisms regulating sexual motivation of virgin female *Drosophila*. Cell Mol. Life Sci. 78, 4805–4819.

Jonsson, T., Kravitz, E.A., Heinrich, R., 2011. Sound production during agonistic behavior of male *Drosophila melanogaster*. Fly 5, 29–38.

Jung, Y., Kennedy, A., Chiu, H., Mohammad, F., Claridge-Chang, A., Anderson, D.J., 2020. Neurons that function within an integrator to promote a persistent behavioral state in *Drosophila*. Neuron 105, 322–333.e325.

Kallman, B.R., Kim, H., Scott, K., 2015. Excitation and inhibition onto central courtship neurons biases *Drosophila* mate choice. elife 4, e11188.

Kamikouchi, A., 2013. Auditory neuroscience in fruit flies. Neurosci. Res. 76, 113–118.

Kamikouchi, A., Inagaki, H.K., Effertz, T., Hendrich, O., Fiala, A., Gopfert, M.C., Ito, K., 2009. The neural basis of *Drosophila* gravity-sensing and hearing. Nature 458, 165–171.

Kerwin, P., von Philipsborn, A.C., 2020. Copulation song in *Drosophila*: do females sing to change male ejaculate allocation and incite postcopulatory mate choice? BioEssays 42, 2000109.

Kerwin, P., Yuan, J., von Philipsborn, A.C., 2020. Female copulation song is modulated by seminal fluid. Nat. Commun. 11, 1430.

Kimura, K., Hachiya, T., Koganezawa, M., Tazawa, T., Yamamoto, D., 2008. Fruitless and doublesex coordinate to generate male-specific neurons that can initiate courtship. Neuron 59, 759–769.

Kimura, K.I., Urushizaki, A., Sato, C., Yamamoto, D., 2019. A novel sex difference in *Drosophila* contact chemosensory neurons unveiled using single cell labeling. J. Neurogenet. 33, 116–124.

Koganezawa, M., Haba, D., Matsuo, T., Yamamoto, D., 2010. The shaping of male courtship posture by lateralized gustatory inputs to male-specific interneurons. Curr. Biol. 20, 1–8.

Koganezawa, M., Kimura, K.-I., Yamamoto, D., 2016. The neural circuitry that functions as a switch for courtship versus aggression in *Drosophila* males. Curr. Biol. 26, 1395–1403.
Kohatsu, S., Yamamoto, D., 2015. Visually induced initiation of *Drosophila* innate courtship-like following pursuit is mediated by central excitatory state. Nat. Commun. 6, 6457.
Kohatsu, S., Koganezawa, M., Yamamoto, D., 2011. Female contact activates male-specific interneurons that trigger stereotypic courtship behavior in *Drosophila*. Neuron 69, 498–508.
Kohlmeier, P., Zhang, Y., Gorter, J.A., Su, C.-Y., Billeter, J.-C., 2021. Mating increases *Drosophila melanogaster* females' choosiness by reducing olfactory sensitivity to a male pheromone. Nat. Ecol. Evol. 5, 1165–1173.
Kowalski, S., Aubin, T., Martin, J.-R., 2004. Courtship song in *Drosophila melanogaster*: a differential effect on male–female locomotor activity. Can. J. Zool. 82, 1258–1266.
Krstic, D., Boll, W., Noll, M., 2009. Sensory integration regulating male courtship behavior in *Drosophila*. PLoS One 4, e4457.
Kulkarni, S.J., Steinlauf, A.F., Hall, J.C., 1988. The *dissonance* mutant of courtship song in *Drosophila melanogaster*: isolation, behavior and cytogenetics. Genetics 118, 267–285.
Kyriacou, C.P., Hall, J.C., 1980. Circadian rhythm mutations in *Drosophila melanogaster* affect short-term fluctuations in the male's courtship song. Proc. Natl. Acad. Sci. U. S. A. 77, 6729–6733.
Kyriacou, C.P., Green, E.W., Piffer, A., Dowse, H.B., 2017. Failure to reproduce period-dependent song cycles in *Drosophila* is due to poor automated pulse-detection and low-intensity courtship. Proc. Natl. Acad. Sci. U. S. A. 114, 1970–1975.
Laturney, M., Billeter, J.-C., 2014. Neurogenetics of female reproductive behaviors in *Drosophila melanogaster*. Adv. Genet. 85, 1–108.
Laturney, M., van Eijk, R., Billeter, J.-C., 2018. Last male sperm precedence is modulated by female remating rate in *Drosophila melanogaster*. Evol. Lett. 2, 180–189.
Lee, S.S., Wu, M.N., 2020. Neural circuit mechanisms encoding motivational states in *Drosophila*. Curr. Opin. Neurobiol. 64, 135–142.
Li, H., 2020. Single-cell RNA sequencing in *Drosophila*: technologies and applications. Wiley Interdiscip. Rev. Dev. Biol. 10, e396.
Li, X., Ishimoto, H., Kamikouchi, A., 2018. Auditory experience controls the maturation of song discrimination and sexual response in *Drosophila*. elife 7, e34348.
Li, F., Lindsey, J.W., Marin, E.C., Otto, N., Dreher, M., Dempsey, G., Stark, I., Bates, A.S., Pleijzier, M.W., Schlegel, P., Nern, A., Takemura, S.-Y., Eckstein, N., Yang, T., Francis, A., Braun, A., Parekh, R., Costa, M., Scheffer, L.K., Aso, Y., Jefferis, G.S.X.E., Abbott, L.F., Litwin-Kumar, A., Waddell, S., Rubin, G.M., 2020. The connectome of the adult *Drosophila* mushroom body provides insights into function. elife 9, e62576.
Liimatainen, J.O., Hoikkala, A., 1998. Interactions of the males and females of three sympatric *Drosophila virilis*-Group Species, *D. montana*, *D. littoralis*, and *D. lummei*, (Diptera: Drosophilidae) in intra- and interspecific courtships in the wild and in the laboratory. J. Insect Behav. 11, 399–417.
Liu, C., Zhang, B., Zhang, L., Yang, T., Zhang, Z., Gao, Z., Zhang, W., 2020. A neural circuit encoding mating states tunes defensive behavior in *Drosophila*. Nat. Commun. 11, 3962.
Lu, B., LaMora, A., Sun, Y., Welsh, M.J., Ben-Shahar, Y., 2012. Ppk23-dependent chemosensory functions contribute to courtship behavior in *Drosophila melanogaster*. PLoS Genet. 8, e1002587.
Luan, H., Diao, F., Scott, R.L., White, B.H., 2020. The *Drosophila* split Gal4 system for neural circuit mapping. Front. Neural Circuits 14, 603397.

Maeda, R.K., Sitnik, J.L., Frei, Y., Prince, E., Gligorov, D., Wolfner, M.F., Karch, F., 2018. The lncRNA *male-specific abdominal* plays a critical role in *Drosophila* accessory gland development and male fertility. PLoS Genet. 14, e1007519.

Manier, M.K., Belote, J.M., Berben, K.S., Novikov, D., Stuart, W.T., Pitnick, S., 2010. Resolving mechanisms of competitive fertilization success in *Drosophila melanogaster*. Science 328, 354–357.

Manning, A., 1967a. Antennae and sexual receptivity in *Drosophila melanogaster* females. Science 158, 136–137.

Manning, A., 1967b. The control of sexual receptivity in female *Drosophila*. Anim. Behav. 15, 239–250.

Manoli, D.S., Foss, M., Villella, A., Taylor, B.J., Hall, J.C., Baker, B.S., 2005. Male-specific *fruitless* specifies the neural substrates of *Drosophila* courtship behaviour. Nature 436, 395–400.

Marie-Orleach, L., Bailey, N.W., Ritchie, M.G., 2019. Social effects on fruit fly courtship song. Ecol. Evol. 9, 410–416.

Markow, T.A., 1987. Behavioral and sensory basis of courtship success in *Drosophila melanogaster*. Proc. Natl. Acad. Sci. U. S. A. 84, 6200–6204.

Markow, T.A., O'Grady, P.M., 2005. Evolutionary genetics of reproductive behavior in *Drosophila*: connecting the dots. Annu. Rev. Genet. 39, 263–291.

Mazzoni, V., Anfora, G., Virant-Doberlet, M., 2013. Substrate vibrations during courtship in three *Drosophila* species. PLoS One 8, e80708.

Mezzera, C., Brotas, M., Gaspar, M., Pavlou, H.J., Goodwin, S.F., Vasconcelos, M.L., 2020. Ovipositor extrusion promotes the transition from courtship to copulation and signals female acceptance in *Drosophila melanogaster*. Curr. Biol. 30, 3736–3748.

Mohr, S.E., Tattikota, S.G., Xu, J., Zirin, J., Hu, Y., Perrimon, N., 2021. Methods and tools for spatial mapping of single-cell RNAseq clusters in *Drosophila*. Genetics 217.

Morley, E.L., Jonsson, T., Robert, D., 2018. Auditory sensitivity, spatial dynamics, and amplitude of courtship song in *Drosophila melanogaster*. J. Acoust. Soc. Am. 144, 734.

Moulin, B., Rybak, F., Aubin, T., Jallon, J.-M., 2001. Compared ontogenesis of courtship song components of males from the sibling species, *D. melanogaster* and *D. simulans*. Behav. Genet. 31, 299–308.

Mulloney, B., Smarandache, C., 2010. Fifty years of CPGs: two neuroethological papers that shaped the course of neuroscience. Front. Behav. Neurosci. 4 (45), 1–8.

Myers, J.G., 1929. Insect Singers: A Natural History of the Cicadas. G. Routledge, London.

Neville, M.C., Nojima, T., Ashley, E., Parker, D.J., Walker, J., Southall, T., Van de Sande, B., Marques, A.C., Fischer, B., Brand, A.H., Russell, S., Ritchie, M.G., Aerts, S., Goodwin, S.F., 2014. Male-specific fruitless isoforms target neurodevelopmental genes to specify a sexually dimorphic nervous system. Curr. Biol. 24, 229–241.

Nilsen, S.P., Chan, Y.-B., Huber, R., Kravitz, E.A., 2004. Gender-selective patterns of aggressive behavior in *Drosophila melanogaster*. Proc. Natl. Acad. Sci. U. S. A. 101, 12342–12347.

Nojima, T., Rings, A., Allen, A.M., Otto, N., Verschut, T.A., Billeter, J.-C., Neville, M.C., Goodwin, S.F., 2021. A sex-specific switch between visual and olfactory inputs underlies adaptive sex differences in behavior. Curr. Biol. 31, 1175–1191.

O'Sullivan, A., Lindsay, T., Prudnikova, A., Erdi, B., Dickinson, M., von Philipsborn, A.C., 2018. Multifunctional wing motor control of song and flight. Curr. Biol. 28, 2705–2717.

Pailette, M., Ikeda, H., Jallon, J.M., 1991. A new acoustic signal of the fruit-flies *Drosophila simulans* and *D. melanogaster*. Bioacoustics 3, 247–254.

Palavicino-Maggio, C.B., Chan, Y.B., McKellar, C., Kravitz, E.A., 2019. A small number of cholinergic neurons mediate hyperaggression in female *Drosophila*. Proc. Natl. Acad. Sci. U. S. A. 116, 17029–17038.

Pan, Y., Meissner, G.W., Baker, B.S., 2012. Joint control of *Drosophila* male courtship behavior by motion cues and activation of male-specific P1 neurons. Proc. Natl. Acad. Sci. U. S. A. 109, 10065–10070.
Partridge, L., Ewing, A., Chandler, A., 1987. Male size and mating success in *Drosophila melanogaster*: the roles of male and female behaviour. Anim. Behav. 35, 555–562.
Perelle, I.B., Ehrman, L., 1983. The development of laterality. Behav. Sci. 28, 284–297.
Pfeiffer, B.D., Ngo, T.-T.B., Hibbard, K.L., Murphy, C., Jenett, A., Truman, J.W., Rubin, G.M., 2010. Refinement of tools for targeted gene expression in *Drosophila*. Genetics 186, 735–755.
Phelps, J.S., Hildebrand, D.G.C., Graham, B.J., Kuan, A.T., Thomas, L.A., Nguyen, T.M., Buhmann, J., Azevedo, A.W., Sustar, A., Agrawal, S., Liu, M., Shanny, B.L., Funke, J., Tuthill, J.C., Lee, W.-C.A., 2021. Reconstruction of motor control circuits in adult *Drosophila* using automated transmission electron microscopy. Cell 184, 759–774.
Phillips, L.H., Konishi, M., 1973. Control of aggression by singing in crickets. Nature 241, 64–65.
Ribeiro, I.M.A., Drews, M., Bahl, A., Machacek, C., Borst, A., Dickson, B.J., 2018. Visual projection neurons mediating directed courtship in *Drosophila*. Cell 174, 607–621.
Riddiford, L.M., 2012. How does juvenile hormone control insect metamorphosis and reproduction? Gen. Comp. Endocrinol. 179, 477–484.
Rideout, E.J., Billeter, J.-C., Goodwin, S.F., 2007. The sex-determination genes *fruitless* and *doublesex* specify a neural substrate required for courtship song. Curr. Biol. 17, 1473–1478.
Rideout, E.J., Dornan, A.J., Neville, M.C., Eadie, S., Goodwin, S.F., 2010. Control of sexual differentiation and behavior by the *doublesex* gene in *Drosophila melanogaster*. Nat. Neurosci. 13, 458–466.
Rings, A., Goodwin, S.F., 2019. To court or not to court—a multimodal sensory decision in *Drosophila* males. Curr. Opin. Insect Sci. 35, 48–53.
Rotenberry, J.T., Swanger, E., Zuk, M., 2015. Alternative reproductive tactics arising from a continuous behavioral trait: callers versus satellites in field crickets. Am. Nat. 185, 469–490.
Rybak, F., Sureau, G., Aubin, T., 2002. Functional coupling of acoustic and chemical signals in the courtship behaviour of the male *Drosophila melanogaster*. Proc. R. Soc. Lond. B 269, 695–701.
Ryner, L.C., Goodwin, S.F., Castrillon, D.H., Anand, A., Villella, A., Baker, B.S., Hall, J.C., Taylor, B.J., Wasserman, S.A., 1996. Control of male sexual behavior and sexual orientation in *Drosophila* by the *fruitless* gene. Cell 87, 1079–1089.
Saarikettu, M., Liimatainen, J.O., Hoikkala, A., 2005. The role of male courtship song in species recognition in *Drosophila montana*. Behav. Genet. 35, 257–263.
Saleem, S., Ruggles, P.H., Abbott, W.K., Carney, G.E., 2014. Sexual experience enhances *Drosophila melanogaster* male mating behavior and success. PLoS One 9, e96639.
Sato, K., Yamamoto, D., 2020. Contact-chemosensory evolution underlying reproductive isolation in *Drosophila* species. Front. Behav. Neurosci. 14, 597428.
Sato, K., Ahsan, M.T., Ote, M., Koganezawa, M., Yamamoto, D., 2019. Calmodulin-binding transcription factor shapes the male courtship song in *Drosophila*. PLoS Genet. 15, e1008309.
Sato, K., Tanaka, R., Ishikawa, Y., Yamamoto, D., 2020. Behavioral evolution of *Drosophila*: unraveling the circuit basis. Genes 11, 157–170.
Scheffer, L.K., Xu, C.S., Januszewski, M., Lu, Z., Takemura, S.-Y., Hayworth, K.J., Huang, G.B., Shinomiya, K., Maitlin-Shepard, J., Berg, S., Clements, J., Hubbard, P.M., Katz, W.T., Umayam, L., Zhao, T., Ackerman, D., Blakely, T., Bogovic, J., Dolafi, T., Kainmueller, D., Kawase, T., Khairy, K.A., Leavitt, L.,

Li, P.H., Lindsey, L., Neubarth, N., Olbris, D.J., Otsuna, H., Trautman, E.T., Ito, M., Bates, A.S., Goldammer, J., Wolff, T., Svirskas, R., Schlegel, P., Neace, E., Knecht, C.J., Alvarado, C.X., Bailey, D.A., Ballinger, S., Borycz, J.A., Canino, B.S., Cheatham, N., Cook, M., Dreher, M., Duclos, O., Eubanks, B., Fairbanks, K., Finley, S., Forknall, N., Francis, A., Hopkins, G.P., Joyce, E.M., Kim, S., Kirk, N.A., Kovalyak, J., Lauchie, S. A., Lohff, A., Maldonado, C., Manley, E.A., McLin, S., Mooney, C., Ndama, M., Ogundeyi, O., Okeoma, N., Ordish, C., Padilla, N., Patrick, C.M., Paterson, T., Phillips, E.E., Phillips, E.M., Rampally, N., Ribeiro, C., Robertson, M.K., Rymer, J.T., Ryan, S.M., Sammons, M., Scott, A.K., Scott, A.L., Shinomiya, A., Smith, C., Smith, K., Smith, N.L., Sobeski, M.A., Suleiman, A., Swift, J., Takemura, S., Talebi, I., Tarnogorska, D., Tenshaw, E., Tokhi, T., Walsh, J.J., Yang, T., Horne, J.A., Li, F., Parekh, R., Rivlin, P.K., Jayaraman, V., Costa, M., Jefferis, G.S.X.E., Ito, K., Saalfeld, S., George, R., Meinertzhagen, I.A., Rubin, G.M., Hess, H.F., Jain, V., Plaza, S.M., 2020. A connectome and analysis of the adult *Drosophila* central brain. elife 9, e57443.

Schretter, C.E., Aso, Y., Robie, A.A., Dreher, M., Dolan, M.J., Chen, N., Ito, M., Yang, T., Parekh, R., Branson, K.M., Rubin, G.M., 2020. Cell types and neuronal circuitry underlying female aggression in *Drosophila*. elife 9, e58942.

Selverston, A.I., 2010. Invertebrate central pattern generator circuits. Philos. Trans. R. Soc. B 365, 2329–2345.

Shirangi, T.R., Stern, D.L., Truman, J.W., 2013. Motor control of *Drosophila* courtship song. Cell Rep. 5, 678–686.

Shirangi, T.R., Wong, A.M., Truman, J.W., Stern, D.L., 2016. Doublesex Regulates the connectivity of a neural circuit controlling *Drosophila* male courtship song. Dev. Cell 37, 533–544.

Shorey, H.H., 1962. Nature of the sound produced by *Drosophila melanogaster* during courtship. Science 137, 677–678.

Shuker, D.M., Burdfield-Steel, E.R., 2017. Reproductive interference in insects. Ecol. Entomol. 42, 65–75.

Silies, M., Gohl, D.M., Clandinin, T.R., 2014. Motion-detecting circuits in flies: coming into view. Annu. Rev. Neurosci. 37, 307–327.

Simon, J.C., Heberlein, U., 2020. Social hierarchy is established and maintained with distinct acts of aggression in male *Drosophila melanogaster*. J. Exp. Biol. 223, jeb232439.

Simpson, J.H., Looger, L.L., 2018. Functional imaging and optogenetics in *Drosophila*. Genetics 208, 1291–1309.

Sirot, L.K., Wolfner, M.F., 2020. "Call and response": a case of behavioral-molecular copulatory dialogue? BioEssays 42, 2000248.

Sitnik, J.L., Gligorov, D., Maeda, R.K., Karch, F., Wolfner, M.F., 2016. The female post-mating response requires genes expressed in the secondary cells of the male accessory gland in *Drosophila melanogaster*. Genetics 202, 1029–1041.

Smith, J.M., 1956. Fertility, mating behaviour and sexual selection in *Drosophila subobscura*. J. Genet. 54, 261–279.

Smith, J.M., Harper, D., 2003. Animal Signals. Oxford University Press.

Spieth, H.T., 1952. Mating behavior within the genus *Drosophila* (Diptera). Bull. Am. Mus. Nat. Hist. 99, 395–474.

Spieth, H.T., 1974. Courtship behavior in *Drosophila*. Annu. Rev. Entomol. 19, 385–405.

Stern, D.L., Crocker, J., Ding, Y., Frankel, N., Kappes, G., Kim, E., Kuzmickas, R., Lemire, A., Mast, J.D., Picard, S., 2017. Genetic and transgenic reagents for *Drosophila simulans, D. mauritiana, D. yakuba, D. santomea*, and *D. virilis*. G3 (Bethesda) 7, 1339–1347.

Stocker, R.F., 1994. The organization of the chemosensory system in *Drosophila melanogaster*: a review. Cell Tissue Res. 275, 3–26.

Stockinger, P., Kvitsiani, D., Rotkopf, S., Tirian, L., Dickson, B.J., 2005. Neural circuitry that governs *Drosophila* male courtship behavior. Cell 121, 795–807.
Strausfeld, N.J., 1976. Atlas of an Insect Brain. Springer-Verlag.
Strausfeld, N.J., 2012. Arthropod Brains: Evolution, Functional Elegance, and Historical Significance. Belknap Press of Harvard University Press.
Sturtevant, A.H., 1915. Experiments on sex recognition and the problem of sexual selection in *Drosophila*. J. Anim. Behav. 5, 351–366.
Sturtevant, A.H., 1921. The North American Species of *Drosophila*. Carnegie Institution of Washington.
Takemura, S.-Y., Bharioke, A., Lu, Z., Nern, A., Vitaladevuni, S., Rivlin, P.K., Katz, W.T., Olbris, D.J., Plaza, S.M., Winston, P., Zhao, T., Horne, J.A., Fetter, R.D., Takemura, S., Blazek, K., Chang, L.-A., Ogundeyi, O., Saunders, M.A., Shapiro, V., Sigmund, C., Rubin, G.M., Scheffer, L.K., Meinertzhagen, I.A., Chklovskii, D.B., 2013. A visual motion detection circuit suggested by *Drosophila* connectomics. Nature 500, 175–181.
Tanaka, R., Higuchi, T., Kohatsu, S., Sato, K., Yamamoto, D., 2017. Optogenetic activation of the *fruitless*-labeled circuitry in *Drosophila subobscura* males iduces mating motor acts. J. Neurosci. 37, 11662–11674.
Tauber, E., Eberl, D.F., 2001. Song production in auditory mutants of *Drosophila*: the role of sensory feedback. J. Comp. Physiol. A. 187, 341–348.
Tauber, E., Eberl, D.F., 2002. The effect of male competition on the courtship song of *Drosophila melanogaster*. J. Insect Behav. 15, 109–120.
Thistle, R., Cameron, P., Ghorayshi, A., Dennison, L., Scott, K., 2012. Contact chemoreceptors mediate male-male repulsion and male-female attraction during *Drosophila* courtship. Cell 149, 1140–1151.
Tobias, J.A., Seddon, N., 2009. Signal jamming mediates sexual conflict in a duetting bird. Curr. Biol. 19, 577–582.
Toda, H., Zhao, X., Dickson, B.J., 2012. The *Drosophila* female aphrodisiac pheromone activates *ppk23* + sensory neurons to elicit male courtship behavior. Cell Rep. 1, 599–607.
Tomaru, M., Oguma, Y., 2000. Mate choice in *Drosophila melanogaster* and *D. sechellia*: criteria and their variation depending on courtship song. Anim. Behav. 60, 797–804.
Tomaru, M., Yamada, H., 2011. Courtship of *Drosophila*, with a special interest in courtship songs. Low Temp. Sci. 69, 61–85.
Tompkins, L., Gross, A.C., Hall, J.C., Gailey, D.A., Siegel, R.W., 1982. The role of female movement in the sexual behavior of *Drosophila melanogaster*. Behav. Genet. 12, 295–307.
Trannoy, S., Kravitz, E.A., 2017. Strategy changes in subsequent fights as consequences of winning and losing in fruit fly fights. Fly 11, 129–138.
Trimarchi, J.R., Schneiderman, A.M., 1994. The motor neurons innervating the direct flight muscles of *Drosophila melanogaster* are morphologically specialized. J. Comp. Neurol. 340, 427–443.
Trott, A.R., Donelson, N.C., Griffith, L.C., Ejima, A., 2012. Song choice is modulated by female movement in *Drosophila* males. PLoS One 7, e46025.
Turner, T.L., Miller, P.M., Cochrane, V.A., 2013. Combining genome-wide methods to investigate the genetic complexity of courtship song variation in *Drosophila melanogaster*. Mol. Biol. Evol. 30, 2113–2120.
Vaughan, A.G., Zhou, C., Manoli, D.S., Baker, B.S., 2014. Neural pathways for the detection and discrimination of conspecific song in *D. melanogaster*. Curr. Biol. 24, 1039–1049.
Venkatasubramanian, L., Mann, R.S., 2019. The development and assembly of the *Drosophila* adult ventral nerve cord. Curr. Opin. Neurobiol. 56, 135–143.
Versteven, M., Vanden Broeck, L., Geurten, B., Zwarts, L., Decraecker, L., Beelen, M., Göpfert, M.C., Heinrich, R., Callaerts, P., 2017. Hearing regulates *Drosophila* aggression. Proc. Natl. Acad. Sci. U. S. A. 114, 1958–1963.

Villella, A., Hall, J.C., 1996. Courtship anomalies caused by *doublesex* mutations in *Drosophila melanogaster*. Genetics 143, 331–344.
von Philipsborn, A.C., 2020. Neuroscience: the female art of saying no. Curr. Biol. 30, R1080–R1083.
von Philipsborn, A.C., Liu, T., Yu, J.Y., Masser, C., Bidaye, S.S., Dickson, B.J., 2011. Neuronal control of *Drosophila* courtship song. Neuron 69, 509–522.
von Philipsborn, A.C., Jörchel, S., Tirian, L., Demir, E., Morita, T., Stern, D.L., Dickson, B.J., 2014. Cellular and behavioral functions of fruitless isoforms in Drosophila courtship. Curr. Biol. 24, 242–251.
von Schilcher, F., 1976a. The behavior of *cacophony*, a courtship song mutant in *Drosophila melanogaster*. Behav. Biol. 17, 187–196.
von Schilcher, F., 1976b. The function of pulse song and sine song in the courtship of *Drosophila melanogaster*. Anim. Behav. 24, 622–625.
von Schilcher, F., 1976c. The role of auditory stimuli in the courtship of *Drosophila melanogaster*. Anim. Behav. 24, 18–26.
von Schlicher, F., Hall, J.C., 1979. Neural topography of courtship song in sex mosaics of *Drosophila melanogaster*. J. Comp. Physiol. A. 129, 85–95.
Wang, K., Wang, F., Forknall, N., Yang, T., Patrick, C., Parekh, R., Dickson, B.J., 2021. Neural circuit mechanisms of sexual receptivity in *Drosophila* females. Nature 589, 577–581.
Watanabe, K., Suzuki, Y., Inami, S., Ohashi, H., Sakai, T., 2018. Light is required for proper female mate choice between winged and wingless males in *Drosophila*. Genes Genet. Syst. 93, 119–123.
Wohl, M., Ishii, K., Asahina, K., 2020. Layered roles of *fruitless* isoforms in specification and function of male aggression-promoting neurons in *Drosophila*. elife 9, e52702.
Yamada, D., Ishimoto, H., Li, X., Kohashi, T., Ishikawa, Y., Kamikouchi, A., 2018. GABAergic local interneurons shape female fruit fly response to mating songs. J. Neurosci. 38, 4329–4347.
Yoon, J., Matsuo, E., Yamada, D., Mizuno, H., Morimoto, T., Miyakawa, H., Kinoshita, S., Ishimoto, H., Kamikouchi, A., 2013. Selectivity and plasticity in a sound-evoked male-male interaction in *Drosophila*. PLoS One 8, e74289.
Yorozu, S., Wong, A., Fischer, B.J., Dankert, H., Kernan, M.J., Kamikouchi, A., Ito, K., Anderson, D.J., 2009. Distinct sensory representations of wind and near-field sound in the *Drosophila* brain. Nature 458, 201–205.
Yu, J.Y., Kanai, M.I., Demir, E., Jefferis, G.S.X.E., Dickson, B.J., 2010. Cellular organization of the neural circuit that drives *Drosophila* courtship behavior. Curr. Biol. 20, 1602–1614.
Zheng, Z., Lauritzen, J.S., Perlman, E., Robinson, C.G., Nichols, M., Milkie, D., Torrens, O., Price, J., Fisher, C.B., Sharifi, N., Calle-Schuler, S.A., Kmecova, L., Ali, I.J., Karsh, B., Trautman, E.T., Bogovic, J.A., Hanslovsky, P., Jefferis, G.S.X.E., Kazhdan, M., Khairy, K., Saalfeld, S., Fetter, R.D., Bock, D.D., 2018. A complete electron microscopy volume of the brain of adult *Drosophila melanogaster*. Cell 174, 730–743.
Zhou, C., Franconville, R., Vaughan, A.G., Robinett, C.C., Jayaraman, V., Baker, B.S., 2015. Central neural circuitry mediating courtship song perception in male *Drosophila*. elife 4, e08477.

CHAPTER FOUR

Communication by substrate-borne mechanical waves in insects: From basic to applied biotremology

Johannes Strauß[a,b,*], Nataša Stritih-Peljhan[c,*], Rachele Nieri[d], Meta Virant-Doberlet[c], and Valerio Mazzoni[e]

[a]AG Integrative Sensory Physiology, Institute for Animal Physiology, Justus-Liebig-Universität Gießen, Gießen, Germany
[b]Center for Mind, Brain and Behavior (CMBB), University of Marburg and Justus Liebig University Gießen, Gießen, Germany
[c]Department of Organisms and Ecosystems Research, National Institute of Biology, Ljubljana, Slovenia
[d]Laboratory of Bio-inspired, Bionic, Nano, Meta Materials & Mechanics, Department of Civil, Environmental and Mechanical Engineering, University of Trento, Trento, Italy
[e]Research and Innovation Centre, Fondazione Edmund Mach, Italy

Contents

*Equal contributors.

Advances in Insect Physiology, Volume 61
ISSN 0065-2806
https://doi.org/10.1016/bs.aiip.2021.08.002

Abstract

Biotremology is a field of study focused on an ancient way of communication by animals endowed with specialized organs for emission and reception of substrate-borne vibrations. Established as a scientific discipline of its own, formally separated from bioacoustics only recently, biotremology is now rapidly spreading throughout the scientific community, offering valuable cooperation also to numerous other scientific disciplines. Insects, make significant use of substrate-borne vibrational signals and cues for inter- and intraspecific communication, to accomplish many different behaviours. In this chapter, after highlighting the main relations, and differences, between insect acoustic and vibrational communication, we provide a detailed, comprehensive overview of the knowledge and ongoing research in the field of insect biotremology. We particularly emphasise the mechanisms of sensory detection of vibrational signals and cues, and the central neural processing of the received information; the latter especially being discussed in relation to insect auditory processing. In addition, we dedicate considerable attention to the behavioural and ecological aspects of insect vibrational communication, to the methods and instruments of research in neurophysiology and behavioural ecology, as well as the use of acquired basic knowledge in practical applications. Whilst some topics in biotremology, such as the neuronal basis of behaviour, need highly increased research attention, the interest in applied biotremology is rising especially quickly for the high potential offered by substrate-borne vibrations as semiophysicals for pest control. Hence, insects are both the elected study model for basic biotremology research and strongly relevant for agricultural and resource management issues.

1. Relation between vibration and sound in insect communication

1.1 Vibration and sound

Vibration and sound are closely related sensory modes, both relating to mechanical disturbances emitted by an oscillating body into the surrounding medium. Such mechanical disturbances, or physical vibration, may transmit as different types of waveforms, often depending on the medium properties, and show a continuum rather than strict physical separation between "sound" and "vibration" *per se* (e.g., Hill, 2009; Markl, 1983). To engineers and physicists, both terms encompass mechanical waves that can be technically described as both vibrational and acoustic. The ultimate categorization of these signals is biological or perceptual in nature, as it depends on the sensory organs excited by these waveforms in the receiver. When we speak of sound, or acoustic waves, we think of far-field, purely longitudinal waves in a homogenous medium, perceived by pressure or pressure-difference receiver organs (i.e., as hearing, in the strict sense). When considering vibration in animal communication, this mostly relates to the use of mechanical waves traveling at the boundary of two media (i.e., as substrate-borne vibration propagated in Rayleigh or bending waves), mainly occurring at the solid-air and water-air interface (Hill and Wessel, 2016). Initially, the term vibration has been applied to two further ways of emitting mechanical energy in biological interactions, including "contact vibration or rhythmic touch," and the "nearfield medium motion" that relates to the medium flow in phase with the movement of the emitter (Markl, 1983). The latter is now categorized as the "near-field sound" and the motion detectors for its perception as hearing organs (e.g., Albert and Göpfert, 2015; Cocroft and Rodriguez, 2005), but the separation in this case appears to be merely a question of definition. Whether research on contact vibration or sense of touch are ever included under the umbrella of "biotremology" remains to be seen, but arguments could be made that these could be incorporated without much change to the basic premises defining either.

The transmission medium imposes several important differences between air-borne and substrate-borne communication. Air is a relatively homogeneous medium, properties are fairly predictable and the speed of transmission of mechanical waves through the air is also more or less constant (Attenborough and Van Renterghem, 2021). On the other hand, substrates are very heterogeneous media, especially when referring to biological systems like plants, which have complicated geometry and are composed of distinct tissues with different mechanical properties. Such heterogeneity introduces a certain level of unpredictability into reliability of transferred information. Transmission properties that influence attenuation and filtering differ depending on the physical properties of the substrate, such as thickness, stiffness and elasticity (Michelsen et al., 1982). Moreover, the speed of signal transmission depends on mechanical properties of the substrate, as well as on signal frequency and, consequently, varies enormously. As a result, the effects of the environment on the efficiency and evolution of air-borne and substrate-borne mechanical signals also differ. Whilst the effects on the air-borne sounds are moderate, these environmental effects on vibrational signals are very strong (Elias and Mason, 2014; Markl, 1983).

1.2 Signal production

Due to the common physical nature of sound and vibration, the production of signals in one modality is inevitably related to production of signals in the other modality. As a consequence, acoustic and vibrational communication in insects (and other animals) are closely interrelated by function and in some lineages also by evolution, with respect to different levels from signal production, to detection, and neural processing (Caldwell, 2014). In insects, loud sound signalling is typically associated with the release of significant amounts of mechanical energy into the substrate, as well, and examples using such co-occurring vibrational signals in communication are known from various Ensifera, such as crickets (Weidemann and Keuper, 1987), bush-crickets (Keuper et al. 1985), mole crickets (Hill and Shadley, 1997, 2001), and from cicadas (Stölting et al., 2002). The reverse relation is generally not the case for insects, however, as their vibrational signalling is often silent or only produces sound signals at low intensities (see Section 1.3). This is one of the reasons why the rich world of animal interactions via the substrate stayed hidden to humans for a long time (Cocroft et al., 2014a).

Four main mechanisms of vibrational signal production have been described for insects: tremulation, percussion, stridulation, and tymbal buckling (Hill, 2012, 2014; Virant-Doberlet and Čokl, 2004). Tremulation (commonly described as jerking, shaking, rocking, etc.) describes oscillation of the body, or some of its parts, without any percussive impact to the substrate, and emits low-frequency narrow-band vibration with no substantial energy release into the air. Stridulation and percussion (the latter also called drumming, or tapping) refer to rubbing of body parts against each other (or against the substrate, also called scraping) and to impacts against the substrate (or between body parts), respectively. These mechanisms produce broadband or high frequency signals that typically include both substrate and airborne components; in the absence of a resonator, the latter often only with very low intensity (e.g., Gogala, 2006). Tymbal buckling (developed specifically in Hemiptera) relates to deformation of a specialized area of abdominal cuticle, which produces a series of "clicks" upon contraction and release and thereby emits low-frequency substrate vibration. Emission of airborne sound by the tymbal, such as is present in cicadas, is only enabled in association with a specialized abdominal resonator chamber (Hoch et al., 2006; Wessel et al., 2014).

Due to physical constraints related to both body size, as well as an impedance difference between the signalling animal and the surrounding medium, insect signalling through the substrate is energetically more efficient then signalling through the air (Bennet-Clark, 1998; Caldwell, 2014). As small radiators, insects can only emit relatively high-frequency sound, which is often conducted in association with resonating structures or chambers that enable emission of loud signals (Bennet-Clark, 1999; Mhatre et al., 2012). Such size- and impedance-related constraints do not exist, or are considerably smaller, for vibrational signalling (Cocroft and De Luca, 2006), and insects are able to emit also very low-frequency vibration efficiently into the substrate without the need of special adaptations for signal amplification.

1.3 Phylogeny and evolution

From the aspect of physics of signal production alone, it is not surprising to find substrate vibrations as the predominant means of mechanical signalling in insects, which largely exceeds and also predates the use of sound in communication (Cocroft and Rodriguez, 2005). By estimation, about 70% of insect species relying on some sort of mechanical signalling use substrate vibrations alone, about 20% of the species use both sound and vibration, and only about 5% of them use sound signals alone (both near- and far-field sound; Cocroft and Rodriguez, 2005). The widespread use of substrate vibrations by insects, and animals in general, has been postulated to be surpassed in phylogenetic presence only by their reliance on chemical signals and cues (Cocroft et al., 2014a). Such common appearance of communication via the substrate is based also on the ubiquity of vibrational sensilla in insects, often occurring in highly specialized vibrosensory organs

(Lakes-Harlan and Strauß, 2014; see also Section 4). This is contrasted by only a subset of insect taxa that are able to detect sound (Strauß and Lakes-Harlan, 2014). In some lineages of the hearing insects, including crickets, tettigoniids and cicadas (at least), auditory signalling appears to have evolved from vibratory signalling (e.g., Bailey, 1991; Hoch et al., 2006; Stumpner and von Helversen, 2001), and hearing in some groups has built up on pre-existing vibrosensory precursor organs (ensiferan Orthoptera: Rössler, 1992; Strauß and Lakes-Harlan, 2009; Strauß et al., 2014; and tachinid and sarcophagid flies: Lakes-Harlan and Lehmann, 2015; Lakes-Harlan et al., 1999). During evolution, a part of the underlying interneuron network, primarily vibrosensory, was also recruited for the newly developed sense of hearing (as indicated for the Ensifera: Stritih and Stumpner, 2009). Consequently, reception and central processing of sound and vibration, not only in these lineages of hearing insects, are tightly interrelated. This has been documented also for insects with hearing organs originating from other types of mechanosensory precursors. In Caelifera (locust and grasshoppers), tympanal ears evolved from abdominal proprioreceptor organs (Meier and Reichert 1990; van Staaden and Römer, 1998) but were found mediating information onto common low-order interneurons with vibrosensory input from the legs (Kühne, 1982b). Such convergence of sound and vibration inputs already at low synaptic levels of processing apparently takes place regardless of the use of these signals in communication (e.g., in locusts), and indicates high importance of combined perception of sound and vibration for a wide range of behavioural contexts.

The close relation of auditory and vibratory senses enables hearing species of insects to rapidly adapt the degree of using sound vs. vibrational signals to the changing environmental pressures. Therefore, the predominant use of sound may be reverted to the predominant use of vibration in communication, following regressive evolution of the auditory system. Such cases of adaptive switch to extensive vibratory signalling are well known from a number of neotropical tettigoniids, which reduced the amount of stridulation and shifted to the use of tremulation in both calling and courtship context as a response to selection by hearing predators (Morris et al., 1994; Rajaraman et al., 2018; Römer et al., 2010; ter Hofstede et al., 2017). A shift to the more extensive vibratory signal use may also be presumed for crickets with reduced acoustic signalling and/or hearing (Otte, 1990; Zuk et al., 2006); although, this change may be less evident behaviourally than in tettigoniids. However, there are exceptional cases of non-hearing Australian crickets that use abdominal drumming signals for both long- and short-distance communication (Evans, 1988). Since these crickets also retain the stridulatory apparatus, they may use the vibratory component of stridulation signals for communication, as well (Otte, 1990). For crickets, signalling by stridulatory substrate vibrations (Broder et al., 2021) is likely adaptive under predation pressure and future behavioural studies can confirm the role of the vibrational channel.

2. Vibrational behaviour and communication

Any vibrational signals or cues of biological relevance are detected by specialized sensory organs, copied and transferred to the central nervous system, analysed for information contents, and integrated with inputs from other receptor organs. Eventually, the sensory inputs are evaluated to initiate necessary behavioural reactions. Local reflexes can do this, by more complex orientations and different sets of responses, or by modulation of behavioural responses. The signals and their behavioural context show great diversity, which would be impossible to summarize here in detail. Instead, we review the main behavioural contexts in which vibrational communication is known to occur.

2.1 Vibrational signals and cues

One central process mediated by substrate vibrations in intraspecific communication is mate detection and pair formation. It requires successive steps of signal analysis to allow for signal detection, species recognition, orientation and localization, courtship, precopula and copula. These processes are well studied in several taxa of Hemiptera (Auchenorrhyncha and Heteroptera) (Cocroft and McNett, 2006; Čokl and Virant-Doberlet, 2003) and Orthoptera (Bell, 1980a; Benediktov et al., 2020; Stritih and Čokl, 2014; Weissman, 2001). Vibrational signals for mate attraction are usually species-specific, thus allowing species identification of potential mates. In Orthoptera, vibrational signals are simultaneously produced with airborne sound by stridulation and can mediate long-distance mate attraction in multimodal signalling (Latimer and Schatral, 1983; Stiedl and Kalmring, 1989). In some taxa, courtship typically takes place in close proximity, or even direct contact, and does not require that signals propagate over longer distances (Brown, 2016; Fianco et al., 2018; Mazzoni et al., 2013). Other examples of intraspecific communication are antagonistic behaviours including territoriality (Benediktov et al., 2020; Dejan et al., 2013;

Guedes et al., 2012), aggression (Hofmann and Schildberger, 2001; Rillich et al., 2009), and post-contest signalling (Bertram et al., 2010). In some relation to these, vibrations are also involved in male rivalry (Fattoruso et al., 2021) and competition avoidance (Evans et al., 2009).

Apart from evolutionarily adapted communication signals, vibrations can also serve as cues to detect and avoid predators. This is probably the most ubiquitous and ancient context of vibrational behaviour in insects, though well-researched case studies are not available from all major taxa. Insects of different developmental stages were shown to detect vibrations caused by predators, highlighting the importance of this behaviour for caterpillars (Castellanos and Barbosa, 2006; Guedes et al., 2012), late instar larvae and pupae (Bacher et al., 1997; Djemai et al., 2001), and adults (Hager and Krausa, 2019; Oberst et al., 2017; reviewed in Virant-Doberlet et al., 2019). Whether or not these cues are also evolutionarily adapted, and therefore allow them to be considered as serving a function in communication along with signals, is a point that requires some resolution. For now, responses to cues are considered as components of vibrational behaviour by biotremologists.

Substrate vibrations, usually of the low-frequency spectrum, can trigger protective responses, such as startle responses, with fast movements of legs, antennae or the body upon reception of vibration stimuli (Friedel, 1999; Takanashi and Kojima, 2021; von Dohlen and Dambach, 1981; see Bullock, 1984, for a neuroethological analysis of startle responses). Such startle responses are understood as preparatory for following locomotion sequences (Takanashi et al., 2016). Response to vibrations can also be defensive stridulation, freezing, or walking in response to vibrations below 1 kHz (Takanashi et al., 2016). Freezing responses stop all body movements upon reception of vibrations (e.g., in the honeybee; see Michelsen et al., 1986a).

In addition, predatory insects can use vibrational cues for prey detection on different media, like terrestrial antlions (Devetak, 2014; Mencinger-Vračko and Devetak, 2008), stink bugs (Pfannenstiel et al., 1995), parasitic wasps (Meyhöfer and Casas, 1999; Otten et al., 2002), or aquatic water striders (Markl and Wiese, 1969). Vibrations caused by wind are even used as indicators of likely plant food availability (Hanrahan and Kirchner, 1994). In mutualistic relationships, vibrational signals can be relevant between species, usually between ants and Lepidoptera (DeVries, 1990; Fiedler et al., 1996; Lin et al., 2019; Travassos and Pierce, 2000) or treehoppers (Cocroft and McNett, 2006; Morales et al., 2008).

2.2 Social signals

Use of social signals is obviously linked to the level of sociality of the respective organisms, and not only the eusocial species of bees, wasps, termites, and ants require communication between conspecifics within groups (Cocroft, 2001; Cocroft and Hamel, 2010). However, for eusocial insects, the division of labour certainly demands the most diverse communication processes, and multiple behaviours or contexts of social signalling (Leonhardt et al., 2016). Communication between members of colonies in social insects depends on several modalities, and even if chemical signalling is perceived as the main signalling channel (Richard and Hunt, 2013), vibrational signals are involved in intracolony communication (Hunt and Richard, 2013). Substrate vibrations play a role in recruitment of conspecifics (Baroni-Urbani et al., 1988), in coordinating colony defence (Kirchner et al., 1994; Markl, 1965; Markl and Fuchs, 1972; Roces and Hölldobler, 1995), in adult-brood communication (Pepiciello et al., 2018), in spatial organization (Pielström and Roces, 2012), and in maintaining the social structure by identifying reproductive individuals (Mignini and Lorenzi, 2015). Recruitment to food sources is present in several ant species (Baroni-Urbani et al., 1988; Masoni et al., 2021; Roces and Hölldobler, 1996). In stingless bees (*Melipona*), thoracic vibrations by returning foragers indicate food profitability without known indications of the direction, whilst the possible information about distance is debated (Hrncir and Barth, 2014).

Substrate vibrations may play a role in the detection of signallers (Nieh and Tautz, 2000), even when other mechanical modalities are used for further information exchange (Michelsen, 2014). Often, comparisons in the signalling behaviour can also indicate different adaptations, e.g., the waggle dance of honeybees includes information about the direction and distance of a food source (Hunt and Richard, 2013), whilst thoracic vibrations occur to adjust the foraging and processing activities (Michelsen et al., 1986a; Nieh, 1993). Recently, vibrations have been found to play a pivotal role in wasps' social structure, by mediating gene expression and thus caste determination (Jandt et al., 2017; Suryanarayanan et al., 2011).

In group living insects, not strictly eusocial, vibrational signals are used for recruitment, defence, and developmental processes. In juveniles of both treehoppers and moths, individuals recruit conspecifics to food sources (Cocroft, 2005; Yadav et al., 2017). In treehoppers (*Umbonia*), the mother guards groups of nymphs, and is activated for defence against diverse predators by vibrations transmitted through the group of nymphs (Cocroft,

1999). The parent also signals by vibrations to the nymphs after the predator encounter. In stink bugs (*Halyomorpha halys*), hatching is triggered by vibrational stimuli (Endo et al., 2019).

2.3 Modulatory signals

The relevance of the vibrational channel for signalling is now established for numerous insect taxa, and these vibrational signals can induce very different behaviours. One fascinating aspect of vibrational signal is a modulatory activity, by which vibrations from conspecifics, e.g., by stridulation, do not elicit a response behaviour specific to such stridulatory signalling, but affect the probability for a certain behaviour to occur (Markl, 1985). Vibrational signals can affect the response threshold for behaviour especially in multimodal signalling (Hölldober, 1999; Markl, 1985), which is especially well described in ants (Hölldober, 1999; Markl and Hölldobler, 1978). Modulation can be analysed at different levels of behaviour, neuronal network activity, or molecular mechanisms. So far, the picture for modulation by vibrational signals is far from complete, and certainly more aspects of modulatory functions need to be addressed. Some established examples of modulation by vibrational signals include learning in stingless bees (Mc Cabe et al., 2015), modification of behavioural signal thresholds in treehopper groups to prevent against false alarms (Hamel and Cocroft, 2012), and the saturation of neuronal activity in the bimodal auditory-vibratory system of bushcrickets (Kalmring et al., 1985).

3. Signal range and active space

The behavioural study of vibrational communication in insects shows that signals and cues are versatile and can be used on a diversity of substrates. Whilst insect signals transmitted by airborne sound can travel over relatively long distances of at least tens of meters (Römer et al., 2010) and occasionally even kilometres (Römer et al., 2014), vibrational signals are usually associated with communication over close-range distances of centimetres up to a few meters (Cocroft and McNett, 2006; Endler, 2019; Römer et al., 2010; Virant-Doberlet and Čokl, 2004). Whilst this may give a general orientation comparing the two signalling channels, a more quantitative analysis is required for signalling ranges by vibrations in relation to different substrates. Vibrations and vibrational signals are transmitted through diverse types of substrates, including sand, soil, plant leaves and stems, water surfaces, spider webs, and other structures built by insects themselves such as termite colonies or bee hives (on honeycombs) (Elias and Mason, 2014; Hill, 2008). Generally, vibrations attenuate over distance in a frequency-dependent manner, and the way of filtering typically is substrate-specific. Together with this, the change in amplitude is not always a linear phenomenon for substrate vibrations. In plants, characterized by low internal damping, vibrations typically reflect at the structures' edges, and form complicated patterns of standing-waves, with amplitude fluctuating with distance in a frequency-dependent manner (Čokl and Virant-Doberlet, 2003; Elias and Mason, 2014; Michelsen et al. 1982; Polajnar et al. 2012).

Different plants have been investigated for their transmission characteristics, which showed highly efficient transfer of vibrational signals (Bell, 1980b; Michelsen et al., 1982; Traue, 1978). Michelsen et al. (1982) established that plants are particularly suited for the transmission of vibrational signals by bending waves, and enable effective communication over 1–2 m if the signaller and the receiver are placed on the same plant (see also Römer et al., 2010). In addition, standing waves form on stems and branches (Čokl and Virant-Doberlet, 2003). The stonefly *Pertinella drymo* showed responses to drumming signals on a wooden rod over 8 m distant from the signaller. This distance was limited by the maximal length of the set-up (Stewart and Zeigler, 1984), suggesting that maximal active space by vibrational signals in plant materials are principally limited by structure dimensions (see Römer et al., 2010).

Insect signallers commonly produce vibrational signals that are 10–30 times above the sensory threshold of receptor organs (Michelsen et al., 1982), which would allow a signal to be effective over some distance. Woody plants allow for better vibration propagation with limited attenuation (Michelsen et al., 1982) than grass, where signals are more strongly attenuated (Bell, 1980b; Keuper and Kühne, 1983). However, the transmission characteristics in plants are complex, and the amplitudes of vibrational signals are not monotonically reduced with increasing distance (Magal et al., 2000; Michelsen et al., 1982; Polajnar et al., 2012; Römer et al., 2010; Velilla et al., 2020b). Due to low attenuation, signals may travel over a plant not only in a single direction, but also up and down several times (Michelsen et al., 1982), and can even become amplified, rather than damped, compared to the input site (Čokl et al., 2007). Bending waves of plants typically transmit with a frequency-specific velocity (Michelsen et al., 1982; Čokl and Virant-Doberlet, 2003). It has been suggested that this frequency dispersion may provide information for receivers to determine the distance to the signaller (Keuper et al., 1985). On plants, due to standing-wave conditions, there is no direct relationship between the amplitude of the signal and the distance to the signaller (De Groot et al., 2011a; Michelsen et al., 1982; Stritih et al., 2000). This makes signalling ranges difficult to predict. In addition, the stems, which are basically one-dimensional substrates (Elias and Mason, 2014), limit the available information for the receiver on the direction of the signals' source (De Groot et al., 2011a).

As the plant dimensions limit the signalling ranges (Römer et al., 2010), some cases are documented where these limitations can be bridged. Stink bugs (*Nezara viridula*) can respond to conspecifics located on a neighbouring plant over 2m away, with plants only coupled by roots and soil (Čokl and Virant-Doberlet, 2003). Vibrational signals of leafhoppers and stink bugs can also transfer between neighbouring plants, when leaves have no direct physical contact, by airborne resonances over a distance of 3–7cm (Eriksson et al., 2011; Kavčič et al., 2013). This finding indicates that vibrational transmission does not necessarily depend on the physical continuity of substrate structures.

Ranges of $\geq$2m are also found for trees (McVean and Field, 1996) and soil (Hill and Shadley, 1997). Other substrates are effective in the range of a few centimetres up to a meter, including soil (Markl, 1967), grass plants (Keuper and Kühne, 1983), shrub plants (Stiedl and Kalmring, 1989), or sand (Devetak, 2014). A study on larval antlions (*Euroleon nostras*), which form pits to trap prey in sand, showed that the responses to vibrations caused by movements of prey depends on the sand particle size (Devetak et al., 2007). The behavioural response was recorded as the antlions tossed sand after detecting vibrations caused by the prey insects. Coarser sand as a substrate resulted in greater reaction ranges for the antlions, whilst finer sand reduced the reaction distances (Devetak, 2014; Devetak et al., 2007). The greatest mean reaction distance was found at $\sim$12cm (Devetak et al., 2007). The basis for these differences is in the stronger damping of signals in the finer sand, especially for relatively high-frequency vibrations (Devetak et al., 2007; Podlesnik et al., 2019). Choice experiments from fine to coarse sand particles showed that antlions prefer the relatively fine sand particles for building traps, as coarser sand hinders larval locomotion and allows prey a greater chance to escape (Devetak et al., 2005; Mencinger-Vračko and Devetak, 2008). Using the preferred particle size, reaction distances also vary with different prey species (Mencinger-Vračko and Devetak, 2008).

Importantly, even the longest signalling ranges are established by the physical parameter of signal transmission and often also by the biological parameter of a behavioural response (e.g., Čokl and Virant-Doberlet, 2003; Stiedl and Kalmring, 1989; Stewart and Zeigler, 1984; Traue, 1978). However, in the concept of active space, these are combined under the perspective of the communicating animals in the habitat. The active space is defined by the signal amplitude generated by the signaller, the attenuation of the signal during transmission, the amplitude of background noise, and the sensory threshold of the receiver (Mazzoni et al., 2014; Römer et al., 2010). By this definition, the active space extends the purely physical propagation of vibrational signals and also requires an understanding of the vibrational sensitivity of the organisms. If the sensitivity cannot be tested in the habitat, signal transmission over distance can be compared to the known sensitivity of interneurons (Keuper and Kühne, 1983) or sensory organs determined in the laboratory. The localization of vibrational signals can use different orientation strategies. It should be noted that vibrational signalling likely cannot only be framed by signal transmission distances: as the vibrational signals are not expanding homogeneously, it has been suggested that marked changes in the signal amplitude that can occur between different parts of a plant, like the stem and petiole, could be important for orientation (Mazzoni et al., 2014). In addition, it is important to note that vibrational communication in several behavioural contexts takes place over relatively short distances (Keuper et al., 1985; Stiedl and Bickmeyer, 1991; Stritih and Čokl, 2012), and it is not obviously adaptive to extend the signal reach further.

For social signals, maximal signalling ranges are reported in only a few cases. Usually, animals are within close range in a group (e.g., Cocroft, 1999; Cocroft and Hamel, 2010), though signalling is especially relevant between individuals with spatially or temporally distinct roles. In colonies, some behavioural observations are reported for bees, ants, and termites. The signals of honeybee queens after emerging on the honeycomb were calculated to extend for about 10 cm, which matches well with the observed distances of 5–10 cm in which queens stop for tooting (Michelsen et al., 1986b). The begging signals of honeybees, produced by thoracic vibrations on the comb, caused displacements that were measured at a few centimetres from the signalling bee (Sandeman et al., 1996). Alarm stridulation in leaf-cutting ants (*Atta cephalotes*) attracts workers up to 8 cm away and buried ants will get help from conspecifics by digging over shorter distances (3–5 cm; Markl, 1967). In termite colonies of the genus *Macrotermes*, alarm signals generated upon disturbance by soldiers through head drumming propagate for about 40 cm (Hager and Kirchner, 2013; Hager et al., 2019). Upon detection of the signals, other individuals within that range also produce drumming signals, effectively extending the signalling range over several meters (Hager and Kirchner, 2013; Röhrig et al., 1999). Other termites respond by retreating into the colony (Röhrig et al., 1999). This social amplification of drumming signals in long-range communication is a unique vibrational signalling system associated with the evolution of larger colonies (Kirchner et al., 1994; Röhrig et al., 1999).

4. Sensory organs involved in vibration detection in insects

4.1 Diversity of vibratory receptors

All behaviours of vibrational communication depend on sensory organs that detect substrate vibrations, and relay the information to the central nervous system for analysis, integration, and the generation of behaviour. Different types of sensilla respond to substrate vibrations, including campaniform sensilla, scolopidial sensilla (forming chordotonal organs), and hair sensilla of different types (Kalmring, 1985; Lakes-Harlan and Strauß, 2014; Römer and Tautz, 1992; Yack, 2016; Fig. 1). These sensilla are primary sensory neurons with direct axonal projections into the central nervous system (Field and Matheson, 1998; Hustert et al., 1981; Keil, 1997; Schmitz et al., 1991; see Section 7). Sensory hairs not only detect tactile stimuli in direct contact, but also vibrations of different media, including vibrations of air (Tautz, 1977) or water surface: in the water strider *Aquarius paludum*, trichobothria located at the ventral tarsus are in contact with the water surface and contribute to vibration detection (Goodwyn et al., 2009).

4.1.1 Campaniform sensilla

Campaniform sensilla are sensilla with cuticular caps (Keil, 1997; Fig. 1A). The orientation of these caps along the leg axes determines the detected directionality of stimuli (Pringle, 1938). Isolated or grouped campaniform sensilla are found in different regions of the leg, including the trochanter, femur, tibia, or tarsal segments (Devetak, 1998; Harris et al., 2020; Hofmann and Bässler, 1982; Mücke, 1991; Zill et al., 1999, 2011). Notably, groups of campaniform sensilla are located at the proximal tibia close to the internal subgenual organ (SGO) and associated scolopidial organs (Eibl, 1978; Lakes and Schikorski, 1990; Mücke, 1991; Schnorbus, 1971; Strauß, 2017a, 2020;

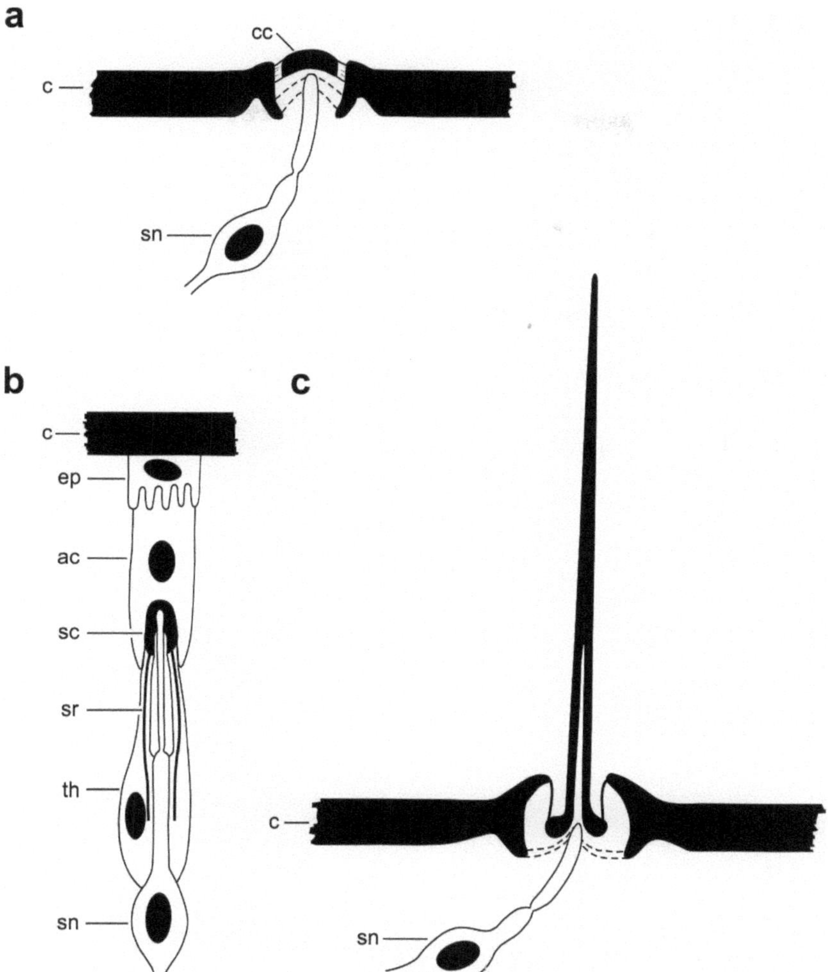

Fig. 1 Schematic of sensilla types responsive to substrate vibrations: (A) campaniform sensillum, (B) scolopidial sensillum, and (C) filiform hair sensillum. Abbreviations: ac, attachment cell; c, cuticle; cc, cuticular cap; ep, epidermis; sc, scolopale cap; sn, sensory neuron; sr, scolopale rod; th, thecogen (scolopale) cell. *Adapted from Keil T.A., 1997. Functional morphology of insect mechanoreceptors. Microsc. Res. Tech. 506–531, with permission from John Wiley and Sons.*

Strauß and Stritih, 2016; Fig. 2A). The campaniform sensilla detect cuticular strain (Pringle, 1938; Schnorbus, 1971). They are often included amongst the vibration receptors, as they can respond to the changes in cuticular strain caused by low-frequency vibrations transmitted over the leg cuticle

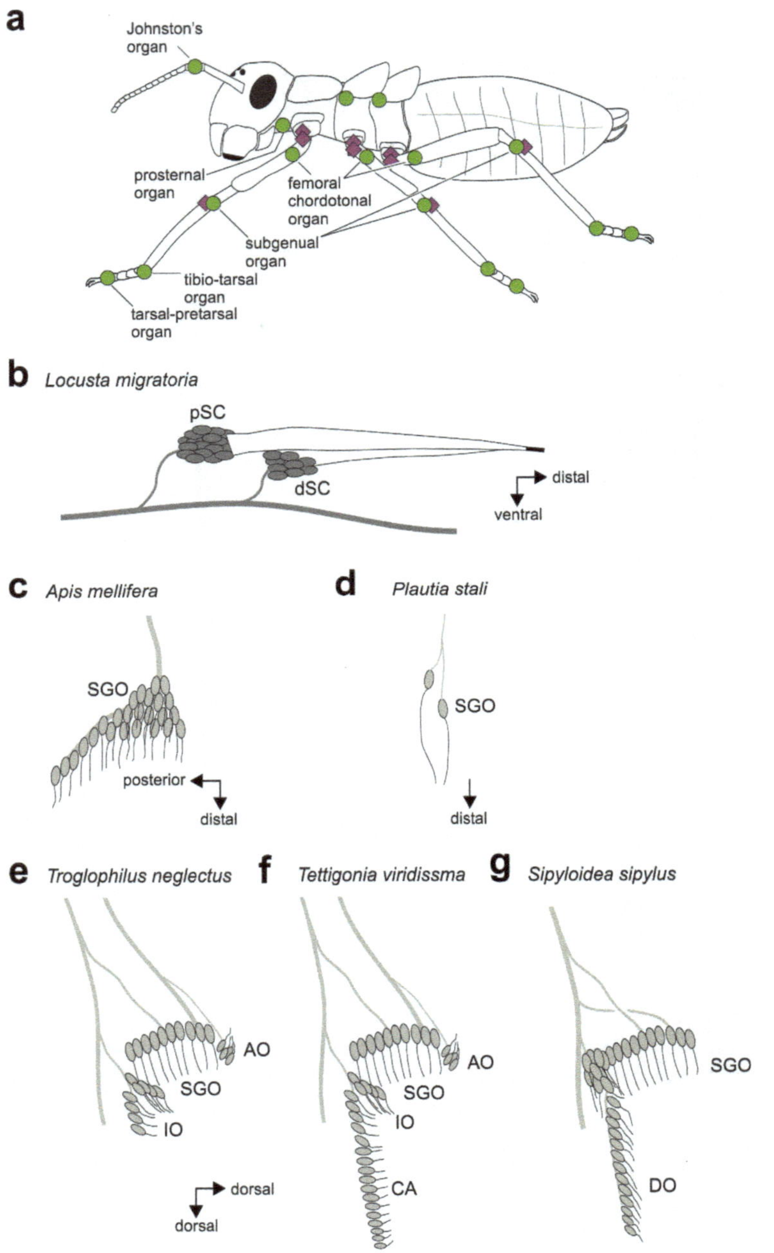

Fig. 2 Chordotonal organs relevant to the detection of substrate vibrations. (A) Overview of the insect body plan indicating confirmed vibration-sensitive chordotonal organs (green circles) and campaniform sensilla (magenta squares). (B) The femoral chordotonal organ of *Locusta migratoria*, midleg, consisting of two scoloparia, *(Continued)*

(Čokl and Virant-Doberlet, 2003; Devetak and Amon, 1997; Kalmring, 1985; Kalmring et al., 1978; Kühne, 1982a; Kühne et al., 1984; Fig. 4A and B). As a result, campaniform sensilla are possibly not very sensitive to vibration stimuli, as they respond at relatively high accelerations, and were not considered as relevant vibration receptors in earlier work (Devetak, 1998; Devetak et al., 2004). Campaniform sensilla show phase-locked responses to sinusoidal vibration stimuli (Kalmring et al., 1978; Kühne, 1982a; Schnorbus, 1971). The physiological responses that remain after ablation of chordotonal organs in the leg are ascribed to the campaniform sensilla (Devetak and Amon, 1997; Schnorbus, 1971). However, extirpation of the campaniform sensilla did not affect vibrational threshold in the cricket (Dambach, 1972a). Physiological experiments in ants support the contribution of campaniform sensilla to the lowest vibrational thresholds, whilst the chordotonal organs are less relevant for vibration detection (Markl, 1970). In this case, campaniform sensilla at the base of the leg (trochanter and proximal femur; Fig. 2A) seem to be the main receptors of substrate vibrations, rather than the SGO. Despite the physiological responses to low-frequency vibrations by campaniform sensilla, there is little evidence of their behavioural role in response to vibrations after exclusion of chordotonal organs. In the field cricket *Gryllus campestris*, campaniform sensilla located at the trochanter-femur joint respond to low frequency vibrations, and mediate the startle response by which crickets stop or interrupt sound production by stridulation and may start moving (von Dohlen and Dambach, 1981).

Fig. 2—Cont'd in lateral view. (C) The subgenual organ of the honeybee. (D) The subgenual organ of the stink bug *Plautia stali*. (E) The subgenual organ complex of *Troglophilus neglectus* (Ensifera: Rhaphidophoridae). (F) The subgenual organ complex of *Tettigonia viridissima* (Ensifera: Tettigoniidae). (G) The subgenual organ complex of *Sipyloidea siyplus* (Phasmatodea). Abbreviations: AO, accessory organ; CA, *crista acustica*; DO, distal organ; dSC, distal scoloparium; IO, intermediate organ; pSC, proximal scoloparium; SGO, subgenual organ. *Panel (B): Adapted from Field L.H., Pflüger H.-J., 1989. The femoral chordotonal organ: a bifunctional orthopteran (*Locusta migratoria*) sense organ? Comp. Biochem. Physiol. A: Physiol. 93, 729–743 with permission from Elsevier. Panel (C): Adapted from Schön, A., 1911. Bau und Entwicklung des tibialen Chordotonalorgans bei der Honigbiene und bei Ameisen. Zool. Jb. Anat. Ontog. Tiere 31, 439–472. Panel (D): Reprinted by permission from Springer-Verlag, Berlin, Heidelberg: Springer Nature: Nishino, H., Mukai, H., Takanashi, T., 2016. Chordotonal organs in hemipteran insects: unique peripheral structures but conserved central organization revealed by comparative neuroanatomy. Cell Tissue Res. 366, 549–572.*

4.1.2 Chordotonal organs

Chordotonal organs are internal mechanoreceptors unique to insects and crustaceans (Field and Matheson, 1998; Moulins, 1976). They consist of scolopidial sensilla, which contain one to four sensory neurons and distinct types of accessory cells around the neuron. The dendrite of the sensory neuron is located in a scolopale rod secreted by the scolopale cell surrounding it (Keil, 1997, 2012; Moulins, 1976; Fig. 1B). In insects, scolopidial sensilla occur in proprioceptive organs, tympanal hearing organs, near-field sound receptor organs, and vibration receptor organs. Therefore, this common sensillum type indicates a great similarity at the cellular level between vibration receptor organs and hearing organs (Faure et al., 2009; Strauß and Stumpner, 2015; Yack, 2004; Yager, 1999). Often, sensory organs can respond to both vibrations and sound depending on stimulus amplitudes (Kalmring et al., 1994, 1997a). The sensory neurons are activated by mechanical forces acting at the dendrite, resulting from stretch or by tilting of the dendrite (Field and Matheson, 1998; Hummel et al., 2016; Keil, 2012). The molecular mechanisms of mechanotransduction in scolopidial sensilla has been investigated thoroughly in *Drosophila melanogaster* with the Johnston's organ as model system (Albert and Göpfert, 2015; Hehlert et al., 2021; Keil, 2012; Liang et al., 2017). Of the receptor organs for substrate vibrations, the femoral chordotonal organ (FeCO) in *Drosophila* was investigated in detail for the molecular components of mechanosensation (McKelvey et al., 2021). The legs of *Drosophila* contain the FeCO and also a chordotonal organ in the distal tibia. Selective activation and depletion of the chordotonal organs by genetic approaches showed that the FeCO has the crucial role in detecting male courtship vibrations by females, which leads to female immobility and eventually copulation. A subset of FeCO sensilla responds to leg vibrations, projecting to the central nervous system with a characteristic arborization pattern (Mamiya et al., 2018) onto interneurons involved in eliciting the female immobility (Agrawal et al., 2020). The mechanosensitive ion channels Nanchung and Piezo were shown to mediate the vibration detection in these FeCO sensilla but these channel proteins are not expressed in the tibial chordotonal organ (McKelvey et al., 2021). These findings also imply that the tibial chordotonal organ in larger flies (*Calliphora*) is not the leg's main vibration receptor, as had been suggested earlier based on combined physiological and ablation experiments (Schneider, 1950). In *Calliphora*, the distal tibial chordotonal organ contains three scolopidial sensilla (Schneider, 1950), as it does in *Drosophila* (McKelvey et al., 2021).

In the following, the molecular physiology will not be reviewed in more detail, and we provide an overview on the neuroanatomy, functional morphology, and mechanical coupling of vibration receptor organs.

Chordotonal organs in insects have been studied in detail for their neuroanatomy, sensory physiology, and functional morphology. Chordotonal organs are the most sensitive type of vibration receptor in insects (e.g., Kühne, 1982a; Fig. 4A), and several organs have been shown to respond to substrate vibrations (Table 1; Fig. 2A). They were suggested in the first half of the 20th century to detect "wave-like" or "transient" movements of both solid and aquatic media (Eggers, 1928, p. 341; "Erschütterungssinn"). The SGO was specifically hypothesized to detect vibrations of the substrate due to its position in the tibia (Eggers, 1928; see also Debaisieux, 1938). Autrum (1941) investigated the physiological responses of leg chordotonal organs to vibrational stimuli, distinguished specific responses to airborne sound and substrate vibrations, and used ablation experiments to locate the vibration receptors. Further applications of electrophysiological recordings from sensory organs, organ parts (scoloparia), or individual sensilla have addressed the sense of vibration detection at different sensory levels.

For chordotonal organs, distinct types have been classified based on their functional morphology. Connective chordotonal organs have endings in connective tissue attached internally to other body parts (Field and Matheson, 1998; Howse, 1968; Moulins, 1976). They play an important role in proprioception and occur at joints. The other chordotonal organs are grouped as non-connective chordotonal organs, including the SGO (see Section 4.1.3). Other terminologies refer more broadly to joint chordotonal organs (Čokl et al., 2006), which also includes sensory organs that attach to a cuticular apodeme. Again, this position and attachment indicates a proprioceptive role.

Insect legs contain several chordotonal organs (Debaisieux, 1938; Field and Matheson, 1998). This position may be particularly suited to detect vibrations transferred through the substrate, and different leg chordotonal organs can detect substrate vibrations. Chordotonal organs show a great variation in the number of scolopidial sensilla, ranging from a single sensillum to several hundred, or even 2000 sensilla. The sensilla of given chordotonal organs can cluster into groups, called scoloparia (Field and Matheson, 1998). Such groups have been found in the FeCO of several orthopteroid insects, including locusts and stick insects, cicadas, stink bugs, and *Drosophila* flies, and the sensilla groups are anatomically separated (Matheson and Field, 1990).

Table 1 Chordotonal organs of insects that are physiologically responding to vibrational stimuli.

Order: Species	Recording method	Frequencies tested	Frequency tuning/Highest sensitivity	References
Summed responses from the intact leg				
Blattodea: *Periplaneta americana*	Tungsten electrodes inserted in femur	30–5000 Hz	1000–2000 Hz, $\sim 10^{-8}$ ms^{-1}, middle leg	Schnorbus (1971)
Coleoptera: *Carabus hortensis*	Platinum wire	50–10,000 Hz	300 Hz, 5700 μm, middle leg	Autrum and Schneider (1948)
Coleoptera: *Geotrupes sylvaticus*	Platinum wire	50–9000 Hz	1500 Hz, 0.48 μm, all leg pairs similar	Schneider (1950)
Coleoptera: *Melolontha*	Platinum wire	50–9000 Hz	1800 Hz, 0.1 μm, all leg pairs similar	Schneider (1950)
Coleoptera: *Pseudophonus* sp.	Platinum wire	50–9000 Hz	750 Hz, 61.0 μm, all leg pairs similar	Schneider (1950)
Coleoptera: *Pterostichus* sp.	Platinum wire	50–9000 Hz	800 Hz, 40 μm, all leg pairs similar	Schneider (1950)
Coleoptera: *Silpha obscura*	Platinum wire	50–9000 Hz	1000 Hz, 40.5 μm, all leg pairs similar	Schneider (1950)
Diptera: *Calliphora erythrocephala*	Platinum wire	50–9000 Hz	300 Hz, 17,400 μm, all leg pairs similar	Schneider (1950)
Diptera: *Eristalis* sp.	Platinum wire	50–10,000 Hz	200 Hz, 6590 μm, middle leg	Autrum and Schneider (1948)

Hemiptera: *Canthophorus (Sehirus) dubius*	Tungsten wire	50–5000 Hz	150–500 Hz, 0.008 ms^{-2}, hind leg	Devetak et al. (1978)
Hemiptera: *Okanagana rimosa*	Tungsten hook electrode	200–8000 Hz	1000 Hz, 1 ms^{-2}	Alt and Lakes-Harlan (2018)
Hemiptera: *Plautia stali*	Tungsten electrode inserted in coxa	50–1000 Hz	50 Hz, ~1 ms^{-2}	Mukai et al. (2020)
Hemiptera: *Rhodnius prolixus*	Platinum wire	50–10,000 Hz	400 Hz, 1300 μm, middle leg	Autrum and Schneider (1948)
Hemiptera: *Sehirus luctuosus*	Tungsten wire	50–5000 Hz	100–200 Hz, 0.009 ms^{-2}, hind leg	Devetak et al. (1978)
Hemiptera: *Tritomegas (Sehirus) bicolor*	Tungsten wire	50–5000 Hz	300–2000 Hz, 0.007 ms^{-2}, hind leg	Devetak et al. (1978)
Hymenoptera: *Acromyrmex octospinosa*	Silver electrode in proximal femur	50–5000 Hz	3000 Hz, displacement 1.8 μm, front leg	Markl (1970)
Hymenoptera: *Andrena nitida*	Platinum wire	50–10,000 Hz	1000 Hz, 18 μm, front leg	Autrum and Schneider (1948)
Hymenoptera: *Apis mellifera*	Platinum wire electrodes	100–10,000 Hz 100–2000 Hz	500 Hz, 60–80 μms^{-1}, hind leg 300–500 Hz, ~70 μms^{-1}, middle leg	Kilpinen and Storm (1997) and Rohrseitz and Kilpinen (1997)
Hymenoptera: *Atta cephalotes*	Silver electrode in proximal femur	50–5000 Hz	1500 Hz, 0.12 ms^{-2}, front leg	Markl (1970)

Continued

Table 1 Chordotonal organs of insects that are physiologically responding to vibrational stimuli.—cont'd

Order: Species	Recording method	Frequencies tested	Frequency tuning/Highest sensitivity	References
Hymenoptera: *Bombus soroeensis*	Platinum wire	50–10,000 Hz	1500 Hz, 32 μm, hind leg	Autrum and Schneider (1948)
Hymenoptera: *Camponotus sp.*	Platinum wire	50–10,000 Hz	2500 Hz, 18 μm, middle leg	Autrum and Schneider (1948)
Hymenoptera: *Vespa crabro*	Platinum wire	50–10,000 Hz	1500 Hz, 36 μm, hind leg	Autrum and Schneider (1948)
Lepidoptera: *Agrotis sp.*	Platinum wire	50–10,000 Hz	1200 Hz, 57 μm, hind leg	Autrum and Schneider (1948)
Lepidoptera: *Pyrameis atalanta*	Platinum wire	50–10,000 Hz	3000 Hz, 0.29 μm, hind leg	Autrum and Schneider (1948)
Lepidoptera: Satyridae indet.	Platinum wire	50–10,000 Hz	2000 Hz, 0.47 μm, hind leg	Autrum and Schneider (1948)
Mantophasmatodea: Austrophasmatidae sp. n. 3	Minuten pin steel electrode	5–2560 Hz	Calculated minimum threshold: 0.000038 ms^{-2}, front leg	Eberhard et al. (2010)
Mantophasmatodea: *Karoophasma biedouwense*	Minuten pin steel electrode	5–2560 Hz	Calculated minimum threshold: 0.00053 ms^{-2}, front leg	Eberhard et al. (2010)
Mecoptera: *Panorpa communis*	Tungsten wire	50–5000 Hz	2000–4000 Hz, 0.005 ms^{-2}, middle leg	Devetak et al. (1978)
Neuroptera: *Chrysoperla carnea*	Tungsten electrodes inserted in coxa	50–2000 Hz	300–2000 Hz, below 0.1 ms^{-2}, hindleg	Devetak and Amon (1997)

Neuroptera: *Hemerobius micans*	Tungsten electrodes inserted in coxa	50–2000 Hz	300–2000 Hz, 01–0.001 ms^{-2}, middle leg	Devetak (1998)
Neuroptera: *Sialis lutaria*	Tungsten electrodes inserted in coxa	50–2000 Hz	50–700 Hz, 1–10 μm displacements, middle leg	Devetak (1998)
Orthoptera: *Ametrus tibialis*	Silver hook electrode	100–2000 Hz	0.028 ms^{-2} front leg	Strauss and Lakes-Harlan (2008a)
Orthoptera: *Comicus calcaris*	Silver hook electrode	100–2000 Hz	~0.1 ms^{-2} front leg	Strauß and Lakes-Harlan (2010)
Orthoptera: *Gampsocleis gratiosa*	Silver hook electrode	50–2000 Hz	1000 Hz, 0.1 ms^{-2} front leg, middle leg	Kalmring et al. (1994)
Orthoptera: *Gryllus bimaculatus*	Copper hook electrode	50–3000 Hz	900 Hz, 0.0011 ms^{-2}, hindleg	Dambach (1972a)
Orthoptera: *Gryllus campestris*	Copper hook electrode	50–3000 Hz	600 Hz, 0.009 ms^{-2}, middle leg	Dambach (1972a)
Orthoptera: *Phasmodes ranatriformis*	Steel hook electrode	30–5000 Hz	200–800 Hz, 0.002 ms^{-2}, hindleg	Lakes-Harlan et al. (1991)
Orthoptera: *Stenopelmatus* spec.	Silver hook electrode	100–2000 Hz	500 Hz, 0.06 ms^{-2}, front leg	Strauß and Lakes-Harlan (2008b)
Orthoptera: *Tachycines asynamorus*	Platinum wire	50–10,000 Hz	1000 Hz, 0.1 μm, middle leg	Autrum and Schneider (1948)
Orthoptera: *Tettigonia cantans*	Platinum wire	50–10,000 Hz	2000 Hz, 0.068 μm, middle leg	Autrum and Schneider (1948)

Continued

Table 1 Chordotonal organs of insects that are physiologically responding to vibrational stimuli.—cont'd

Order: Species	Recording method	Frequencies tested	Frequency tuning/Highest sensitivity	References
Femoral chordotonal organ				
Diptera: *Drosophila melanogaster*	Microelectrode elgiloy afferent calcium imaging	0.2–500 Hz 100–2000 Hz	Front leg Front leg	Lakes-Harlan and Lefèvre (2012) and Mamiya et al. (2018)
Hemiptera: *Tritomegas (Sehirus) bicolor*	Tungsten wire	50–5000 Hz	100–200 Hz, 0.009 ms^{-2}, front leg	Devetak et al. (1978)
Neoptera: *Chrysoperla carnea*	Tungsten electrodes	50–5000 Hz	2000 Hz, ~0.1 ms^{-2}, hind leg	Devetak et al. (2004)
Orthoptera: *Locusta migratoria*, proximal scoloparium	Tungsten needle	10–300 Hz	300 Hz, 40 μm, tibial displacement middle leg	Field and Pflüger (1989)
Phasmatodea: *Cuniculina impigra*	Intracellular recordings	10–5000 Hz	200–800 Hz, 0.3 ms^{-2}, middle leg	Stein and Sauer (1999)
Subgenual organ complex				
Phasmatodea: *Sipyloidea sipylus*	Silver hook electrode	50–5000 Hz	700 Hz, 0.09 ms^{-2}, front leg	Strauß and Lakes-Harlan (2017)
Subgenual organ (spanning haemolymph channel in the proximal tibia)				
Blattodea: *Periplaneta americana*	Copper wire	1.57 kHz	1.57 kHz, 0.2 nm, hind leg	Shaw (1994b)

Hemiptera: *Nezara viridula*	Extracellular, glass microcapillary electrodes	30–5000 Hz	~0.01 ms^{-2}	Čokl (1983) and Čokl et al. (2006)
Isoptera: *Zootermopsis angusticollis*	Nerve cord recordings	25–2075 Hz	1150–1200 Hz, 10 μm, displacement	Howse (1964)
Neuroptera: *Chrysoperla carnea*	Tungsten electrodes inserted in coxa	50–2000 Hz	1500–200 Hz, 0.03 ms^{-2}, hind leg	Devetak and Amon (1997)
Orthoptera: *Decticus verrucivorus*	Glass micropipettes	30–5000 Hz	200–1000 Hz, <0.01 ms^{-2}	Kühne (1982a)
Orthoptera: *Gampsocleis gratiosa*	Glass microelectrodes	30–4000 Hz	600–800 Hz, 0.05 ms^{-2}, middle leg	Kalmring et al. (1994)
Orthoptera: *Gryllus bimaculatus*	Copper hook electrode	50–3000 Hz	400–500 Hz, ~0.0015 ms^{-2}, hind leg	Dambach (1972a)
Orthoptera: *Gryllus campestris*	Glass micropipettes	30 5000 Hz	500–1000 Hz: 0.005–0.01 ms^{-2} (front/middle legs) 100–300 Hz: 0.01 ms^{-2} (hind legs)	Kühne et al. (1984)
Orthoptera: *Hemideina femorata*	Silver hook electrodes, in the femur	20–3000 Hz	20 Hz, 0.012 ms^{-2}, middle leg	McVean and Field (1996)
Orthoptera: *Locusta migratoria*	Glass micropipettes	30–5000 Hz	200–1000 Hz, <0.01 ms^{-2}	Kühne (1982a)

Continued

Table 1 Chordotonal organs of insects that are physiologically responding to vibrational stimuli.—cont'd

Order: Species	Recording method	Frequencies tested	Frequency tuning/Highest sensitivity	References
Orthoptera: *Polysarcus denticauda*	Glass microelectrodes	Not specified	Not specified	Kalmring et al. (1996)
Orthoptera: *Troglophilus neglectus*	Glass microelectrodes	100–5000 Hz	200–700 Hz, 0.04 ms^{-2}, front leg, middle leg	Čokl et al. (1995)
Nebenorgan (posterior position in the proximal tibia)				
Blattodea: *Periplaneta americana*	Tungsten electrodes	30–5000 Hz	63 Hz, 3.1×10^{-6} ms^{-1}, middle leg	Schnorbus (1971)
Accessory organ (posterior position in the proximal tibia)				
Orthoptera: *Troglophilus cavicola*	Intracellular recording	30–4000 Hz	50–300 Hz, 0.22 ms^{-2}, front leg	Stritih-Peljhan et al. (2019)
Intermediate organ				
Orthoptera: *Gampsocleis gratiosa*	Glass microelectrodes	30–4000 Hz	600–1000 Hz, 0.01–0.03 ms^{-2}, front leg, middle leg	Kalmring et al. (1994)
Orthoptera: *Polysarcus denticauda*	Glass microelectrodes	Not specified	Not specified	Kalmring et al. (1996)
Orthoptera: *Troglophilus neglectus*	Glass microelectrodes	100–5000 Hz	700–2000 Hz, 0.01 ms^{-2}, front leg, middle leg	Čokl et al. (1995)

Tympanal organ				
Orthoptera: *Teleogryllus commodus*	Tungsten hook electrode, neck connective	35–36,000 Hz	4 kHz, front leg	Ball and Hill (1978)
Orthoptera: *Teleogryllus oceanicus*	Tungsten hook electrode, neck connective	35–36,000 Hz	5 kHz, front leg	Ball and Hill (1978)
Crista acustica				
Orthoptera: *Gampsocleis gratiosa*	Glass microelectrodes	30–4000 Hz	2000 Hz, 0.05–0.1 ms^{-2}, front leg, middle leg	Kalmring et al. (1994)
Orthoptera: *Polysarcus denticauda*	Glass microelectrodes	Not specified	7000–8000 Hz?, middle leg	Kalmring et al. (1996)
Tibio-tarsal chordotonal organ				
Blattodea: *Periplaneta americana*	Tungsten needles inserted in tibia	Not specified	Not specified, middle leg	Young (1970)
Tarsal chordotonal organ				
Hemiptera: *Aquarius paludum*	Glass suction electrode	50 Hz	Middle leg, hind leg	Goodwyn et al. (2009)
Hemiptera: *Notonecta glauca*	Steel wire	10–1000 Hz	100–150 Hz, 0.5 μm displacement, middle leg	Wiese (1972)

Continued

Table 1 Chordotonal organs of insects that are physiologically responding to vibrational stimuli.—cont'd

Order: Species	Recording method	Frequencies tested	Frequency tuning/Highest sensitivity	References
Prosternal organ				
Diptera: *Emblemasoma auditrix*	Tungsten electrode	100–10,000 Hz	500 Hz, 20 ms^{-2}	Lakes-Harlan et al. (1999)
Diptera: *Sarcophaga bullata*	Tungsten electrode	100–10,000 Hz	100 Hz, 0.4 ms^{-2}	Lakes-Harlan et al. (1999) Stölting et al. (2007)
Wing chordotonal organ (located at the base of the wing)				
Orthoptera: *Gryllus campestris*				Möss (1971)
Orthoptera: *Locusta migratoria*				Pearson et al. (1989)

Note that the sensitivity is given for acceleration, velocity or displacement, according to the reference.

Not all types of vibration receptors are morphologically identified. Locusts and tettigoniids have a receptor type that is most sensitive to low-frequency vibrations, with thresholds following a constant acceleration (Kühne, 1982a; Fig. 4A). This type was compared to the chordotonal organs located at leg joints (Devetak and Amon, 1997; Kühne, 1982a), specifically the FeCO (Field and Pflüger, 1989), but also to multipolar sensilla (Lakes-Harlan and Strauß, 2014). The multipolar sensilla are placed at, or close to, joints and occur in the femur, tibia, and tarsi (Mücke, 1991). In *Drosophila* larvae, they are involved in the detection of stretch and touch (Liang et al., 2017).

4.1.3 Subgenual organ and the subgenual organ complex

The subgenual organ (SGO) is present in the proximal tibia of most insects except for beetles and flies (Autrum and Schneider, 1948; Debaisieux, 1938; Field and Matheson, 1998; Fig. 2). It is located in the haemolymph channel, and is considered the most sensitive receptor organ detecting substrate vibrations. The term subgenual organ complex (SGOC) is used for orthopteroid insects, where additional organs occur close to the SGO (Dambach, 1972a; Eberhard et al., 2010; Strauß and Lakes-Harlan, 2013). These additional organs can include the auditory sensilla in tympanal organs that are developed, e.g., in crickets (Dambach, 1972a). In the SGOC of orthopteroid insects, there is always at least one more chordotonal organ besides the SGO, and up to four different chordotonal organs overall (Eberhard et al., 2010; Strauß and Lakes-Harlan, 2013; Fig. 3). Outside the orthopteroid groups, sensory structures were described only in dragonflies, in addition to the SGO in the position of the orthopteran distal organ (Debaisieux, 1938).

Subgenual organ. The SGO is usually positioned within the proximal tibia. In some species of Neuroptera, the SGO attachment has a different position in the tibia, as it can be found in the middle of the hind leg tibia (Devetak, 1998). The SGO is located in the dorsal haemolymph channel and attaches to the inner cuticle (Eberhard et al., 2010; Lin et al., 1995; Strauß et al., 2017, 2021) and also locates close to the leg tracheae. It shows notable morphological differences amongst Orthoptera, where the sensory neurons form a hemi-circle at the dorsal tibia, Hymenoptera, where the organ is club-shaped, and Lepidoptera, where it is a diffuse structure with branches into the haemolymph channel (Autrum and Schneider, 1948). The number of sensilla in the SGO ranges between 20 and 70 in Hymenoptera (Autrum and Schneider, 1948; Menzel and Tautz, 1994; Fig. 2C), in Orthoptera

Fig. 3 Phylogeny of *Polyneoptera* matched with the sensory organs of the subgenual organ complex. Whilst the subgenual organ is present in all taxa investigated, additional organs are documented from selected lineages, with different combinations of organs. Note that Ensifera have not a homogenous combination of sensory organs, with different organs present besides the subgenual organ in specific lineages. Abbreviations: AO, accessory organ; CA, *crista acustica*; CAH, *crista acustica* homolog; DO, distal organ; IO, intermediate organ; NO, Nebenorgan; SGO, subgenual organ; TO, tympanal organ. *Adapted from Wipfler B., et al., 2019. Evolutionary history of Polyneoptera and its implications for our understanding of early winged insects. Proc. Natl. Acad. Sci. USA 116, 3024–3029.*

(Lin et al., 1994, 1995; Nishino and Field, 2003; Rössler, 1992; Strauss and Lakes-Harlan, 2008a; Fig. 2E and F), in cockroaches (Shaw, 1994b), and in stick insects (Strauß et al., 2021; Fig. 2G). There are also several insects with very few scolopidial sensilla in the SGO, like cicadas and stink bugs, which have two sensilla (Alt and Lakes-Harlan, 2018; Autrum and Schneider, 1948; Michel et al., 1983; Nishino et al., 2016; Fig. 2D) and the lacewing *Chrysoperla carnea* with three sensilla (Devetak and Pabst, 1994). A single sensillum was described in the SGO of the common scorpionfly, *Panorpa communis* (Debaisieux, 1938). In the ant *Atta cephalotes*, 10–15 sensilla form the SGO (Markl, 1970). The differences in the functional morphology of the SGO amongst different taxa has raised the question whether these organs are indeed homologous, or not (Lakes-Harlan and Strauß, 2014). At least amongst Polyneoptera, the position, orientation and attachment of SGO sensilla is consistent. Notably, in this group the SGO is expanded into the SGOC by additional chordotonal organs (Fig. 3).

The high sensitivity of the SGO to vibrational stimuli was indicated by comparisons of vibration sensitivity amongst insects with and without this organ (Autrum and Schneider, 1948) and the recording of compound action potentials with selective ablation of organs (Autrum, 1941; Dambach, 1972a; Devetak and Amon, 1997; Schnorbus, 1971; Fig. 4B). Intracellular recordings also support the sensitivity of the SGO

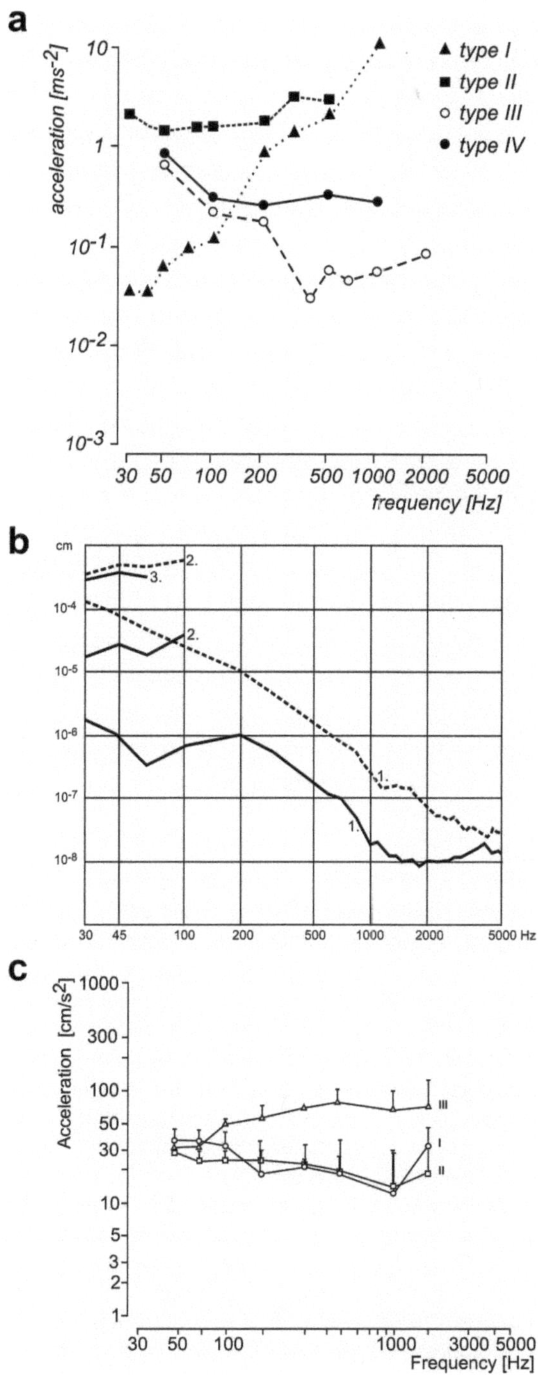

Fig. 4 See figure legend on next page.

(Kalmring et al., 1994, 1996; Kühne, 1982a; Fig. 4A and B), though during electrophysiological recordings, the precise localization of a receptor or receptor type needs to be confirmed (see Section 6). The SGO usually has broad tuning with responses over a frequency spectrum from 50 up to 4000 Hz (e.g., Čokl et al., 1995). It also responds to airborne sound at relatively low frequencies (Kalmring et al., 1994; Shaw, 1994a; Stumpner, 1996).

Intermediate organ/distal organ. The intermediate organ is present in some groups of ensiferan Orthoptera, including weta, tettigoniids, splay-footed crickets, and cave crickets (Nishino and Field, 2003; Strauß and Lakes-Harlan, 2009; Strauß et al., 2014). The intermediate organ is placed distally to the SGO. Amongst the organs in the SGOC, the physiological role of the intermediate organ is probably least understood. It usually has fewer sensilla than the SGO and these sensilla may form a proximal and distal group. They can respond to both sound and vibration (Čokl et al., 1995; Kalmring et al., 1994; Stölting and Stumpner, 1998). With respect to vibration detection, the sensilla can be grouped into different functional classes (Čokl et al., 1995; Kalmring et al., 1994). The vibrational threshold curves partly overlap with those from the SGO, but generally have the highest sensitivity at higher frequencies than the SGO (but see Kalmring et al., 1994).

The distal organ is likely homologous to the intermediate organ (Strauß and Lakes-Harlan, 2013). For the distal organ, the physiology has not been

Fig. 4 (A) Electrophysiological thresholds recorded from individual vibrosensitive sensilla in the locust and bushcricket allow to distinguish different sensilla types. (B) Electrophysiological thresholds from summed responses in the cockroach leg. Numbers give responses from (1) the subgenual organ or (2 + 3) campaniform sensilla located at the tibia (2: proximal tibial sensilla; 3: distal tibial sensilla). Broken line give responses from free-standing legs, solid lines give responses from legs with tarsi attached to the substrate. Response type (3) was only recorded in the latter setting. (C) Summed electrophysiological thresholds from the subgenual organ complex of the legs of a bushcricket (*Gampsocleis gratiosa*), showing a broad and relatively flat tuning. *Panel (A): Reprinted from Kühne R., 1982. Neurophysiology of the vibration sense in locusts and bushcrickets: response characteristics of single receptor units. J. Insect Physiol. 28, 155–163, with permission from Elsevier. Panel (B): Reprinted from Schnorbus, H., 1971. Die subgenualen Sinnesorgane von* Periplaneta americana*: Histologie und Vibrationsschwellen. Z. vergl. Physiol. 71, 14–48, by permission from Springer-Verlag, Berlin, Heidelberg, Springer Nature. Panel (C): Adapted from Kalmring, K., Rössler, W., Unrast, C., 1994. Complex tibial organs in the forelegs, midlegs, and hindlegs of the bushcricket* Gampsocleis gratiosa *(Tettigoniidae): comparison of the physiology of the organs. J. Exp. Zool. 270, 155–161, with permission from John Wiley and Sons.*

studied in detail (e.g., identification of functional groups of sensilla) in those insect species where a sensitivity to vibration stimuli was demonstrated, as in locusts (Kühne, 1982a), cockroaches (Schnorbus, 1971), or stick insects (Strauß and Lakes-Harlan, 2017; Fig. 2G). In crickets, the intermediate organ is the putative precursor organ that was adapted into the tympanal hearing organ (Strauß and Lakes-Harlan, 2013).

Accessory organ/Nebenorgan: This organ is relatively small with few scolopidial sensilla (4–17) (Strauß, 2017b; Fig. 2E and F). It is only present in some groups of orthopteroid insects like cockroaches, heelwalkers, and in some Ensifera (Fig. 3). This organ was described in few representatives of relatively species-rich taxa like cockroaches (Schnorbus, 1971) and tettigoniids (Strauß et al., 2016), and it is likely common in other members of these groups as well. The organ is found at the posterior side of the tibia, close under the cuticle beside the SGO (Nishino and Field, 2003; Schnorbus, 1971; Strauß, 2017a; Strauß and Stritih, 2016). This sensory organ is usually associated with one or few campaniform sensilla (Strauß, 2017b). The central projection of afferents to a mechanosensory neuropil (mVAC) supports the vibroreceptive function (Nishino and Field, 2003; Stritih-Peljhan et al., 2019), though the physiological data on accessory organs and Nebenorgans are so far very limited (Strauß, 2017b). The tuning to low frequency vibrations is supported from electrophysiological recordings in cockroaches (Schnorbus, 1971; Nebenorgan) and cave crickets (Stritih-Peljhan et al., 2019; accessory organ). In most cases, the orientation of dendrites in a proximal or dorsal direction is different to the SGO sensilla (Strauß, 2017b). This, together with the position underneath the cuticle, might provide the anatomical basis for a different tuning to the SGO sensilla.

Crista acustica/crista acustica homolog: The *crista acustica* contains the sensilla responding to airborne sound in the tympanal organs in forelegs of tettigoniids, weta, and grigs (Lakes and Schikorski, 1990; Mason, 1991; Nishino and Field, 2003; Strauß et al., 2014, 2017). The sensilla are arranged linearly along the proximo-distal axis of the tibia (Fig. 2F), which correlates with tuning characteristics to sound of increasing frequencies (Montealegre-Z et al., 2012; Sarria et al., 2017; Stölting and Stumpner, 1998; Stumpner and Nowotny, 2014). The *crista acustica* exists in all leg pairs of these insects. Notably, the *crista acustica* homolog also exists in some taxa of Ensifera that lack tympanal membranes in the tibia and further modifications for sound detection in all leg pairs (Strauss and Lakes-Harlan, 2008a,b, 2009). It has been speculated that this *crista acustica* homolog is involved in the detection of vibration stimuli rather than airborne sound, with either distinct frequency tuning or orientation to vibrational planes (Lakes-Harlan and Strauß, 2014). The ancestrally atympanate organization is likely indicated by the presence of the *crista acustica* or *crista acustica* homolog in all leg pairs. Intracellular recordings from the *crista acustica* of the tettigoniid *Gampsocelis gratiosa* showed responses to vibrational stimuli in fore- and midlegs (Kalmring et al., 1994). In the ground weta *Hemiandrus pallitarsis*, the *crista acustica* homolog is jointly coupled with the intermediate organ by connective tissue to the inner tibial cuticle, and likely detects vibrations transferred over the cuticle (Strauß et al., 2017). However, physiological characteristics of any *crista acustica* homolog from an atympanate species are still to be documented.

4.1.4 Femoral chordotonal organ

The femoral chordotonal organ (FeCO) is located in the femur (Fig. 2A). The exact position can vary between different insect groups from a position in the proximal femur in orthopteroid insects (Field and Pflüger, 1989; Füller and Ernst, 1973; Nishino, 2003; Theophilidis, 1986) and the distal half or third of the femur (Alt and Lakes-Harlan, 2018; Devetak et al., 2004; Michel et al., 1983; Nishino et al., 2016). The FeCO is proximally placed at the dorsal hypodermis of the femur and is organized usually into two scoloparia of sensory neurons in orthopteran and hemipteran insects. The two scoloparia are homologous in several insect groups (proximal and distal scoloparia in locusts, and the dorsal and ventral scoloparia in stink bugs, crickets, and stick insects) (Nishino et al., 2016; Fig. 2B). No such distinction into two scoloparia is found in beetles (Takanashi et al., 2016). The organ links distally to the cuticular receptor apodeme that extends to the proximal tibia. Therefore, the FeCO is considered as a non-connective chordotonal organ (Field and Matheson, 1998, p. 27). With this connection to the tibia, the FeCO is involved in the postural control of legs. In stink bugs, the attachment of the organ is more complex, and the scoloparia connect to two ligaments that link to a tibial extensor muscle or its apodeme (Michel et al., 1983; Nishino et al., 2016).

Similar to the SGO, the FeCO shows great differences in the number of sensory neurons, from 12 in *N. viridula* (Michel et al., 1983), 20–24 in other stink bugs and cicadas (Alt and Lakes-Harlan, 2018; Nishino et al., 2016), to ~180 in weta (Nishino, 2003) and ~240 in stick insects (Füller and Ernst, 1973). In insect species where both the FeCO and the SGO have been investigated, the former always had a higher number of sensory neurons.

The two scoloparia of the FeCO in orthopteroid insects (Fig. 2B) are not equally involved in the detection of limb movements (Büschges, 1994; Field and Pflüger, 1989; Kittmann and Schmitz, 1992). The proximal/dorsal group is likely involved in the detection of substrate vibrations, which is also supported by the different central projection of sensory afferents (Field and Pflüger, 1989; Nishino, 2003; Nishino and Sakai, 1997). Responses of this organ to vibration stimuli are likely recorded after cutting at the femur-tibia joint in crickets (Dambach, 1972a), or ablation of the SGO (Devetak and Amon, 1997). Physiological recordings find that the FeCO is less sensitive to substrate vibrations than the SGO (Devetak and Amon, 1997; Devetak et al., 1978).

4.1.5 Connective chordotonal organs

Other joint chordotonal organs with attachment by connective tissue are located at or connected between parts of the leg (tibia-tarsus and tarsus-pretarsus). Such organs occur in all insects (Debaisieux, 1938). They can respond to sinusoidal vibrations when parts of these segments are bent up- or downwards. As a result, they respond best to rather low vibration frequencies, and likely with a relatively low vibration sensitivity compared to SGO and SGOC (Kühne, 1982a). Ablation experiments have repeatedly found no influence of such organs on the overall sensitivity recorded from summed thresholds (Eberhard et al., 2010; Strauß and Lakes-Harlan, 2017). Therefore, the connective chordotonal organs may respond to substrate vibrations, but at higher stimulus amplitudes, which covers their contribution in recordings of summed responses.

A chordotonal organ in the distal tibia connects to a retractor tendon, the tibial distal chordotonal organ in *N. viridula* (Čokl et al., 2006) or distal tibial chordotonal organs in *Carausius morosus* (Godden, 1972) and *Sipyloidea sipylus* (Strauß and Lakes-Harlan, 2017). This organ consists of six to seven sensilla. Distal tibial organs had been suggested as the main receptor organs for vibrations in flies and beetles that lack the SGO (Schneider, 1950). The tibio-tarsal chordotonal organ in cockroaches responds to upwards and downwards flexion of the tarsus but also responds to vibrations, though these responses were not systematically analysed (Young, 1970). Similar organs occur in locusts with about 30 sensilla (Mücke, 1991), in tettigoniids with 10–13 sensilla (Lakes and Schikorski, 1990) and in Mantophasmatodea with 8–10 sensilla (Eberhard et al., 2010).

The tarsal scolopidial organ or tarsal-pretarsal organ occurs in the tarsi of many insects (Čokl et al., 2006; Debaisieux, 1938; Field and Matheson, 1998). They are usually involved in measuring the raising and flexion of tarsal and pretarsal elements. Such movements are relevant for aquatic insects like water striders moving on the water surface (Ma et al., 2020; Murphey, 1971), and tarsal scolopidial organs have been identified as vibration receptors in the water striders *Notonecta glauca* (Wiese, 1972) and *A. paludum* (Goodwyn et al., 2009). The chordotonal organs consist of two groups of sensilla in the distal tarsomere (*Notonecta glauca*; Wiese and Schmidt, 1974) or three groups of sensilla (*A. paludum*; Goodwyn et al., 2009). These differ in their position in *Notonecta glauca* with a proximo-lateral (three sensory neurons) and a median-distal group (five sensory neurons) (Wiese and Schmidt, 1974). The sensilla also differ in size (Wiese, 1972). In *A. paludum*, three groups of a few sensilla occur in the two tarsomeres (Goodwyn et al., 2009). The tarso-pretarsal chordotonal organ in the stink bug *N. viridula* was also assumed to respond to substrate vibrations (Čokl et al., 2006).

4.1.6 Johnston's organ

This chordotonal organ is located in the second antennal segment, the pedicel (Field and Matheson, 1998). It is best studied in *Drosophila* as a receptor for sound, wind and gravity (Albert and Göpfert, 2015; Kamikouchi and Ishikawa, 2016). In *Drosophila* hearing, the third antennal segment, the funiculus, which also holds the feathery arista, is inserted in the pedicel, and moves against the pedicel at the attachment site of the Johnston's organ (Albert and Göpfert, 2015; Zannini et al., 2014). The Johnston's organ in *Drosophila* contains ~200 scolopidia with two to three sensory neurons, and ~480 sensory neurons in total (Kamikouchi and Ishikawa, 2016; Matsuo and Kamikouchi, 2013). The distal segments of the antenna can also be directly stimulated by sinusoidal vibrations. In the stink bug *N. viridula*, the antennae are obviously used for orientation towards vibrational signals (Ota and Čokl, 1991), indicating the Johnston's organ as a possible receptor organ. Stimulating the antenna directly at the proximal flagellum gave responses to vibrations, though the leg receptor organs are more sensitive (Jeram and Čokl, 1996). The Johnston's organ in *N. viridula* contains 45 scolopidia and a further seven scolopidia in the antennal central organ, both with three or one to two sensory neurons (Jeram and Pabst, 1996). The antenna with Johnston's organ likely also mediates pair formation in the whirligig beetle *Dineutes discolor* by the detection of water surface vibrations, as shown with antennal ablations (Kolmes, 1985).

4.1.7 Prosternal chordotonal organ

Vibration receptor organs are not necessarily restricted to a location on the appendages. The prosternal chordotonal organ is present in the ventral prothorax of Diptera (Fig. 2A) and consists of ~35 scolopidial sensilla in *Sarcophaga bullata* (Stölting et al., 2007) and 50–55 scolopidial sensilla in *Phormia regina* (Lakes-Harlan et al., 2007). It is located in the prothorax between the sternal apodeme and the lateral prosternum (Lakes-Harlan et al., 1999; Stölting et al., 2007). The prosternal organ responds to vibrational stimulation with highest sensitivity to frequencies below 2 kHz, but extending to at least 10 kHz (Lakes-Harlan et al., 1999). The prosternal organ has been adapted into tympanal hearing organs in two dipteran lineages of Sarcophagidae and Tachinidae (Lakes-Harlan and Lehmann, 2015; Lakes-Harlan et al., 1999; Strauß and Lakes-Harlan, 2014). Notably, the sensilla in the tympanal organ of *Emblemasoma auditrix* (Sarcophagidae) still respond to direct vibration stimulation of the organ, but with a lower sensitivity over the entire frequency range tested (Lakes-Harlan et al., 1999).

4.2 Mechanical coupling and functional morphology of vibratory receptor organs

With multiple potential receptor organs, the detection of vibration in insects represents a complex situation. Especially, the specific roles of the multiple vibrosensory organs are not well understood with respect to generating vibrational behaviours. Likely, the SGO as well as the FeCO are the main receptors of substrate vibrations in the legs. The diversity of receptor organs is related to different types of mechanical couplings of these organs in the leg, which provide different input pathways of substrate vibrations to the sensory organs. Such differences in coupling may cause the different best frequencies for individual organs, as it appears in the SGOC for the SGO, intermediate organ, and accessory organ (Čokl et al., 1995; Kalmring et al., 1994; Schnorbus, 1971; Stritih-Peljhan et al., 2019). As previously mentioned, the SGO is generally located in the dorsal haemolymph channel and attaches internally to the leg cuticle. However, in orthopteroid insects, the subgenual sensilla insert into a tissue structure perpendicular to the leg's main axis (Eberhard et al., 2010; Lakes and Schikorski, 1990; Lin et al., 1994; Nishino and Field, 2003; Strauß et al., 2017, 2021); in honeybees and a parasitoid wasps (*Pimpla turionellae*), the SGO has the shape of a hollow cone

(Kilpinen and Storm, 1997; Otten et al., 2002). The dendrites orient along the sides of this cone in honeybees (Kilpinen and Storm, 1997; Schön, 1911). Based on the position in the haemolymph channel, the SGO can be displaced by haemolymph movements caused by vibrations transmitted along the tibia. The perpendicular position of the SGO in the haemolymph channel and the coverage of a large portion of the channel make it well placed to respond to pressure changes in the haemolymph. The biomechanics of the SGO have been demonstrated in bees, where its movements could be visualized by applying a vital stain (Kilpinen and Storm, 1997). During vibrational stimulation, the SGO is moved within the haemolymph with the same frequency as the stimulus, and these organ movements lag behind those of the leg. The biomechanics of the bee SGO correspond to an overdamped mass-spring system and its displacement is affected by both the haemolymph movements and the leg oscillations (Kilpinen and Storm, 1997). In orthopteroid insects, the SGO contacts the cuticle or elaborated trachea, which can provide sensitivity also to airborne sound. Notably, cockroaches have no tympanal membranes of thinned cuticle in the tibia, which usually indicate auditory adaptations in insects. The detection of airborne sound can also be behaviourally relevant, as in the alarm response described in cockroaches (Shaw, 1994a).

In Hemiptera, the SGO is connected to a ligament running into the distal tibia. In the stink bug *N. viridula*, this ligament is flat and broadened, and extends longitudinally in the tibia between the anterior and posterior side. At the sensory neurons, the proximal end of the ligament is not fixed to the tibia. Distally, the ligament divides into two strands partly fixed at the dorsal cuticle and the two tibial nerves (Michel et al., 1983). In cicadas, the SGO also links to a long ligament extending into the distal tibia (Alt and Lakes-Harlan, 2018; Nishino et al., 2016). The role of these ligaments and their broad structure is not understood mechanistically, e.g., for a possible role in tuning to relatively low vibration frequencies. Since it locates in the leg in perpendicular orientation to the SGO in Orthoptera, it likely has a different direction of displacements in the tibia.

Other organs in the SGOC like the intermediate organ or the distal organ have a different suspension compared to the SGO. They are also located in the haemolymph channel, with connections to the leg trachea and the inner cuticle (Eberhard et al., 2010; Lin et al., 1994; Strauß et al., 2017, 2021). In the ground weta, *H. pallitarsis*, the intermediate organ is attached closely to the cuticle, and only slightly narrows the diameter of the haemolymph channel. Based on this attachment, it was suggested that this organ responds to vibrations transmitted over the leg rather than in the haemolymph, which is detected by the SGO (Strauß et al., 2017). The accessory organ and the Nebenorgan in several species are located at the hypodermis (Strauß, 2017b). These organs preferentially respond to low-frequency vibrations, which may be due to the tight coupling to the cuticle (Nishino and Field, 2003; Schnorbus, 1971; Stritih-Peljhan et al., 2019). The presence of several receptor organs may also improve the analysis of vibrational stimuli transferred over different structures of the leg (Stritih-Peljhan and Strauß, 2018).

The vibration sensitivity of the FeCO is based on the detection of fast and small displacements by sinusoidal stimuli. This is an important distinction to the slow and large-amplitude leg movements detected at joints during locomotion (Stein and Sauer, 1999). The FeCO could detect vibration stimuli either by movements of the entire tibia via the apodeme attachment, or by vibrations transferred over the cuticle via the femur to the sensory neurons. Notably, a group of sensilla in many insects is found closely under the cuticle (Nishino, 2003; Nishino and Sakai, 1997). For stick insects, a response of the FeCO was shown upon direct vibrational stimulation of the tibia (Stein and Sauer, 1999), whilst the proximal scoloparium in locusts responds to sinusoidal vibration of the receptor apodeme (Field and Pflüger, 1989). The best sensitivity to vibrational stimuli for the FeCO is found for lower frequency ranges than for the SGO (Čokl et al., 2006). This difference in responses is supposedly found for other joint chordotonal organs as well.

5. Physiological methods for investigating vibratory receptor organs

Vibrational physiology has some specific difficulties in investigating sensory detection and processing in comparison to the auditory system. These include the mechanical isolation of preparations during physiological recordings to avoid vibrational artefacts in recordings, but also different methods of vibrating parts of the insect body. This section discusses some aspects relevant to the recording of sensory responses to vibration stimuli and to the identification of vibration-sensitive organs (as well as vibratory neuron elements at subsequent synaptic levels).

5.1 Experimental delivery of vibration stimuli

For experimental stimulations, vibrations are usually delivered as sinusoidal oscillations. Vibrational stimuli require a direct mechanical coupling of a stimulator or a transmitting substrate to the insect body, e.g., one or several legs. This is in contrast to studying auditory systems and the stimulation with sound stimuli, which can be delivered to preparations of insects as far-field sound during experiments in precisely defined amplitudes and angles (e.g., Schöneich and Hedwig, 2010). Sound stimuli are picked up by resonant tympanal membranes or auditory spiracles that transfer the sound energy to the receptor organs.

For vibrations, the insect body is linked to a generator of vibrational stimuli. Such stimuli can be delivered as oscillations via a mini-shaker (Čokl, 1983; Eberhard et al., 2010; Kalmring and Kühne, 1980; Kalmring et al., 1994; Kühne, 1982a; Strauß and Lakes-Harlan, 2010, 2017). The vibrations are often transferred to the insect leg by a rod attached to the mini-shaker. Alternative generators are a loudspeaker membrane on which the insect rests (Autrum and Schneider, 1948; Devetak, 1998; Devetak and Amon, 1997; Devetak et al., 1978; Markl, 1970; Wiese, 1972) or the vibrating surface on which the animal is placed (Autrum, 1941; Dambach, 1972a; Schnorbus, 1971). The direct coupling between the insect and the stimulator could potentially interfere with the biomechanics of appendages relevant for the stimulus transmission. Therefore, it is also important to consider the types of attachments to the stimulator (Schnorbus, 1971). In addition, the travelling vibration stimuli may cause rebounds from the body (Stritih-Peljhan and Strauß, 2018).

Such direct stimulation always raises the question of the specificity of sensory responses to vibrations. Obviously, several sensory organs on the body or in the appendages can potentially respond to vibrations, and the specific origin of responses needs to be confirmed further by neuronal staining, by ablation experiments (see Section 5.5, for details), or by physiological characteristics. In one study on the cricket *Teleogryllus commodus*, the leg receptor organs were stimulated by contact vibrations at frequencies typical for sound stimuli between 3.5 and 35 kHz (Ball and Hill, 1978). Recorded neuronal responses from interneurons in the central nervous system ascending to the brain were attributed to the auditory sensilla of the tympanal organ rather than the vibrosensitive sensilla in the SGO based on sensitivity ranges, strictly ipsilateral processing in connectives, and similarities in coding properties between sound and vibrational stimulation (Ball and Hill, 1978; Dambach, 1972b). Given the differences in mechanical activations by airborne sound and substrate vibrations of the sensory organs, this finding suggests that the tympanal organ can function as a vibration detector. For the auditory sensilla of tympanal organs, it is assumed that they become mechanically isolated from other displacements caused, e.g., by substrate vibrations or body movements during the evolutionary adaptation to sound detection (Hoy and Yack, 2009; Shaw, 1994a). Therefore, the vibrational activation of auditory sensilla would be interesting from both a physiological and an evolutionary perspective.

Depending on the preparations, the delivery of the stimuli can be more or less controlled for, e.g., the angle of the leg to the stimulator. Usually, the legs are set to resemble a natural position (Eberhard et al., 2010; Kalmring et al., 1978; Kilpinen and Storm, 1997; Kühne, 1982a; Kühne et al., 1984; Lakes-Harlan et al., 1991; Schneider, 1950), or into a controlled femur-tibia angle of 90° (Čokl, 1983; Dambach, 1972a; Lakes-Harlan and Lefèvre, 2012; Strauß and Lakes-Harlan, 2017; Stritih-Peljhan et al., 2019) or 70° (Alt and Lakes-Harlan, 2018). Some studies stimulated more than one leg simultaneously, attaching several legs to the stimulator (Kühne, 1982b). This approach obviously makes it almost impossible to keep all relevant legs in a natural standing position. However, simultaneous stimulation of different legs is necessary especially in physiological studies of interneurons. This has been done (due to spatial constraints of the set-up) with usually one, or at the most two, stimulators. However, even with two stimulators it is almost impossible to keep all relevant legs in a natural standing position. In the Orthoptera, thus, most studies focused on stimulation of fore- and mid-legs only. Only Kühne et al. (1984), Čokl et al. (1985), and Bickmeyer (1991) stimulated all six legs simultaneously. Consequently, vibratory interneurons of Orthoptera originating from the metathoracic segment stayed so far largely unexplored (see Section 7.2). Simultaneous stimulation is still easier to be achieved with smaller insects, however. In the study of the stink bug *N. viridula*, Prešern et al. (2018) used a plexiblock extension from each mini-shaker that provided a support for the set of legs on each body side, which were attached in close to natural position.

5.2 Mechanical isolation of the set-up

Given the highly sensitive receptor organs in insects, it is important to isolate the preparation from accidental vibrations or other sources of vibrations that could affect the recordings. Such mechanical isolation is, of course, important for any physiological (and especially intracellular) preparation, also with respect to other sensory systems, but is especially important in studies of the vibratory system since it directly affects its sensory input. Different support structures are used in sophisticated set-ups to prevent such vibrational artefacts. Usually, the preparation (insect or isolated leg) is mounted on a holder placed on a solid metal plate (Dambach, 1972a; Devetak and Amon, 1997; Kilpinen and Storm, 1997; Schnorbus, 1971) that can also serve as the basis of a Faraday cage resting on air-filled rubber sacks (Strauss and Lakes-Harlan, 2008a,b; Wiese, 1972). The plate can be placed on a desk, or in sand containers resting on air-filled rubber sacks (Devetak and Amon, 1997), or a concrete plate (Dambach, 1972a; Kilpinen and Storm, 1997). Schnorbus (1971) employed a set-up where the preparation was placed on a metal plate with air-filled rubber cushions on a desk, and the desk was suspended from the ceiling of the laboratory with expander rubber.

The choice of the set-up obviously depends on the size of the animal, and on the method of recording that is used, and therefore on the anatomical dissections required to gain access to sensory nerves, e.g., for inserted electrodes or hook electrodes.

5.3 Electrophysiological recordings of neuronal activity

An important aspect to study vibration receptor organs is to understand their response characteristics over a range of frequencies and amplitudes. For vibration receptors, different types of preparations for electrophysiological recordings have been developed. This is especially interesting in comparison to recordings from auditory nerves/neurons. In the auditory pathway, recording from peripheral nerves and central neurons requires intact tympanal membranes and sound-transmitting structures like tracheae. For recordings with vibrational stimulation, it is probably most notable that in insects they can be carried out on the sensory nerves of isolated legs (Autrum, 1941; Devetak et al., 1978; Goodwyn et al., 2009; Howse, 1962; Sandeman et al., 1996; Schneider, 1950; Schnorbus, 1971; Shaw, 1994b; Strauß and Lakes-Harlan, 2017; Wiese, 1972; but see Markl, 1970). Other approaches are leg preparations of insects fixed in the set-up (Alt and Lakes-Harlan, 2018; Čokl et al., 1995; Eberhard et al., 2010; Kalmring et al., 1994; Stein and Sauer, 1999; Strauss and Lakes-Harlan, 2008a,b), and from permanently implanted electrodes in mobile animals at the base of the leg or the prothorax (Dambach, 1972a), or the neck membrane (Dambach, 1972b).

Accordingly, different types of recording electrodes can be employed. These can be chosen depending on the recording of compound action potentials from sensory nerves, or the recording from single neurons. Nerve recordings can be achieved after opening of the leg cuticle to expose the sensory nerve, which is placed on a hook electrode of silver or tungsten (Alt and Lakes-Harlan, 2018; Field and Pflüger, 1989; Kalmring et al., 1994; Strauß and Lakes-Harlan, 2010, 2017). A less invasive method is the insertion of recording electrodes through the cuticle close to the sensory nerve in the femur, trochanter, or coxa. Such electrodes can be either sharpened electrodes or metal wires like tungsten (Alt and Lakes-Harlan, 2018; Devetak and Amon, 1997; Field and Pflüger, 1989; Lakes-Harlan et al., 1999; Schnorbus, 1971), platinum (Kilpinen and Storm, 1997; Rohrseitz and Kilpinen, 1997), Elgiloy (Lakes-Harlan and Lefèvre, 2012), or minuten pin steel electrodes (Eberhard et al., 2010; Sandeman et al., 1996).

Intracellular recordings from single receptor cells can usually be achieved from axons of vibration-sensitive sensilla running in the sensory nerve. Recordings are usually carried out in the nerve close to the nerve entrance to the ganglion (Büschges, 1994), in the ganglion close to the nerve entrance (Stritih-Peljhan et al., 2019), or in the ganglion/neuromere in the termination site of sensory afferents (Stölting et al., 2007). These recordings used glass microcapillary electrodes, which can also be filled with stain for intracellular staining of neurons (see Section 5.4). This technique can also be used for recordings from interneurons in the central nervous system. These methods are identical also for studying auditory receptor neurons and interneurons.

Some studies used glass capillaries for extracellular recordings from single receptor neurons (Čokl, 1983; Kalmring et al., 1994; Kühne, 1982a). This method also used glass micropipettes for intracellular recordings from single cells, placed at the entrance of the leg nerve to the ganglion or from the sensory neuropil.

5.4 Identifying sensory units: Staining of sensilla during electrophysiological recordings

Physiological recordings require identifying the origin and location of the vibration-sensitive units. This is especially complicated in the vibrosensory system by the occurrence of more than one vibration-sensitive organ in the legs of insects, or even in the tibial SGOC. Therefore, the origin of neuronal potentials in response to vibration stimuli can be difficult to identify (Kühne, 1982a; Markl, 1970; Sandeman et al., 1996; Wiese, 1972). To link the recorded responses, their origin should be traced to a specific sensillum, sensillum group/type, or a receptor organ. Based on the identification of the location of different sensory organs, e.g., in the leg, during recordings from single cells, a dye (e.g., solutions of Lucifer yellow, neurobiotin, or cobalt chloride) can simultaneously be injected to stain the neuronal structures intracellularly such as the cell body and the afferents to the central nervous system (ganglion). Staining of the cell body allows us to identify its position in the organ responding to the stimuli (Stölting et al., 2007). However, the cell body located in the periphery can be more difficult to stain, depending on the site of recording and the resulting distance to the soma that needs to be bridged by the dye (e.g., Stritih-Peljhan et al., 2019). The stained pattern of the central projection allows us to compare the investigated cell with the known central projection known from other organs (Büschges, 1994; Stein and Sauer, 1999; Stritih-Peljhan et al., 2019), and to classify the recorded neurons into distinct subtypes based on their physiology, combined with neuroanatomical characteristics (e.g., Ahi et al., 1993).

5.5 Identifying sensory units: Ablation experiments

Another approach to identifying the relevance of a specific sensory organ for vibration detection is the selective ablation of receptor organs, coupled with behavioural or physiological experiments. The position of relevant receptor organs on the legs or on the body can be identified by ablations during behavioural experiments, e.g., testing for responses and orientation (Devetak, 1985; Markl and Wiese, 1969; Murphey, 1971; Murphey and Mendenhall, 1973; Rupprecht, 1968). Several physiological studies used ablation experiments to distinguish the role of different chordotonal organs, and revealed the importance of the SGO (Autrum, 1941; Schnorbus, 1971). The ablations can be carried out by different means: immobilization of joints and associated sensory organs (Wiese, 1972), the amputation of parts of the leg (Autrum, 1941; Schneider, 1950; Strauß and Lakes-Harlan, 2017) or antenna (Eriksson et al., 2011), micro-operations on afferents of specific sensory organs (Takanashi et al., 2016), or excluding different organs with a thermocauter (Dambach, 1972a; Devetak et al., 1978; Markl, 1970; Schnorbus, 1971). Schnorbus (1971) cautioned use of heat extirpation because of the denaturation of the haemolymph, which could affect the mechanical properties of the sensory organ in the leg; although, such effects have not been investigated systematically. More subtle methods include destroying different organs mechanically by extirpation with a needle (Dambach, 1972a; Devetak, 1998; Devetak and Amon, 1997) or by heat with a needle (Devetak, 1998; Devetak and Amon, 1997; Strauß and Lakes-Harlan, 2017). To account for potential problems due to ablation procedures, several studies used more than one method for comparison to evaluate the efficiency of a specific type of ablation (Dambach, 1972a; Devetak, 1998; Devetak and Amon, 1997; Strauß and Lakes-Harlan, 2017).

Such control experiments are especially relevant when using amputations, as these shorten the leg and inevitably influence the mechanical properties of the leg, especially when the leg joints are affected. Such ablations may even result in an increased sensitivity, when the cutting of the tibia-tarsus joint leads to a direct stimulation of the tibia (Schneider, 1950). In several insects, the chordotonal organ in the distal tibia or tarsus was ablated to record only responses from the SGO/SGOC (Autrum, 1941; Devetak and Amon, 1997; Eberhard et al., 2010; Sandeman et al., 1996; Strauß and Lakes-Harlan, 2017). Schneider (1950) investigated flies and beetles, which have no SGO and approximated the position of the relevant receptor organ by successively shortening the leg from the tarsi and distal tibia. Markl (1970) and Dambach (1972a) cut the tibia between the femur-tibia joint and the SGO in the proximal tibia, and replaced the tibia with a prosthesis made from a thin insect needle. Eberhard et al. (2010) cut off different parts of the leg including the distal femur and proximal tibia, thereby severing the sensory afferents of receptors located distally in the leg, and glued the isolated leg back with insect wax, to prevent mechanical influences from shortening the leg. For the FeCO in locusts, selective ablation of one of the two sensilla subgroups (scoloparia) were conducted by cutting the nerve to the respective group (Field and Pflüger, 1989). In the legs of bees (meso- and metathoracic legs), Kilpinen and Storm (1997) isolated the SGO by cutting the nerve close to the organ, thus abolishing neuronal responses to vibrations previously recorded in the femur. Takanashi et al. (2016) micro-operated on all six legs from beetles to remove the FeCO, which resulted in abolishing behavioural responses to vibration stimuli.

6. Sensory physiology of vibratory receptor organs

6.1 The adequate stimulus

Sinusoidal stimuli used for physiological experiments can be characterized by different parameters of displacement, velocity, and acceleration of the sinus wave, following a frequency-dependent relationship (Griffin, 1990). Accordingly, the physiological responses can be analysed for any of these different parameters. Depending on the frequency, chordotonal organs can respond to a similar acceleration or displacement over part of the tested frequencies, which has often been considered an indication of sensitivity to this particular parameter of the stimulus (Čokl, 1983; Devetak and Amon, 1997; Kalmring et al., 1978; Kühne, 1982a; Schnorbus, 1971). Individual sensilla in the FeCO in stick insects and locusts respond to different parameters like position, velocity, and also acceleration (Büschges, 1994; DiCaprio et al., 2002; Hofmann et al., 1985). For the SGO, the acceleration component of the sine wave is the effective stimulus (Schnorbus, 1971) and for this reason it has been compared to an accelerometer (Kilpinen and Storm, 1997; Schnorbus, 1971). Chordotonal organs are very versatile in responding to different mechanical stimuli, and their physiological function depends on the mechanical coupling. Other chordotonal organs are also adapted to the detection of airborne sound, near-field sound, or gravity (Albert and Göpfert, 2015; Field and Matheson, 1998; Römer and Tautz, 1992; Yack, 2004).

6.2 Peripheral coding

Electrophysiological recordings have been used to investigate the physiological properties of chordotonal organs and other types of sensilla sensitive to substrate vibrations. Vibrosensory sensilla and organs may show broad responses over frequency ranges up to several kHz (e.g., Kalmring et al., 1994; Kühne, 1982a; Fig. 5A). The SGO and SGOC, like the intermediate organ and the accessory organ, especially contribute to a broad tuning and high sensitivities (Čokl and Virant-Doberlet, 2009; Dambach, 1972a,b; Devetak and Amon, 1997; Eberhard et al., 2010). Responses of chordotonal organs to vibrational stimuli can have different physiological characteristics with tonic, phasic, or phasic-tonic responses (Field and Matheson, 1998). SGO and intermediate organ sensilla in locust, bushcricket, and cave cricket were characterized by tonic or slightly phaso-tonic responses, the latter indicative of adaptation (Čokl et al., 1995; Kalmring et al., 1978; Kühne, 1982a). Other chordotonal units, which respond in a tonic way, may code, for example, for static position of joints and have proprioceptive functions (Field and Matheson, 1998; Field and Pflüger, 1989). Proprioceptive campaniform sensilla and sensilla in the FeCO typically respond to vibratory stimuli in a phase-locked fashion and thus directly code for stimulus frequencies below ~200 Hz (Kühne, 1982a; Schnorbus, 1971; Stein and Sauer, 1999). In the FeCO of the stick insect *Cuniculina impigra*, stimuli at a frequency of 50 Hz elicited phase-locked responses but not every stimulus was found to trigger an action potential (Stein and Sauer, 1999). The proximal scoloparium of the FeCO in locusts contains tonic, phasic, and

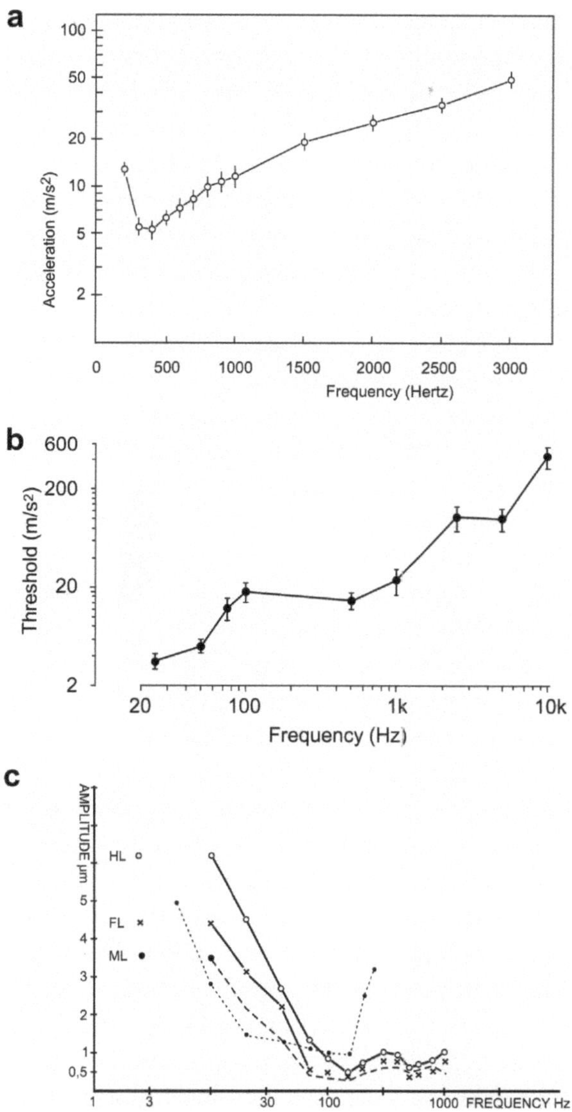

Fig. 5 Behavioural thresholds of (A) the freezing response in honeybee, (B) startle response and stridulation in the beetle *Monochamus alternatus*, (C) thresholds in the water strider *Notonecta glauca* with thresholds for prey capture (small circles, dotted line) and sensory thresholds from the front leg (FL), middle leg (ML), and hind leg (HL). *Panel (A): Reprinted by permission from Springer-Verlag, Berlin, Heidelberg, Springer Nature: Michelsen, A., Kirchner, W.H., Lindauer, M., 1986a. Sound and vibrational signals in the dance language of the honeybee,* Apis mellifera. *Behav. Ecol. Sociobiol. 18, 207–212. Panel (B): Reprinted from Takanashi, T., Fukaya, M., Nakamuta, K., Skals, N., Nishino, H., 2016. Substrate vibrations mediate behavioural responses via femoral chordotonal organs in a cerambycid beetle. Zool. Lett. 2, 18, Publisher Springer Nature, published under the Creative Commons CC BY license. Panel (C): Reprinted by permission from Springer-Verlag, Berlin, Heidelberg, Springer Nature: Wiese, K., 1972. Das mechanorezeptorische Beuteortungssystem von Notonecta. I. Die Funktion des tarsalen Scolopidialorgans. J. Comp. Physiol. 78, 83–102.*

phaso-tonic units (Field and Pflüger, 1989). Phase-locked responses allow direct encoding of the stimulus frequency, whilst coding of higher frequencies can only be extracted by the topography of sensilla projections (place principle; see Field and Matheson, 1998). Whilst somatotopic projections of different organs are found in the central nervous system (Nishino, 2003; Nishino and Field, 2003), "vibrotopic" encoding of frequencies is probably not very specific in vibrosensory systems, as only a rough frequency representation seems to exist. Given such overlaps of projections and the broad tuning of sensilla, only a rough discrimination of "low" vs. "high" frequency may be possible, based on the place principle (see Section 7.1 on central projections).

In addition, the neuronal activity transmitted over peripheral nerves encodes several parameters of the vibrational signal. Responses increase with increasing stimulus amplitude and saturate within about 20–30 dB range (Field and Matheson, 1998; Kühne, 1982a; Stritih, 2006; Stritih-Peljhan et al., 2019). Similar dynamic ranges are also characteristic for auditory receptors in tympanal hearing organs located in the legs of crickets and tettigoniids (e.g., Imaizumi and Pollack, 2001; Oldfield, 1984; Oldfield et al., 1986; Stumpner and Nowotny, 2014).

The latency denotes the response time between the onset of the stimulus and the first action potential. The latency in the vibratory response of the leg nerve (e.g., Alt and Lakes-Harlan, 2018) and also vibratory interneurons (e.g., Zorović et al., 2008) decreases with increasing stimulus amplitude independent of the carrier frequency. For vibration receptor neurons, similar latencies suggest similar distance to transduction sites, and for interneurons a similar synaptic distance to receptors. Shortest latencies of individual receptors tuned to different frequencies range between 10 and 13 ms, as shown by data from the few studies available (Buh, 2011; Stritih, 2006; Zorović, 2005).

Joint chordotonal organs can be considered mainly proprioceptive, but may respond to vibrational stimuli when those are transferred into movements of leg joints, typically with large displacements that occur with low frequency stimuli. The tarso-pretarsal chordotonal organ in *Notonecta glauca* contains phaso-tonic and phasic units in the proximal and the distal scoloparium, respectively. The phaso-tonic units respond to velocity and position of the claw, and the phasic units respond to movement of the joint (Wiese, 1972; Wiese and Schmidt, 1974).

6.3 Sensory and behavioural thresholds

For sensory systems, the threshold is an important characteristic and documents the sensitivity of a sensory organ. The sensory threshold can be determined as the lowest stimulus amplitude that can still elicit a spike discharge in the sensory neurons. Usually, stimuli of different amplitudes are tested for a specific frequency, and a given amplitude is tested repeatedly for a consistent response. Testing over a frequency range allows one to determine a threshold curve from the highest sensitivity at different frequencies. Importantly, the sensitivity can be determined physiologically for individual sensilla or sensory organs, but also for the organism by the threshold to stimuli at which a behavioural response is initiated.

In vibrosensory organs, the individual scolopidial sensilla are usually tuned to a characteristic best frequency where they show the lowest threshold, whilst thresholds of both lower and higher frequencies rise (Kalmring et al., 1994; Kühne, 1982a; Fig. 4). However, certain other sensilla may lack such a tuning to a best frequency (Kühne, 1982a). In summed recordings, the combined responses of several sensilla form the threshold curve. Here, the recorded thresholds can be relatively homogenous over a frequency spectrum and not show a clear best frequency (Kalmring et al., 1994; Strauss and Lakes-Harlan, 2008a).

Particularly interesting is the absolute sensitivity for vibrational stimuli, both for different sensory organs found in one species, and between different species, to reveal specific sensory adaptations. The absolute threshold of insect sensory organs to vibration stimuli has been a contentious topic, as some studies found responses to notable low amplitudes at a fraction of a nanometre (Autrum and Schneider, 1948; Schnorbus, 1971). This sensitivity was addressed in a rigorous methodological study by Shaw (1994b), which corrected the neurophysiological thresholds from the SGO in the cockroach (*Periplaneta americana*) hindleg to a minimum displacement of 0.2 nm.

Comparing thresholds between insect species shows that the vibration sensitivity varies largely (Table 1). The possible influence of leg dimensions, body morphology, or body mass on vibrational sensitivity can be studied systematically. For the absolute sensitivity of sensory organs, the stimulus direction is a relevant parameter: when compared between different directions of stimulation, the stimulation in the longitudinal direction of the tibia (or vertical direction) correlates with higher sensitivity of vibration receptor organs over horizontal stimulation (Rohrseitz and Kilpinen, 1997; Strauß and Lakes-Harlan, 2017). This finding is plausible from a more efficient energy transmission along the longitudinal leg axis, as well as the position of insects placed on a substrate. Usually the insect body axis is oriented in parallel to the substrate, whilst the bending waves transmitted over leaves are perpendicular to the insect body, and the tibia of the standing leg is at least slightly tilted in the vertical axis or is standing in parallel to the vertical axis (Čokl and Virant-Doberlet, 2003; Stritih-Peljhan and Strauß, 2018; Tsubaki et al., 2014).

One aspect that is largely lacking in investigations so far is the test for sensitivity to vibrational signals in the natural habitat. This requires preparations with electrodes for electrophysiological recordings from sensory organs inserted into the leg, whilst insects are placed or can even move on soil or plants. The detection of vibrational signals transmitted on plant leaves and stems has been confirmed by such recordings only for two ensiferan species (McVean and Field, 1996; Römer et al., 2010).

With respect to behaviour, a high sensitivity is adaptive both for intraspecific communication like mate detection, and for predator or prey detection. In the context of predator detection, insects can show a freezing response or startle response (Fig. 5). So far, behavioural thresholds are reported only for few insect species, and mostly cover freezing and startle responses. Such threshold curves have been reported in honeybees (Michelsen, 2014; Michelsen et al., 1986a; Fig. 5A), locusts (Friedel, 1999), and beetles (Takanashi et al., 2016; Tsubaki et al., 2014; Fig. 5B). For startle and freezing responses, the behavioural threshold curves usually show particular responsiveness to low frequencies (<500 Hz) (Friedel, 1999; Michelsen et al., 1986a; Takanashi et al., 2016; Tsubaki et al., 2014; von Dohlen and Dambach, 1981), depending on the acceleration parameter (Michelsen et al., 1986a). Freezing responses in bees occur for vibrations of the comb at lowest thresholds of $5\,m/s^2$, and adequate amplitudes cause all bees placed on the comb to remain motionless (Michelsen et al., 1986a; Fig. 5A). In the beetle *Monochamus alternatus*, the behavioural thresholds for the freezing response increase with frequency. Whilst the lowest reactions occur in response to 25 Hz stimuli at $3.5\,m/s^2$, thresholds over 50 Hz raise above 10 to even $>100\,m/s^2$ (Takanashi et al., 2016; Fig. 5B). In the carpenter ant, *Camponotus herculeanus*, sine wave vibrations can induce the insects to stop or to run, depending mainly on the stimulus amplitude (Fuchs, 1976). These behaviours are triggered most effectively by stimuli between 250 and 750 Hz, with only higher amplitudes effective in other frequency ranges. The stop reaction was shown in response to stimuli between 15 and 1000 Hz at thresholds of $0.05–10\,m/s^2$, whilst the run reaction was shown in response to any frequency between 5 and 10,000 Hz at extremely high thresholds (larger than $50\,m/s^2$; Fuchs, 1976). Notably, the stimuli evoking freezing responses are detected by different sensory structures (SGO, FeCO, campaniform sensilla), and often have relatively high thresholds compared to physiological sensitivities of the SGO in particular (Michelsen et al., 1986a; Takanashi et al., 2016, but see Dambach, 1989).

The different effectiveness between vertical and horizontal direction of vibration has also been shown in the freezing response of honeybees, which is consistent with the higher sensitivity to vertically applied stimuli (Rohrseitz and Kilpinen, 1997). In the longicorn beetle *Paraglenea fortunei*, different vibrations of biotic and abiotic origin were presented to beetles in combinatory settings on a horizontal or vertical substrate and whilst walking or resting (Tsubaki et al., 2014). The thresholds for walking and resting beetles were similar in tuning, and the lowest behavioural thresholds were generally found at frequencies of 100 Hz or lower. However, the absolute sensitivities differed, and on both substrate orientations, the threshold was lower in walking beetles. This was interpreted as a sequence of freezing and startle responses due to intermittent vibration detection (Tsubaki et al., 2014). The thresholds were lowest for walking on the horizontal substrate at $\sim 2\,m/s^2$ (100 Hz) (Tsubaki et al., 2014), whilst for the vertical substrate and the resting on the horizontal substrate, thresholds did not drop below $10\,m/s^2$. For these combinations threshold curves were similar. The different frequency profiles of wind, or conspecifics walking and landing, possibly allow beetles to distinguish biotic and abiotic signals (Tsubaki et al., 2014; see also Guedes et al., 2012 for vibration discrimination in caterpillars).

Vibration detection can also be indicative for resources like food availability. Tenebrionid beetles from the Namib Desert can detect wind over the sand surface. Wind leads to their emergence from the sand as it exposes detritus on which the beetles feed. The vibration sensitivity was also higher (more sensitive) in desert species than in species from a river bed habitat, which had more vegetation and less wind (Hanrahan and Kirchner, 1994).

Onymachris plana showed movement responses to substrate vibrations between 100 and 500 Hz, with highest sensitivity of $0.1\,m/s^2$ at 500 Hz (Hanrahan and Kirchner, 1994). Antlions detect prey by substrate vibrations caused from movements (reviewed in Devetak, 2014; Scharf et al., 2011) with lowest sensitivities below $0.1\,m/s^2$ between 200 and 500 Hz in *Myrmeleon formicarius* (Devetak, 1985). Arena experiments using *Myrmeleon inconspicuous* with artificial vibration stimuli documented the antlions' behavioural responses to very low amplitude displacements at 0.1 nm (Martinez et al., 2020).

Thresholds for social signals are documented from only a few species, but are especially well studied in termites (Kirchner, et al., 1994;

Röhrig et al., 1999). In the damp-wood termite, *Zootermopsis nevadensis*, acceleration thresholds are rather uniform at ~1–2 m/s^2, but displacement thresholds decrease with frequency below 0.01 μm p-p at 5 kHz (Kirchner et al., 1994). In the fungus-growing termites, *Macrotermes bellicosus*, *M. subhyalinus* and *M. natalensis*, the behavioural responses show much higher sensitivity to acceleration amplitudes with thresholds at 0.012–0.02 ms^2 at 500 Hz (Hager and Kirchner, 2013; Röhrig et al., 1999).

Ideally, both sensory and behavioural thresholds can be compared, which enables an insight into physiological mechanisms and adaptive significance of the behavioural response by sensitivity matching between these. For vibration detection, this was done for the water strider, *Notonecta glauca* where prey capture depends on the tarsal chordotonal organ (Wiese, 1972). For both prey capture and the sensory threshold of this chordotonal organ, the threshold between 50 and 200 Hz are very similar. Notably, for vibration frequencies between 10 and 50 Hz, the behavioural threshold is slightly lower than that of the sensory organs measured in any of the three leg pairs (Wiese, 1972; Fig. 5C). This could be due to input summation from different receptors detecting vibrations in the low-frequency range. A similar mismatch between physiological and behavioural thresholds was found in the termite, *Zootermopsis angusticollis*, where physiological thresholds were recorded from the leg and behavioural thresholds were obtained for the avoiding reaction, in which the insect reverses direction upon stepping from a non-vibrating to a vibrating surface (Howse, 1964). Generally, physiological thresholds are known to be extremely sensitive to vibrations, and more studies should address the relationships to behavioural outputs. Comparing previously recorded sensory thresholds (Autrum and Schneider, 1948) with the behavioural thresholds for the stop reaction of carpenter ants revealed behavioural responses to be more sensitive than physiological responses by a factor of 10 (Fuchs, 1976). Importantly, however, these comparisons are only relevant when the same, or at least similar, experimental settings are used across studies (Strauß et al., 2019).

The vibration receptors in insects are notably sensitive (Shaw, 1994b). Often, behavioural thresholds are higher in stimulus amplitudes than the thresholds recorded from receptor units (Hanrahan and Kirchner, 1994), as indicated above. In certain very sensitive insect species, behavioural thresholds at displacement of 1–2 nm were noted (Röhrig et al., 1999), or even below 1 nm (Eriksson et al., 2011; Martinez et al., 2020). The discrepancy between behavioural and physiological thresholds may depend on the integration of sensory inputs from six legs in the central nervous system, which decreases the threshold due to summation. Certainly, the behavioural and physiological responses will be relevant to compare under controlled settings, and would benefit from including threshold responses of vibratory interneurons, as well.

7. Central neural processing of vibrational information

From the vibrosensory organs in the legs, axons of receptor neurons project into the segmental ganglia of the ventral nerve cord, where they connect synaptically with interneurons. There are different neuropile regions associated with synaptic integration and processing of vibrosensory information in thoracic ganglia. Their output is directed both locally to sensory-motor neuropiles influencing behaviour via reflex pathways, and via the ascending pathway for more complex evaluation in the brain.

After rapid advances in understanding insect vibrosensory processing (initially studied along with hearing) at the end of the previous century, only two insect species remained under investigation for various aspects of the vibratory sense, including central processing: the cave cricket, *Troglophilus neglectus* (Orthoptera: Rhaphidophoridae), and the stink bug, *N. viridula* (Hemiptera: Pentatomidae). The former is thought to belong to ancestrally atympanate Orthoptera (Ensifera; Song et al., 2020; Strauß et al., 2014; Stritih and Stumpner, 2009) and is therefore also relevant as a comparative system to study evolution of acoustic-vibratory systems in this insect lineage. In contrast to *T. neglectus* that only exhibits simple, short-range vibratory signalling by males (Stritih and Čokl, 2012), *N. viridula* relies on elaborate bidirectional vibrational communication in long-range attraction, mate choice, and rivalry. This makes it ideal for studying various aspects of vibratory processing involved in communication, from filtering of conspecific temporal patterns to encoding of directionality (Čokl, 2008; Čokl et al., 2014). This section describes the central findings of these studies in Orthoptera and stink bugs, attempts to draw some generalizations, and postulates hypotheses, often in the light of similarities and differences to the closely related sense of hearing. The overall scarcity of studies being conducted currently calls for increased study efforts focusing on central processing of vibratory information in insects.

7.1 Central sensory projections and organization of input areas

The morphological and physiological diversity of insect vibrational sensilla (see Section 4.1) also relates to termination of their sensory neurons in different neuropile regions of the ventral nerve cord. Neuropiles in insect ganglia are the centres of synaptic interactions between neurons, and different neuropiles are generally associated with different functional specializations (e.g., Smarandache-Wellmann, 2016). Campaniform sensilla, hair sensilla, and joint chordotonal organs, such as the distal/ventral scoloparium of the femoral chordotonal organ (FeCO) in Orthoptera, project primarily into ventral- and lateral-most neuropile regions of the segmental ganglia (Field and Pflüger, 1989; Hustert et al., 1981; Mücke and Lakes-Harlan, 1995; Nishino, 2000; Pflüger et al., 1988). These sensilla are primarily proprioceptive, and their function in vibration detection may be mainly associated with reflex responses to low-frequency vibration of rather high intensity (Friedel, 1999; Stritih and Čokl, 2014; Takanashi et al., 2016). The subgenual organ (SGO) or the subgenual organ complex (SGOC) in orthopteroid insects, and at least part of the FeCO in some taxa (like its dorsal/proximal scoloparium in Orthoptera) are considered specialized detectors of substrate vibration. These project segmentally into the neuropile called the medial ventral association centre (mVAC; Field and Pflüger, 1989; Nishino, 2000, 2003; Nishino and Field, 2003; Strauss and Lakes-Harlan, 2008a,b; Strauß et al., 2014; Stritih and Stumpner, 2009; Fig. 6). This neuropile area at the intermediate position in the ganglion may be generally associated with processing of mechanosensory information related to complex behaviours. It also contains projections of tympanal (auditory) organs and their atympanate (often vibration-sensitive) chordotonal precursors (Boyan, 1993; Lakes-Harlan et al., 1999; Strauss and Lakes-Harlan, 2008a,b; Stölting et al., 2007). Thus, in cases where data on the sensory physiology of an atympanate CO are lacking, projections in the mVAC are considered highly indicative of its vibrosensory function (as is the case with the dorsal FeCO scoloparium in crickets and weta; Field and Pflüger, 1989; Lakes-Harlan and Strauß, 2014; Nishino, 2000, 2003).

Vibrosensory receptors of the SGO/SGOC may also have terminations laterally to and posteriorly to the morphologically well delimited mVAC area (Kalmring et al., 1996; Nishino and Field, 2003; Stritih-Peljhan et al., 2019), formed of dense receptor and interneuron arborizations. In this regard, two adjacent neuropile regions have been described as targets of FeCO and SGO projections in weta; the mVAC anteriorly, and the more diffuse "posterior medial neuropile" (Nishino, 2003; Nishino and Field, 2003). In the atympanate *T. neglectus*, this entire semi-circular area, entered by axons of the SGOC, shows no obvious morphological delimitation between the anterior and posterior parts (Stritih and Stumpner, 2009; Fig. 6D) and has been termed the "ring tract" (RT, after Tyrer and

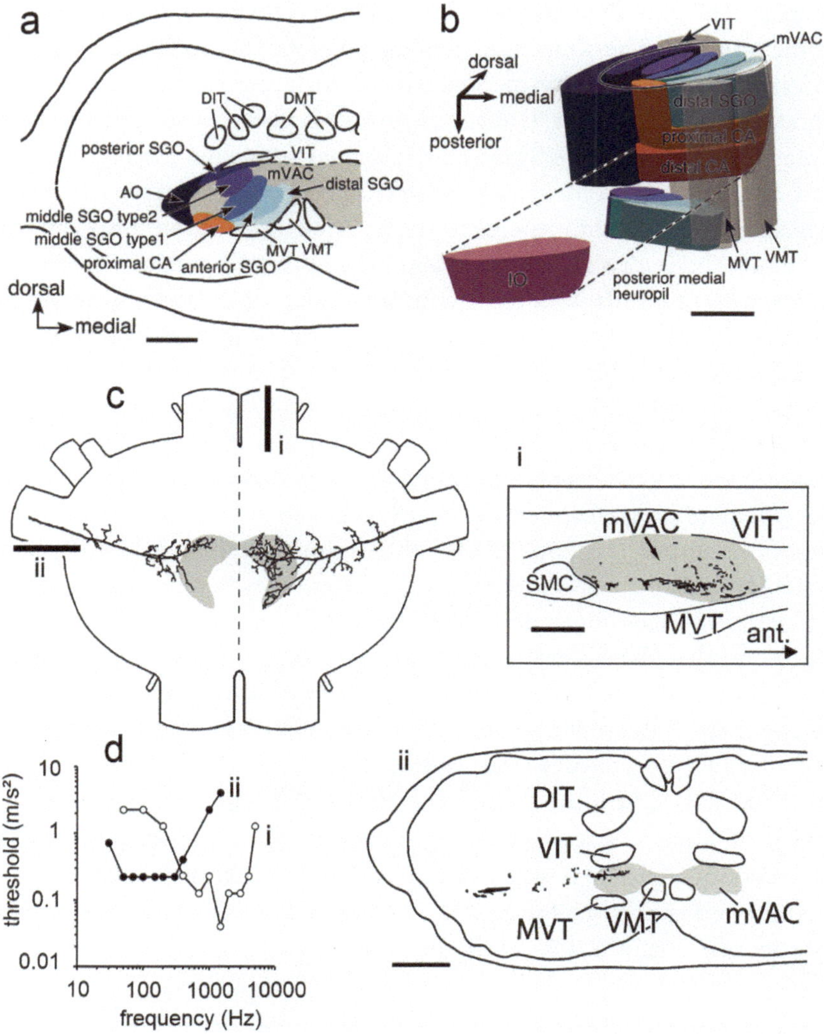

Fig. 6 Sensory projections into the medial ventral association centre in Ensifera. Projection areas of different receptor neuron subgroups from the complex tibial organ of the weta, *Hemideina femorata* (A) in the transverse section of prothoracic ganglion at the level of the medial ventral association centre (mVAC; level as in C-i), and (B) in the three-dimensional reconstruction of the mVAC. Subgenual organ and accessory organ in graded blue, intermediate organ and *crista acustica* projections in graded red. Except for *crista acustica*, other receptors detect substrate vibration. (C) Central projections of a high-frequency and a low-frequency vibratory receptor neuron in the cave cricket, *Troglophilus neglectus*, in the wholemount view of the prothoracic ganglion and in the neuropile in the longitudinal section (i; with the detail of the mVAC) and in the transverse section (ii), as indicated in the wholemount. (D) Threshold tuning of the same two

(Continued)

Gregory, 1982), with its anterior-most, midline-crossing part corresponding to the mVAC. The functional homogeneity of this region in vibration processing is indicated by the dendritic arborizations of many first-order vibratory interneurons in *T. neglectus*, covering the entire range of receptor projections (Stritih, 2009; Stritih and Stumpner, 2009; see also Fig. 7i, ii, iv).

Characteristically, the sensory projections terminating in the mVAC are spatially segregated with respect to their peripheral location, providing the anatomical basis for a functional subdivision of this neuropile. The division has been most clearly associated with processing of different stimulus frequencies in the auditory system of bushcrickets, based on the peripheral tonotopic organization of the sensilla within the linearly arranged sensory organ (Stumpner, 1996; Stölting and Stumpner, 1998). As shown in crickets and weta, also the axons of the vibration-sensitive FeCO, SGO, and the bimodal sound- and vibration-sensitive intermediate organ (IO) receptors are topographically mapped in the mVAC and its posteriorly associated sensory neuropile (Nishino, 2000, 2003; Nishino and Field, 2003). Small subsets of sensory cells associated with different structures and tissues in the periphery have been shown to have different central arrangements, suggesting that these sensilla may also differ physiologically in frequency and threshold sensitivity (Nishino and Field, 2003; Fig. 6A–C). Morphological and physiological data on individually recorded and stained vibration receptors in cave crickets generally support this hypothesis (Stritih and Čokl, 2014; Stritih-Peljhan et al., 2019; Fig. 6D), but definitive evidence of soma location in the periphery by complete staining of vibrosensory neurons is still lacking. A similar functional subdivision of receptor projections in the mVAC has been recently shown for various chordotonal organs (COs) of the body and legs of stink bugs and cicadas

Fig. 6—Cont'd receptors, presumed to originate in the intermediate and accessory organ, respectively. Abbreviations: AO, accessory organ; CA, *crista acustica*; CTO, complex tibial organ; DIT, dorsal intermediate tract; DMT, dorsal medial tract; IO, intermediate organ; mVAC, medial ventral association centre; MVT, medial ventral tract; SGO, subgenual organ; SMC, supramedian commissure; VIT, ventral intermediate tract; VMT, ventral medial tract. *Panels (A, B): Reprinted from Nishino, H., Field, L.H., 2003. Somatotopic mapping of chordotonal organ neurons in a primitive ensiferan, the New Zealand tree weta* Hemideina femorata*: II. Complex tibial organ. J. Comp. Neurol. 464, 327–342, with permission from Wiley. Panels (C, D): Adapted from Stritih-Peljhan, N., Rühr, P.T., Buh, B., Strauß, J., 2019. Low-frequency vibration transmission and mechanosensory detection in the legs of cave crickets. Comp. Biochem Physiol. A 233, 89–96 and new data analysis.*

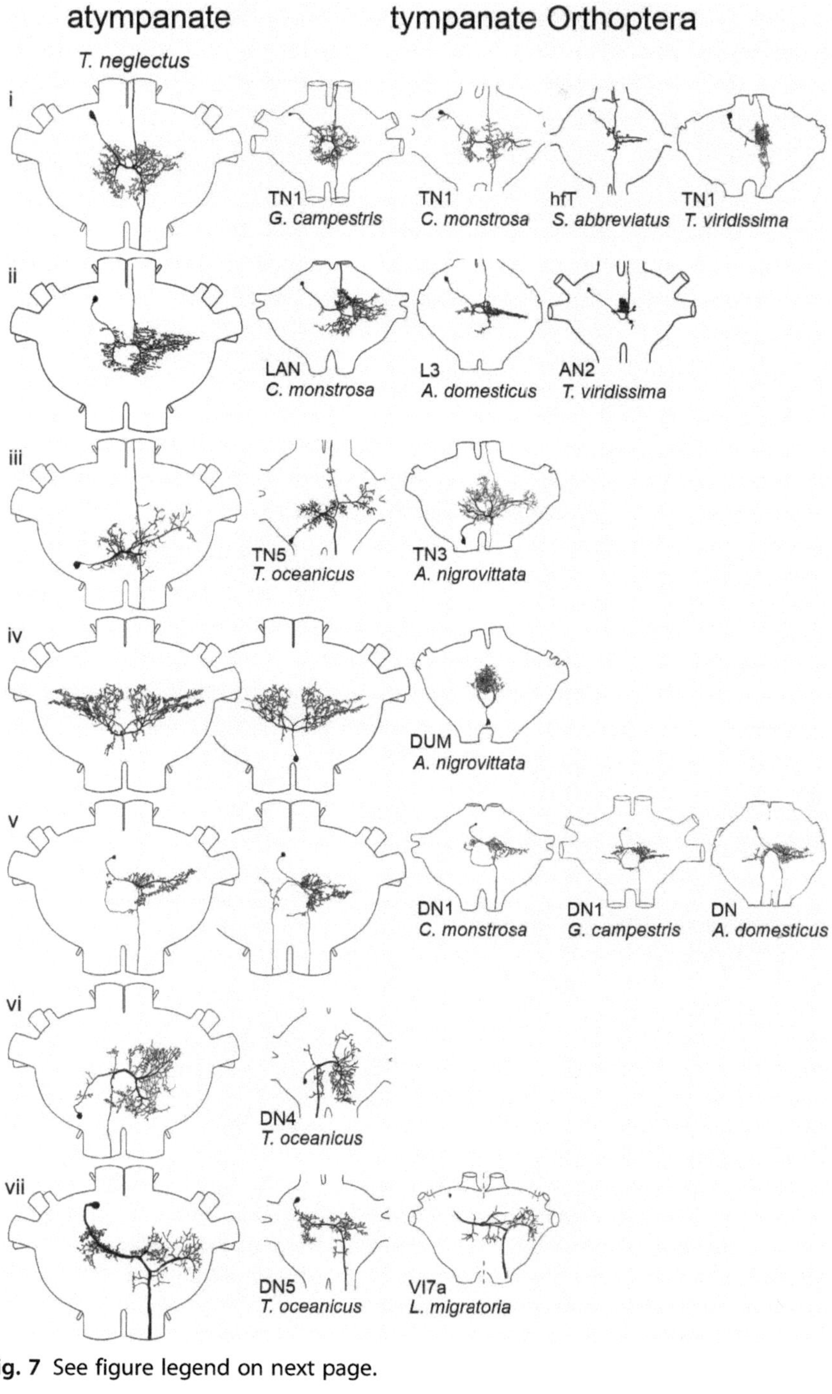

Fig. 7 See figure legend on next page.

(i.e., FeCO, SGO, Johnston's organ, tymbal CO, pleural CO, and ventral CO of the abdomen; Nishino et al., 2016). In crickets, weta, and cicadas, the projections of vibratory (FeCO/SGO) and auditory organ afferents show almost no spatial overlap; auditory receptors terminate most ventrally, whereas vibratory inputs terminate more dorsally or laterally in/to the mVAC (Nishino, 2000, 2003; Nishino and Field, 2003; Nishino et al., 2016; Fig. 6A–C). The most dorso-lateral parts of mVAC were suggested to serve processing the stimuli of lowest frequency (Nishino et al., 2016), which is consistent with the anatomical location of low-order vibratory interneurons and sensory receptors tuned to different frequencies in *T. neglectus* (Stritih, 2009; Stritih and Stumpner, 2009; Stritih-Peljhan et al., 2019; Fig. 6D). These data suggest a rather conserved organization of the insect RT/mVAC neuropile with respect to vibratory processing, which seems to include at least a rough "vibrotopic" organization; dorsal parts seem to be related to more low-frequency and ventral parts to more high-frequency vibration processing. Such subdivision of inputs not only enables separate processing and perception of spectrally diverse signals, but also provides an anatomical basis for differential adaptation to such signals in interneurons that integrate information over a wide range of frequencies. Such "stimulus-specific adaptation" of individual interneurons was demonstrated in the auditory system of Ensifera, and is based on cellular processes restricted to specific parts of the dendritic segment (Prešern et al., 2015; Schul et al., 2012). It would be interesting to investigate whether such adaptation also takes places in the insect vibratory system.

Fig. 7 Homology of sound- and/or vibration-sensitive prothoracic interneurons of Orthoptera. Putative homologues between the atympanate cave cricket *Troglophilus neglectus* (left, larger ganglia) and various species of tympanate Orthoptera (right, smaller ganglia, with neurone labels and species affiliation) are shown in the same line. The sets of homologues in (i), (ii) and (iv) indicate evolutionary changes in morphology (progressively from left to right) following auditory specialization from a vibrosensitive precursor neuron. Other sets of neurons are morphologically, physiologically and most likely also functionally preserved in the different species to a high degree (in the tympanate species they are only weakly responsive to sound and/or respond to vibratory stimuli, as well). *Gryllus campestris*, *Acheta domesticus* and *Teleogryllus oceanicus* are crickets (Gryllidae), *Scapteriscus abbreviatus* is a mole cricket (Gryllotalpidae), *Cyphoderris monstrosa* is a hump-backed cricket (Propalangopsidae), *Tettigonia viridissima* and *Ancistrura nigrovitatta* are bushcrickets (Tettigoniidae; all Ensifera) and *Locusta migratoria* is a grashopper (Acrididae; Caelifera). *Adapted from Stritih, N., Stumpner, A., 2009. Vibratory interneurons of the non-hearing cave cricket indicate evolutionary origin of sound processing elements in Ensifera. Zoology 112, 48–68.*

7.2 Vibratory interneuron types, numbers and interspecific homologies

The thoracic vibrosensory network appears to be a complex system with a large number of constituent interneurons. This may be especially true for orthopteroids, which possess complex vibrosensory organs containing a large number of sensory cells. In general, various local (intrasegmental) and intersegmental interneurons are found in the ventral nerve cord of insects, conveying information within and between segments, the latter in the anterior, posterior, or both anterior and posterior directions (i.e., ascending, descending and T-shaped interneurons, respectively). We are far from knowing, probably, even the majority of its constituent interneuron elements, which, however, may be considered to exist as homologue neurons in at least closely related taxa (Fig. 7).

In the tympanate Orthoptera, at least ten different physiological interneuron types with a considerable excitatory vibratory input are described ascending towards the brain. Five of them responding preferentially to vibratory stimuli (V1-V5) and five responding in a similar degree to vibratory and acoustic stimulation (VS1-VS5 neurons) are known from *Tettigonia cantans*, *Decticus verrucivorus* and *Locusta migratoria* (Kühne, 1982b; Kühne et al., 1980; Silver et al., 1980). Kalmring et al. (1997b) reported at least six functional types of V neurons ascending towards the head ganglia of *Polysarcus denticauda*, and Čokl et al. (1985) recorded in *L. migratoria* from several VS neurons that did not correspond to any of the previously described types. The variability in physiological responses, together with testing several units in the same preparation, often indicated that more than one neuron of each type was present at the same side of the ventral nerve cord, but firm conclusions could not be made based on a physiological basis alone. For V3 neurons, for example, response variability suggested at least two or three coexisting units (Čokl et al., 1985; Kalmring et al., 1997a,b; Kühne, 1982b), whilst additional morphological data indicated at least five distinct ascending elements of the V3 type from different segments of the ventral nerve cord (Bickmeyer, 1991; Sickmann, 1997).

Only a fraction of these interneurons, mostly with pro- and mesothoracic origins, has been identified morphologically in the hearing species of Orthoptera (note that hindleg stimulation was only rarely included in the experimental settings; see Section 5.1, for details). Considering morphological data at least from the segment hosting the soma and the main dendritic region of a neuron, seven vibration-sensitive neurons are known from bushcrickets, with four from the prothoracic (including one ascending, descending, local and T-shaped neuron, each) and three from the mesothoracic ganglion (with one T-neuron and two ascenders; Römer, 1985; Rössler et al., 2006; Sickmann, 1997). From crickets, the morphology of only one vibration-sensitive prothoracic neuron is known (Kühne et al., 1984; together with two specialized auditory neurons inhibited by vibration; Kühne et al., 1984; Wiese, 1981). Seven further neurons were described for locusts, including four with prothoracic (i.e., one T-shaped and three ascending neurons), two with mesothoracic (one T-neuron and one ascender; Bickmeyer, 1991; Bickmeyer et al., 1992) and one with metathoracic origin (i.e., an ascending neuron confined to the thorax; Čokl et al., 1985).

In search of a presumed homology between the vibratory and auditory systems of Ensifera at the level of central neuronal processing, 26 morphologically distinct interneurons with soma and dendritic region in the prothoracic ganglion were described in the primitively non-hearing *T. neglectus* (most likely representing at least 30 coexisting neurons; Stritih, 2006, 2009; Stritih and Stumpner, 2009). Importantly, the study connected several of the vibration- and/or sound-sensitive interneurons known from different species of Orthoptera as putative homologues, or at least as originating from homologous neuron clusters, to a set of neurons now being described from a single species (Stritih and Stumpner, 2009; Fig. 7). In addition to indicating the origin of ensiferan auditory interneurons from the pre-existing vibratory system, the analysis of homologous neurons revealed a high degree of morphological and physiological similarity of the counterparts in both hearing and non-hearing Orthoptera, also indicative of their similar implementation in the network (Fig. 7). Responsiveness to wind of some of these vibration- and/or sound-sensitive neurons (tested by gently blowing at cerci; Stritih and Stumpner, 2009) and/or light (visual) stimuli (Atkins and Pollack, 1987), and morphological resemblance of some of them to multimodal thoracic interneurons of the cockroach escape system (Ritzmann et al., 1991), raises the question of neuronal specialization, which so far has not been tested systematically.

Morphology and physiology of a considerable number of vibratory interneurons has also been described in stink bugs (Čokl and Amon, 1980; Prešern, 2007; Prešern et al., 2018; Zorović, 2011; Zorović et al., 2008). These include at least nine distinct neuron types from the prothoracic ganglion (with five T-neurons, 3 descenders and one local neuron) and 12 from the fused meso- and metathoracic neuromeres of the central ganglion (with eight ascending and four local neurons; see Fig. 8, for some examples). Notably, despite having many fewer receptor elements in the specialized vibrosensory organs in the legs of stink bugs (Michel et al., 1983; Nishino et al., 2016) compared to Orthoptera, stink bugs possess a complex central neural network to process vibration stimuli. Comparison of these elements with the sound- and/or vibration-activated interneurons of orthopteroids does not reveal a high degree of morphological similarity indicative of neuronal homology (but see Fig. 8, for a likely example). This raises the question of the common origin of the insect vibratory systems in general (see also Lakes-Harlan and Strauß, 2014). Alternatively, the differences could reflect different degrees of specialization of homologous circuits diversified by adaptive radiation of these relatively distant insect taxa. Nevertheless, there are local and intersegmental neurons with similar topology of input and output regions in Orthoptera and stink bugs, reflecting similar functional adaptations of their vibratory neuron networks (Fig. 8A and F).

7.3 Vibratory processing mechanisms

7.3.1 Intra- and intersegmental input integration

In the ventral nerve cord, important pre-processing of information with integration of input from various legs and segments takes place before information ascends to the brain. In ascending interneurons passing the cervical connectives of tympanate Orthoptera, all six legs usually contribute to the response, but the characteristic sensitivity peak and response pattern is primarily a result of the input from a particular leg or leg pair. This typically includes both excitatory and inhibitory inputs (Čokl et al., 1977, 1985; Dambach, 1972a,b; Kühne, 1982b; Kühne et al., 1980, 1984; Virant-Doberlet et al., 2006). The dominant input relates in most cases to the specific functional topography of neurons, i.e., the location of the dendritic (postsynaptic) region in the segment or hemiganglion providing the strongest input, as later confirmed at the level of morphologically identified neurons both in Orthoptera and stink bugs (Bickmeyer, 1991; Bickmeyer et al., 1992; Prešern, 2007; Sickmann, 1997; Stritih, 2009; Stritih and Stumpner, 2009). Contributions from other legs, despite being quantitatively minor, are, however, crucial for the qualitative change in the response, reflecting amongst others the timing of leg stimulation and thus the direction of incoming stimuli (Čokl et al., 1985; Prešern, 2007; Prešern et al., 2018; see also Section 7.3.2).

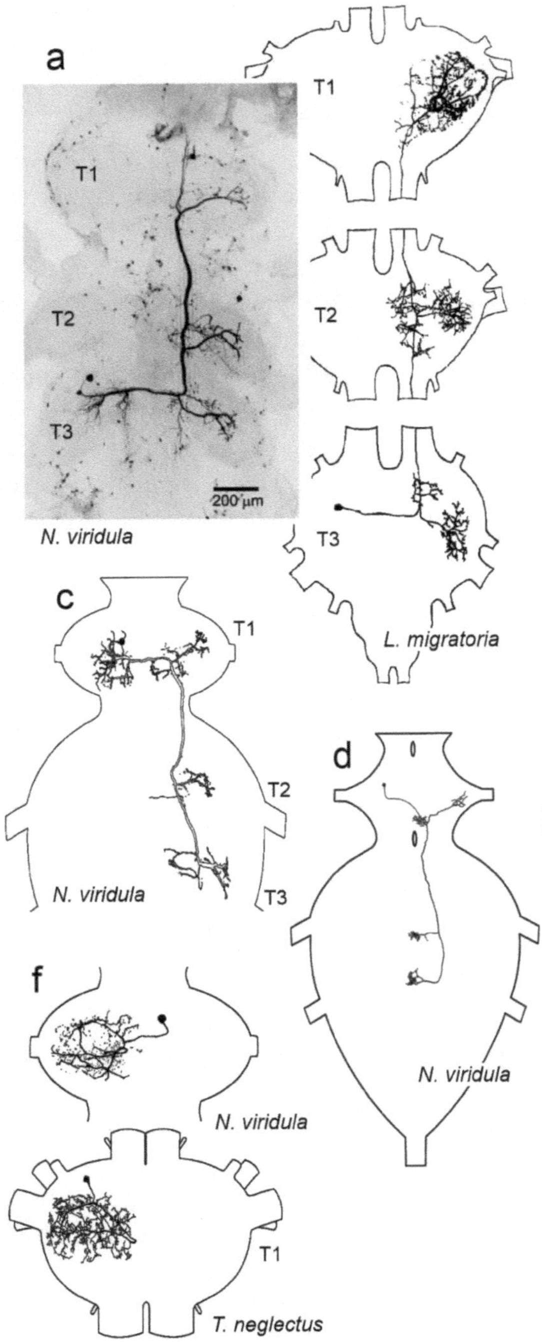

Fig. 8 See figure legend on opposite page.

The largest number of neurons, morphologically identified from one neuromere, have been examined for separate ipsi- and contralateral inputs in the cave cricket *T. neglectus*. The strength of excitatory and (frequency-specific) inhibitory inputs received via each leg was roughly correlated to the extent of dendritic branches in a given hemiganglion (Stritih, 2006, 2009; Stritih and Stumpner, 2009). This suggests that intrasegmental integration

Fig. 8 Functional topology of vibratory interneurons; examples from *Nezara viridula* with some interspecific comparison. (A, B) Ascending neurons, (C, D) descending neurons, (E) T-shaped neurons, (F, G) local neurons. In (A) and (F), neurons of similar functional typologies to that of *N. viridula* are shown from the orthopterans *Locusta migratoria* and *Troglophilus neglectus*. Whilst the former could be homologous (due to similar soma location and primary structure), the latter are clearly non-homologous (note the different soma location), though functionally highly similar neurons, connecting dorsal and ventral sides of the hemiganglion, also with a similar frequency sensitivity in both species. In (A) and (C), location of dendritic and axonal branches indicates input from the soma-ipsilateral leg and extensive output contralaterally in the dorso-lateral (i.e., sensory-motor) regions of all thoracic segments, whilst the two neurons transmit information in the opposite directions. An ascending neuron in (B) integrates input bilaterally from mid- and hindlegs (note the dendritic extensions towards individual legs), and the descending (D) and the T-shaped neuron (E) from one and both forelegs, respectively, whilst connecting to other vibratory interneurons intersegmentally (as indicated by their medial axonal arborizations in the thorax). The local neuron in (G) mediates information from the soma-ipsilateral (dendritic) to the soma-contralateral (axonal) side of the same segment. Abbreviations: T1, T2, T3: prothoracic, mesothoracic, and metathoracic ganglion or segment of the fused ganglia. *Panels (A) (*N. viridula*), (D), and (G): from Zorović, M., Prešern J., Čokl, A. 2008. Morphology and physiology of vibratory interneurons in the thoracic ganglia of the Southern Green Stinkbug* Nezara viridula *(L.). J. Comp. Neurol. 508, 365–381, with permission from Wiley. Panel (A) (*L. migratoria*): Reprinted by permission from Springer: Čokl, A., Otto, C., Kalmring, K., 1985. The processing of directional vibratory signals in the ventral nerve cord of* Locusta migratoria. *J. Comp. Physiol. A 156, 45–52. Panel (B): From Zorović, M., 2005. Morphological and physiological properties of vibrational neurones in thoracic ganglia of the stinkbug* Nezara viridula *(Heteroptera: Pentatomidae). Doctoral dissertation, University of Ljubljana, Ljubljana, Slovenia with permission. Panels (C) and (F) (*N. viridula*): from Prešern, J., 2007. Neurobiological fundamentals of orientation of the stink bug* Nezara viridula *(L.) (Heteroptera: Pentatomidae) towards the source of vibration signal. Doctoral dissertation, University of Ljubljana, Ljubljana, Slovenia with permission. Panel (E): Adapted from Prešern, J., Polajnar, J., de Groot, M., Zorović, M., Virant-Doberlet, M., 2018. On the spot: utilization of directional cues in vibrational communication of a stink bug. Sci. Rep. 8, 5418, published under the Creative Commons CC BY license. Panel (F) (*T. neglectus*): from Stritih, N., 2009. Anatomy and physiology of a set of low-frequency vibratory interneurons in a nonhearing ensiferan (*Troglophilus neglectus*, Rhaphidophoridae). J. Comp. Neurol. 516, 519–532 with permission from Wiley.*

of excitatory inputs in the vibratory system is strictly ipsilateral and thus any commissural neurons with a clear lateralization of input and output sites are inhibitory. Such putatively inhibitory local and bisegmental neurons in the fused meso- and metathoracic neuromeres of the central ganglion have been identified in *N. viridula* (Zorović et al., 2008; Fig. 8G).

In a comparative study of the combined auditory-vibratory system of bushcrickets and locusts, based on a large number of individually recorded interneurons ascending to the brain, similar functional types of predominantly vibration-sensitive neurons ("V"), and both vibration- and sound-sensitive ("VS") neurons, were found in the two insect groups (Kühne, 1982b). However, neurons of the same type were found to differ in their connectivity to vibrational receptors from different legs; in locusts, most neurons received dominant inputs from the midlegs, whilst in bushcrickets most input came from the ipsilateral foreleg (thereby suggesting that structurally different, non-homologous neurons were considered in the two groups). Whilst this difference in input to bimodal VS neurons is clearly related to the different organization of the auditory systems in the two groups, with prothoracic vs. meso- and metathoracic receptor projections, it is unclear why the difference in the auditory input topography should also influence the input to (purely) vibratory neurons. Indeed, Čokl et al. (1985) later found in *L. migratoria* V-type interneurons with input from various leg pairs, and data from identified neurons in Orthoptera suggested no such obvious taxon-related differences in the weighting of vibratory inputs to different nerve cord segments.

In fact, the largest number of ascending vibratory interneurons (in all taxa) can be expected to originate from the metathoracic neuromere, presumably transmitting information mostly in the anterior direction towards other thoracic segments and the brain. However, since hindlegs most often were not stimulated in studies of Orthoptera, only a single neuron from the metathoracic segment has been morphologically identified to date (Čokl et al., 1985; Fig. 8A—*L. migratoria*). In the stink bug *N. viridula*, on the other hand, several morphological types of neurons are known to ascend from the metathoracic neuromere of the central ganglion into the prothoracic ganglion, the subesophageal ganglion, and/or the brain (Prešern et al., 2018; Zorović, 2011; Zorović et al., 2008). In particular, in those that integrate information bilaterally, the dendrites typically extend across both meta- and mesothoracic neuromeres, consistent with mid- and hindleg input, and apparently an adaptation to the fusion of the respective segments (Fig. 8B).

At the same time, of the identified prothoracic vibratory interneurons, only a small proportion send information anteriorly towards the brain, another indication of the importance of intersegmental input integration in the thorax. Of the 22 distinct intersegmental neurons described in the cave cricket, only four are ascending neurons, whilst others have an additional or an exclusively posteriorly projecting axon (6 and 12 neurons, respectively; Stritih, 2006, 2009). Similarly, amongst the 10 morphological types of intersegmental prothoracic interneurons known from the stink bug, there are some T-shaped but no exclusively ascending units (Čokl and Amon, 1980; Prešern, 2007; Zorović et al., 2008). Most of these neurons show terminations of descending axons in both meso- and metathoracic neuromeres (Fig. 8C–E).

Several other candidate neurons for intersegmental integration, with dendritic segments in one thoracic neuromere or hemiganglion and with extensive axonal terminations in other neuromeres, were described morphologically amongst Orthoptera (Bickmeyer et al., 1992; Čokl et al., 1985; Sickmann, 1997) and stink bugs (Zorović, 2011; Zorović et al., 2008; Fig. 8). In the locust, *L. migratoria*, simultaneous recordings from different neuronal sites together with neuronal anatomy showed two low-order vibratory interneurons to be connecting vibrational receptors with ascending interneurons in the aRT (i.e., mVAC) of the same and other thoracic segments (Bickmeyer et al., 1992). In Orthoptera, in general, neuronal anatomy demonstrated presynaptic (output) branches not only in the mVAC but frequently also dorso-laterally in the sensory-motor neuropile areas, suggesting an involvement of the neurons (also) in local reflex responses to vibration (Sickmann, 1997; Stritih, 2009; Stritih and Stumpner, 2009). The bushcrickets *D. verrucivorus* and *D. albifrons* have axonal terminations in the aRT (mVAC) for both vibration- and sound-sensitive interneurons, but lateral projections in the motor regions only for neurons receiving vibratory input (Sickmann, 1997; Rössler et al., 2006). This was highlighted as a characteristic functional differentiation between the auditory and vibratory systems. Although data on neuronal anatomy are lacking for *N. viridula*, their morphology also indicates information transmission, both into the medial parts of the neuropile, presumably to other vibratory neurons, and/or by extensive outputs dorso-laterally into (the motor regions of) the same and/or other neuromeres (Fig. 8).

7.3.2 Directional processing

Directional information in the vibratory system is based on sensory inputs to six spatially separated sets of vibratory sensilla of each leg, which in principle could provide more accurate information for directional processing than inputs from paired (or even unpaired) insect ears. However, heterogeneous and highly variable substrates underlying vibrational signal transmission often render orientation cues in this modality unreliable, or near the detection limit (reviewed in Virant-Doberlet et al., 2006). On plants, due to complex signal interference, time delays of signals arriving at different legs were shown to provide the only reliable cue to the signal source (Prešern et al., 2018; Fig. 9A and B). In insect auditory systems, the specific morphology of sound-transmitting structures can greatly enhance the interaural intensity difference, whilst a comparable peripheral mechanism for amplifying directional vibratory cues is unknown. However, the mechanical response of the entire insect body may contain directional information that could underlie directionality in very small insects, where sufficient time or intensity differences between inputs to different legs are not available (Cocroft et al., 2000; Miles et al., 2001).

The ability to localize the source of vibrations has been demonstrated for several insect taxa (reviewed in Virant-Doberlet et al., 2006), whilst only two studies to date have examined the neuronal basis of vibrational directionality in more detail. The first involved the migratory locust (Čokl et al., 1985; for which vibrational directionality in behaviour has not been documented), and the second was conducted with the stink bug *N. viridula* (Prešern, 2007; Prešern et al., 2018), where the male searches the female on a plant following her vibratory calls (Ota and Čokl, 1991; Prešern et al., 2018).

In the locust, the time delay between the onset of stimulation applied to different pairs of legs was determined to be the dominant directional cue, but in some interneurons the directional change in response (evaluated as spike count) was enhanced when delay was combined with stimulus attenuation (Čokl et al., 1985). For most interneurons, stimulation from ahead (i.e., with the forelegs first) elicited a stronger response than from behind, indicating the highest capacity for direction encoding during the approach to the vibration source. Remarkably, only the neurons with dominant vibratory input ascending towards the brain changed their response as a function of the pattern and delay of leg stimulation, whereas directional sensitivity was not found in bimodal sound- and vibration-sensitive neurons (despite a large number of neurons tested; Čokl et al., 1985). This suggests that in hearing Orthoptera directional auditory and vibratory information is largely processed separately in the ventral nerve cord and the output is first compared in the brain to facilitate orientation to the common source of auditory

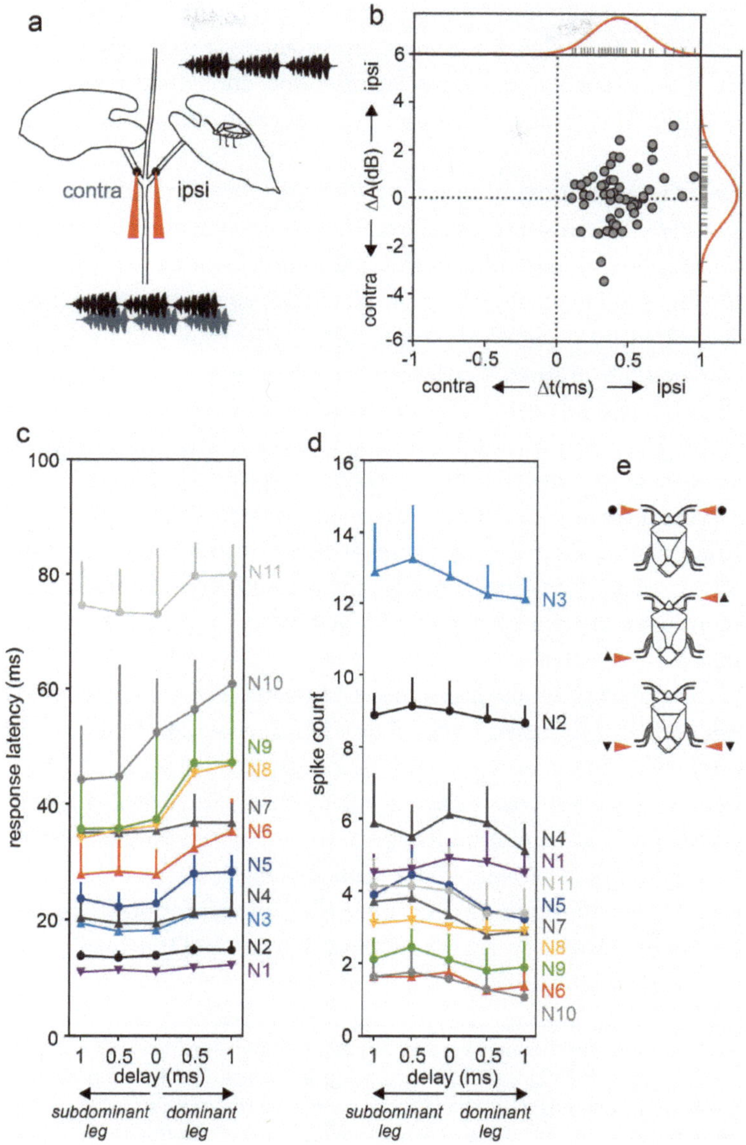

Fig. 9 Directional vibratory cues and encoding in the central nervous system of *Nezara viridula*. (A) Schematic representation of the setup to measure transmission of female vibratory signals over the stem-stalk node of a bean plant, a point of decision for the searching male. Signals of the female calling from a leaf were measured at the indicated points (red) on the same (ipsi) and the opposite leaf stalks (contra) with two laser vibrometers. (B) Distribution of time differences (Δt) and RMS amplitude difference (ΔA) of vibrational signals ($n=55$) recorded simultaneously from the indicated points.

(Continued)

and vibratory stimuli (see Latimer and Schatral, 1983; Weidemann and Keuper, 1987). Such separation in the early stages of processing may reflect different environmental transmission of sound and vibration, which may require different neural mechanisms for direction encoding between the systems.

Low-order vibratory interneurons with a restricted input from one leg or leg pair, as well as transmitting information to other segments, are necessary presynaptic elements to evoke directional responses in higher-order ascending neurons. Such neurons, which alone did not show directional responses, were described by Čokl et al. (1985) in *L. migratoria*, and Prešern (2007) showed that such units in *N. viridula* reliably encode stimulus onset delay as short as 0.5 ms. In a set of higher-order neurons tested in different configurations of leg stimulation in *N. viridula*, such small time delays significantly affected neuronal activity in terms of spike count and/or response latency in a direction-dependent manner (Prešern et al., 2018; Fig. 9C–E). Particularly striking was the sharp increase in the first-spike latency of up to about 10 ms, evoked already by the 0.5 millisecond delay of leg stimulation, in a set of neurons at the highest synaptic distances to receptors (Fig. 9C).

The underlying neural mechanism responsible for such sharp latency shifts has not been discussed in detail by Prešern et al. (2018), but in general, inhibitory interactions between inputs to different legs (i.e., lateral inhibition) have so far been considered mandatory for vibrational directionality. Contralateral and ipsilateral inhibition as a mechanism facilitating temporal and intensity contrast between the stimuli arriving at different legs have been proposed by several model neuron networks for arthropods (Brownell and van Hemmen, 2001; Hergenröder and Barth, 1983; Murphey, 1973).

Fig. 9—Cont'd Values below 0 on y axis indicate signals having higher amplitude on the contralateral stalk. Only the difference in signal arrival time to the male legs straddled across the junction is a reliable directional cue to the calling female. Below is the effect of time delay in the onset of leg stimulation on (C) the response latency and (D) spike count of eleven thoracic vibratory interneurons of *N. viridula*. Symbols relate to the applied configuration of leg stimulation, which is schematised in (E). Neurons (N) are numbered subsequently by their response latency, and are indicated by the same symbol and colour in both diagrams. Dominant is the side with the strongest influence on the response. *Adapted from Prešern, J., Polajnar, J., de Groot, M., Zorović, M., Virant-Doberlet, M., 2018. On the spot: utilization of directional cues in vibrational communication of a stink bug. Sci. Rep. 8, 5418, published under the Creative Commons CC BY license.*

However, it is possible that combined excitatory and inhibitory inputs to neurons received via the same leg(s) also play an important, or even a predominant, role in vibratory directionality. In particular, the leading inhibition that delays the first action potential in the excitatory response could explain the sharp increase in response latency caused by small input delays. This mechanism, known from auditory temporal processing, has the ability to strongly affect first spike latency (Krahe et al., 2002; Ronacher and Stumpner, 1988; Stumpner and Ronacher, 1991) and has been suggested to be potentially important for directionality encoding, as well (Rheinlaender, 1984; Römer et al., 1981; c.f. Stumpner and Ronacher, 1991). Preceding inhibition has been demonstrated amongst ascending vibratory interneurons of *N. viridula* (Zorović, 2005; Zorović et al., 2008), including one of the neurons largely increasing response latency by delayed leg stimulation (neuron N8 after Prešern et al., 2018; CG-AC-2 after Zorović et al., 2008).

In fact, evidence for (contra-) lateral inhibition at the level of thoracic vibratory interneurons is scarce. Even when inputs to different legs were examined systematically for a large number of neurons, mostly a single example with significant contralateral inhibitory effect could be found (in the cricket: Kühne et al., 1984, locust: Bickmeyer, 1991; stink bug: Prešern, 2007; and cave cricket: Stritih, 2006, 2009). On plants, inhibition between inputs from different legs would even counteract directionality encoding. Here, the changes in vibrational signal amplitude with distance are non-monotonous and amplitude is often higher at sites distant from the signaller (Polajnar et al., 2012; Prešern et al., 2018; Fig. 9B). Because neuronal responses to higher-intensity stimuli typically have shorter latencies, such amplitude cues—and especially when enhanced by lateral inhibition—would most likely override directional information encoded in the (small) time delays. With leading inhibition, which keeps response latency fairly constant at different stimulus intensities (e.g., Stumpner and Ronacher, 1991; Zorović et al., 2008), such counteracting amplitude effects are not expected to occur. This proposed mechanism for directional coding in the insects' vibratory systems requires further investigation.

7.3.3 Frequency processing

The capacity for frequency and intensity processing and discrimination at the central neural level depends essentially on the response properties of receptor organs/sensilla, and the number of sensilla mediating the information, on specific frequency/intensity ranges into the central nervous system. In this respect, it is particularly interesting to compare the central vibrosensory interneuron network of Heteroptera and Orthoptera, as these taxa differ greatly in the structural complexity of their vibrosensory SGOs, as well as the FeCOs, which are much more elaborated in the Orthoptera. As a starting point for the Orthoptera, the following discussion uses data from *T. neglectus*, where the largest number of neurons of one segment has been tested so far for frequency sensitivity. For Orthoptera, emission, reception, neural processing, and behavioural discrimination of vibrations of different frequencies have been reviewed previously by Stritih and Čokl (2014).

Low-frequency processing: Heteroptera and Orthoptera appear to be especially similar in the processing of vibration frequencies below 400 Hz. In the stink bug *N. viridula* and the cave cricket *T. neglectus*, two classes of interneurons were recorded with best frequencies in this range. One neuron class expresses sharp tuning to 50 Hz with intermediate to low sensitivity (thresholds between 0.01 and 1 m/s^2), following an equal-displacement line at the high frequency side (Stritih, 2009; Zorović et al., 2008; Fig. 10A). Such physiological neuron types (V4 and VS1) were also described in the combined auditory-vibratory system of tympanate Orthoptera (Bickmeyer, 1991; Kühne, 1982b; Kühne et al., 1980; Sickmann, 1997). Dendritic branches of these neurons are typically located outside the mVAC neuropile (Bickmeyer, 1991; Stritih, 2009), suggesting input is from less specialized vibration receptors, such as are campaniform sensilla, hair sensilla, and ventral FeCO scoloparium in Orthoptera. These interneurons probably represent the most ancient elements of an insect vibrosensory system and could be functionally most similar amongst taxa. For Orthoptera, this idea is supported by a particularly high degree of morphological and physiological similarity between the low-frequency tuned neurons amongst the recognized homologues (Stritih and Stumpner, 2009; Fig. 8), and also by reflex reactions elicited by nearly identical ranges of low-frequency stimuli in their hearing and non-hearing representatives (Stritih, 2009; Stritih and Čokl, 2014).

The second class of neurons in *T. neglectus* and *N. viridula* is tuned to frequencies between 70 and 400 Hz, and contains units with steep inclination of threshold curves and responses limited up to ca. 1000 Hz (V1 functional neuron type of tympanate Orthoptera would correspond to this group; Kühne, 1982b), and more broadly tuned neurons with responses evoked at least up to 2000 Hz, often with a considerable secondary response peak at high frequencies in *T. neglectus* (Čokl and Amon, 1980; Stritih, 2009; Stritih and Stumpner, 2009; Zorović, 2011; Zorović et al., 2008; Fig. 10A). These neurons, most at low synaptic distances from receptors,

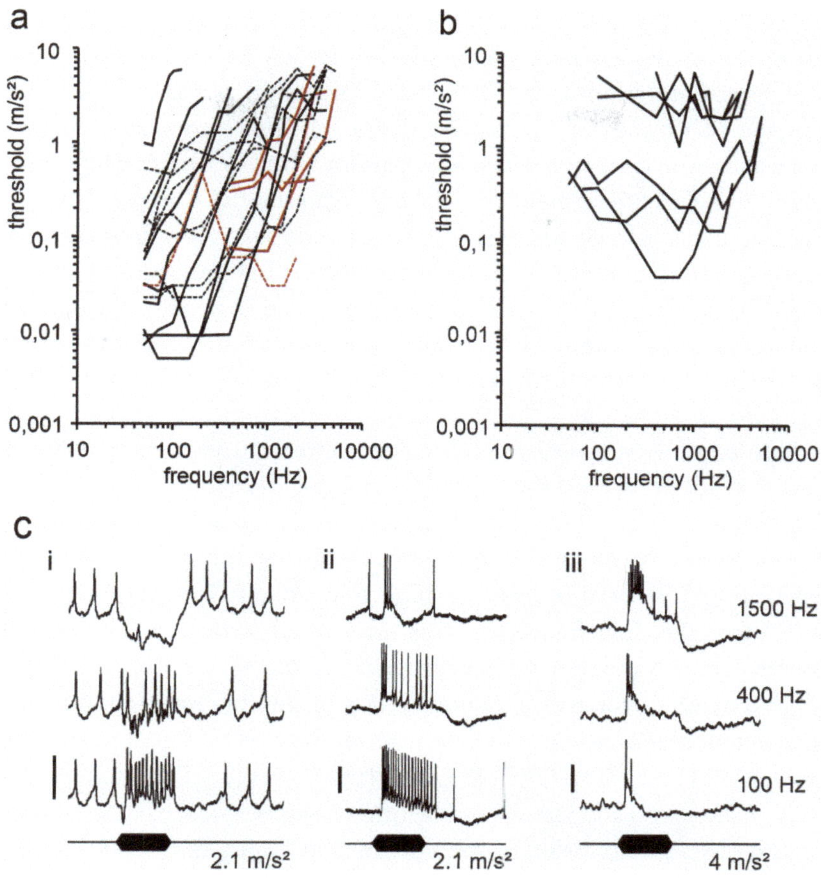

Fig. 10 Frequency sensitivity of prothoracic vibratory interneurons in *Troglophilus neglectus*. Threshold tuning curves of 24 interneuron types, arranged in two classes (black—excitatory thresholds, red—inhibitory thresholds in spontaneously active neurons): (A) neurons with highest sensitivity below 400 Hz ($N=17$; full lines—following equal-displacement at the high-frequency side; dashed lines—broader tuning to low frequencies), (B) neurons with highest sensitivity above 400 Hz ($N=7$). (C) Response traces from three neurons with mixed (frequency-dependent) excitatory and inhibitory inputs, shown for tree frequencies at a high suprathreshold intensity (as indicated; scale bars 5, 20 and 10 mV from left to right). All neurons are tuned to low frequencies at threshold; left and middle example with inhibition at high frequencies, whilst the right example receives inhibition at low frequencies that markedly shifts its tuning as intensity increases. *Adapted from Stritih, N., 2009. Anatomy and physiology of a set of low-frequency vibratory interneurons in a nonhearing ensiferan (*Troglophilus neglectus, *Rhaphidophoridae). J. Comp. Neurol. 516, 519–532, with permission from Wiley.*

include most highly sensitive units (with thresholds between 0.001 and 0.01 m/s^2) with dendrites confined to the mVAC neuropile (Stritih, 2009; Stritih and Stumpner, 2009). Such location is consistent with inputs from the SGO and, in Orthoptera, from its closely associated accessory organ, and/or from the proximal scoloparium of the FeCO. Candidate neurons in *T. neglectus* to receive the accessory organ and FeCO inputs have dendritic segments dorsally in the mVAC, in turn suggesting low-frequency sensitivity of these organs, with demonstrated central projections but scarce physiological data. In the cave cricket, the large majority of descending and local neurons are tuned to frequencies below 400 Hz, which may indicate the importance of low-frequency information (also) in reflex sensory-motor processing.

High-frequency processing: High frequency interneurons with best sensitivity above 400 Hz are in the minority amongst the identified elements in *T. neglectus* (Stritih, 2009; Fig. 10B) and in *N. viridula* (Zorović, 2011; Zorović et al., 2008). In the stink bug, this is not surprising, as only one leg vibration receptor is tuned to high frequencies (700–1000 Hz; Čokl, 1983) and the low-frequency bias in central processing reflects the low-frequency nature of vibratory signals of the species. In *T. neglectus*, by contrast, the strong low-frequency bias in interneurons sensitivity was unexpected, given the extensive input from the SGOC tuned to the range between 400 and 1500 Hz (Čokl et al., 1995).

In bushcrickets and locusts, on the other hand, physiological types with a prominent sensitivity peak, or with broadband tuning in the 400–2000 Hz range (i.e., V2 and V3, and VS2-VS4, respectively) predominate amongst the ascending interneurons from the thorax (Kühne, 1982b; Kühne et al., 1980). In crickets, different types of ascenders with fewer high-frequency units and with more pronounced frequency selectivity have been described (Kühne et al., 1984). At the physiological basis alone, however, the actual ratio of neurons devoted to processing different vibration frequencies cannot be assessed from these studies, and neither can the extent in which the recordings may have been obtained from homologous neurons in different taxa. In addition, these data are not directly comparable to those from *T. neglectus*, as by using different methods these studies have largely investigated different neuronal populations. These constraints highlight the need to study the system at the level of identified neurons.

Frequency-specific inhibitions: In the large majority of neurons in *T. neglectus* and *N. viridula* with processes in/at the location of the mVAC, responses are shaped by frequency-specific inhibition (i.e., inhibitory inputs received along with excitation via stimulation of the same leg). These inputs, clearly visible in intracellular recordings as distinct IPSPs, and/or gaps in the response spike discharge (Fig. 10C), are received either at side frequencies or at intensities above (but also below) the excitatory threshold in most of the neuron's response range. In *N. viridula*, inhibition is received mostly at high intensities of stimuli below 200 Hz in the low-frequency neurons, which results in the optimum-type of intensity-response functions (exhibiting a clear peak at a specific intensity). It may be received also as preceding inhibition throughout the response range of a neuron (Zorović, 2005; Zorović et al., 2008), possibly playing a role in temporal and directional processing. By contrast, in *T. neglectus*, inhibitory inputs are often received as a clear side frequency band, apparently shaping the neurons' frequency tuning (Fig. 10C), a mechanism well known from insect auditory systems (e.g., Stumpner, 1998, 2002). In the cave cricket, such inputs are received in most cases at high frequencies (Fig. 10A and C-i, ii), typically exceeding excitation first at 400 Hz and suggesting this frequency as a behaviourally important border for discrimination. In addition, some of the low-frequency tuned neurons (at threshold) receive inhibition at low frequencies, which markedly shifts their sensitivity towards higher frequencies as stimulation intensity increases (Stritih, 2009; Fig. 10C-iii). Inhibitory side-bands, above and below 200 Hz, have also been detected in the V1 (low-frequency) and V2 (high-frequency) physiological types of bushcricket and locust neurons, as the most sharply tuned vibratory ascenders (Kühne, 1982b; Nebeling, 2000).

The main candidates to provide such frequency-specific inhibitory inputs in the vibratory system are unpaired median neurons (DUM neurons), predominantly local neurons with bilaterally symmetric structure and branches in the mVAC (see Figs. 7iv, and 8B). Recently, the existence of between 15 and 50 auditory DUM interneurons was demonstrated in the bushcricket *Ancistrura nigrovitatta*; by showing a wide range of response selectivity for carrier frequencies and temporal patterns of sound (Lefebvre et al., 2018; Stumpner et al., 2019), as well as containing the inhibitory neurotransmitter GABA, they are the prime candidates for shaping frequency and temporal selectivity of ascending auditory neurons (Stumpner et al., 2020). A few examples of DUM neurons found in the vibratory network of stink bugs and cave crickets (Stritih, 2009; Stritih and Stumpner, 2009; Zorović et al., 2008; Figs. 7iv and 8B) presumably serve such functions, as well, possibly as representatives of a much larger neuronal population. Their homology to auditory DUM neurons may be assumed at least at the basis of descent from the common neuronal precursor (Stritih and Stumpner, 2009).

Behavioural correlates of frequency processing: In the Orthoptera that communicate acoustically on vegetation, neuronal discrimination of high- vs. low-frequency vibration was usually related to the need to distinguish the vibratory component of stridulatory signals (typically evoking rather high- or broadband vibration on plants; e.g., Keuper and Kühne, 1983) from predator or alarming signals (e.g., Dambach, 1989). As pointed out by Kühne et al. (1984), such distinction may be less obvious for ground-living species, such as are crickets and mole crickets, due to strong filtering of high-frequency vibration on soil or sand (e.g., Elias et al., 2004; Hill and Shadley, 2001), and since most crickets also use low-frequency tremulation signals at close communication distances (reviewed in Stritih-Peljhan and Virant-Doberlet, n.d.). Such categorical, frequency-based discrimination of signals appears to be less relevant for insects such as stink bugs and cave crickets, which rely exclusively on low-frequency signals for communication. These species must therefore employ additional neuronal mechanisms for differentiation of signals and cues of various biological significance, based on temporal pattern and co-processing of context-related signals of different modalities.

So far, behavioural experiments to demonstrate differential perception of signals of different significance based on frequency composition are scarce. Responses to low-frequency vibration such as startle and freezing responses are known from various Orthoptera and other insects, and may be related to predator detection (Dambach, 1989; Friedel, 1999; Stritih and Čokl, 2014; see also Section 6.3). In caterpillars, differential behavioural responses have been demonstrated to vibration of different biotic and abiotic origins with similar fundamental frequencies but different higher-frequency composition (Castellanos and Barbosa, 2006; Guedes et al., 2012; see also Section 6.3). The ability of frequency discrimination by the vibrosensory system may thus relate to the need to distinguish signals of different biological significance, despite their being of similar fundamental frequencies. Frequency discrimination may also facilitate directionality, based on the dispersive nature of vibration transmission on plants (reviewed in Virant-Doberlet et al., 2006), and frequency filtering of vibration during transmission may enable estimation of distance to the signal source, similarly as applied by auditory systems (Hedwig and Stumpner, 2016). Yet, behavioural evidence showing either of these cues to be used for orientation or distance estimation is so far lacking.

7.3.4 Temporal processing

The ability of the vibrosensory neuron network to copy and/or filter temporal signal parameters has been scarcely investigated. Most studies into that issue included temporal processing in the combined auditory-vibratory system of acoustically communicating Orthoptera. In most sound- and vibration-sensitive interneurons of bushcrickets *D. verrucivorus* (Kühne et al., 1980) and *T. cantans* (Kalmring and Kühne, 1980; Kalmring et al., 1983), and to a lesser extent amongst the neurons of *P. denticauda* (Kalmring et al., 1997a,b), as well as in the only bimodal neuron recorded from crickets (Kühne et al., 1984), interactions of vibratory and auditory inputs enhanced the response, reduced habituation, and improved coding of syllable and/or chirp structure of conspecific song(s). As noted by Kühne (1982b), this usually resulted not from a simple summation but from qualitative response changes due to complex, cross-modal interactions of excitatory and inhibitory inputs by combined stimulation.

Similar mechanisms underlying temporal processing based on input convergence may generally be expected to take place also in the vibratory system. So far, this issue has been addressed by only one study in *N. viridula*, testing a wide range of combinations of pulse durations and interval durations in four types of ascending vibratory interneurons of the central ganglion (Zorović, 2011). Responses, evaluated as time-averaged spike counts and peak instantaneous spike rates, revealed selectivity of all neurons for short pulse duration and little or no selectivity for interval duration, with the former matching pulse duration of the conspecific vibrational signals of the male. One of the neurons exhibited regular post-stimulus oscillations of membrane potential that were most pronounced with short pulse durations, indicating a possible resonance-based mechanism for temporal selectivity in the thoracic circuitry. Besides that, inhibitory inputs from local thoracic neurons, such as DUM neurons, may underlie temporal selectivity at low levels of vibratory processing in *N. viridula*, and in insects in general.

Due to intersegmental signal integration, the insect vibratory system may exhibit a greater degree of temporal processing already at the thoracic level, when compared to auditory systems, but the major role in signal feature recognition may be considered to take place in the brain. These ultimate levels of vibratory processing by brain neurons have so far remained unexplored.

7.3.5 Noise reduction mechanism

A particular challenge for the insect vibratory system appears to be the task of detecting relevant signals in the presence of noise. Although the vibrational signalling channel has been traditionally considered "private" and thus less influenced by environmental noise than the acoustic channel (e.g., Römer et al., 2010), it can also be highly noisy in plant environments, such as meadows (Šturm et al., 2019, Virant-Doberlet et al., 2019; see also Section 8.1). Whilst the number of simultaneous signallers within the active space of a species may still be smaller compared to the auditory channel, in the vibrational channel the signals of most species are emitted in the range below 1000 Hz and thus largely overlap, spectrally. These signals also interfere with the low-frequency abiotic noise from wind and rain, which on plants is particularly strong (Šturm et al., 2019, Virant-Doberlet et al., 2019). In addition, in the plant environment vibrational signals from different emitters not only become spectrally increasingly similar with increasing transmission distance (due to frequency filtering), but also merge on (literally one-dimensional) stems and stalks as if they had a common origin, which strongly facilitates the masking effect. In addition, the challenges of dealing with self-induced noise due to motion can be considered much greater for vibratory than for auditory systems. Thus, it would be particularly interesting to investigate the neural mechanisms of noise reduction and selective attention in the vibratory system. There are no completed studies on these mechanisms, but the topic is currently under investigation in stink bugs.

8. Ecological context of vibrational communication

Whilst it is now accepted that vibrational signalling is one of the oldest forms of communication and prevalent amongst insects (Cocroft and Rodriguez, 2005; Cocroft et al., 2014a), surprisingly little is known about the natural vibrational environment where this communication takes place and evolves (Šturm et al., 2019; Virant-Doberlet et al., 2019). Neuroethological investigations on orthopterans put air-borne sound signalling at the forefront of the research on insect mechanical communication and until the beginning of the 21st century studies on vibrational communication have often not only replicated the approaches used in the orthopteran research, but also used the same model species (e.g., Kalmring, 1985; Kühne, 1982a,b; Kühne et al., 1984). Taking into account also technical challenges associated with biotremological research, it is not surprising that vibrational communication has long been studied almost exclusively in the laboratory (Cocroft and Rodriguez, 2005). As a result, the reality that the evolution of vibrational communication did not take place on the vibration-damped table in a sound-proof room was long neglected (Šturm et al., 2019).

Animal communication systems should evolve to ensure efficient and reliable transfer of information between the signaller and the receiver (e.g., Endler, 1993). When widening our view from the laboratory to the

Fig. 11 Vegetation in a hay meadow. Plants of different species are growing closely together and are intertwined. *Photo courtesy of Rok Šturm.*

field situation, it first becomes clear that the habitat structure is usually quite different from the simplified environment used in the laboratory. Whilst studies determining plant transmission properties are usually done on single, freely moving plants (e.g., Čokl et al., 2007; Michelsen et al., 1982; Polajnar et al., 2012), in the meadows plants of different species are growing closely together and are also intertwined (Fig. 11). How such complex habitat structure affects signal transmission through the habitat is not known; however, whilst under such circumstances signal active space is not limited to the plant on which an insect is signalling, it is likely that vibrational signals are not transmitted equally well and equally far in all directions (Šturm et al., 2021).

Humans can access the vibratory world where insects live only with the help of specialized equipment (Cocroft and Rodriguez, 2005; Nieri et al., in press; see also Section 10) and we lack an intuitive understanding of the structure of the natural vibrational environment. Crucially, how can we relate the characteristics of the natural vibrational environment where insects live and behave to their sensory mechanisms? At present, on the one hand, we have very limited insight into the natural vibroscape composition and complexity, and on the other hand only a handful of species for which information on vibratory sensory mechanisms is available.

8.1 Vibroscape

Vibroscape has been defined as "the collection of biological, geophysical and anthropogenic vibrations emanating from a given environment to create unique vibrational patterns across a variety of spatial and temporal scales" (Šturm et al., 2019). Field recordings show that the natural vibrational environment in hay meadows is primarily dominated by frequencies below 1 kHz (Šturm et al., 2019, 2021). The studies of the natural vibrational environment show that regardless of the environment studied, geophysical vibrations induced by light wind are nearly always a present component of the natural vibroscape (Barth et al., 1988; Cocroft and Rodriguez, 2005; McNett et al., 2010; Šturm et al., 2021; Tishechkin, 2007; Velilla et al., 2020a,b). Such constant wind-induced background vibrational noise is characterized by frequencies below 50 Hz (Fig. 12). Stronger wind gusts generate high amplitude vibrations in the frequency range up to 5 kHz, characterized by rapid, unpredictable short term variations in the amplitude (Cocroft and Rodriguez, 2005; McNett et al., 2010).

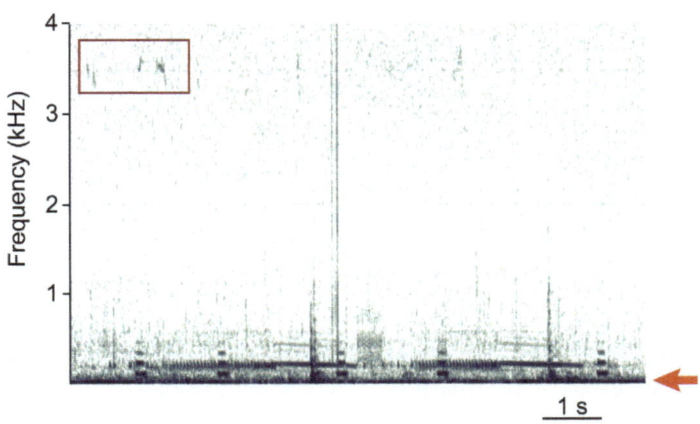

Fig. 12 Visualization in the form of a spectrogram of vibroscape recorded on a clover plant in a hay meadow. Vibroscape includes overlapping vibrational signals species of two unknown insect species and vibrational component of a bird air-borne song (brown frame). Red arrow indicates the background vibrational noise induced by wind. *Recording courtesy of Rok Šturm.*

The main source of biological vibrations is vibrations emitted by animals, not only during intraspecific vibrational communication, but also as a by-product of other activities (e.g., incidental vibrations induced by body movement) (Šturm et al., 2019). In general, incidental vibrations are characterized by a random pattern of short pulses with broadband frequency structure (up to 20 kHz) and dominant frequencies between 0.6 and 8 kHz (e.g., Barth et al. 1988; Potamitis et al., 2009; Zorović and Ćokl, 2015).

Vibrational signals used in intraspecific communication are species- and sex-specific with predictable temporal and spectral characteristics. The frequency range of insect vibrational signals recorded in the field or in the laboratory so far spans the range between 70 and 7000 Hz (e.g., Cocroft et al., 2010; Čokl, 2008; Derlink et al., 2014; Eberhard et al., 2010; Henry et al., 2013; Stritih and Čokl, 2012; Velilla et al., 2020a,b). Whilst some species rely on signals with narrow-band frequency characteristics, others produce broadband vibrational signals; however, regardless of the overall frequency range, the dominant frequencies in most cases appear below 1 kHz. Two cases on the extreme are the bushcricket, *Docidocercus gigliotosi*, whose tremulatory signals have most energy limited to frequencies below 20 Hz (Römer et al., 2010), and the leafhopper, *Megophthalmus scanicus*, that emits vibrational signals with dominant frequencies above 1.5 kHz (Šturm et al., 2021).

Shifting our view from the laboratory to the field reveals that in nature vibrational communication occurs in a group of several conspecific and heterospecific individuals, where each signal can be received by several receivers and each receiver can receive signals from several signallers (Šturm et al., 2019; Virant-Doberlet et al., 2014, 2019). Vibroscape recordings from a single plant in a hay meadow showed that the vibrational communication channel is, on a single day, shared by many species (Šturm et al., 2021). Taking into account that vibrational signals of these species share the same frequency range, incidental masking from overlapping heterospecific signals should be common (Fig. 12). Biotic noise can also originate from signalling conspecifics and such a situation is likely to be common in species that live in aggregations.

Vibroscape recordings also showed that vibrational components of bird air-borne sounds is a common source of biological vibrations in the natural vibrational environment. Such vibrations occur at frequencies above 3 kHz (Fig. 12) and in particular in spring can constitute a majority of recorded biological vibrations (Šturm et al., 2021). Air-borne sounds emitted by insects can also contribute to the vibroscape, either directly by body movements during song production and/or indirectly via transfer of mechanical waves between two adjacent media (Caldwell, 2014).

8.2 Perceived world (Umwelt)

The concept of subjective perception of the world has been developed by Uexküll and Kriszat (1934). The perceived world in which animals live and behave depends on species-specific sensory and neuronal mechanisms that ultimately trigger the appropriate behavioural response (Barth, 1986; Conchou et al., 2019; Uetz and Roberts, 2002). In general, one should expect that components in the natural vibroscape that are relevant to insect behavioural decisions are detected by vibroreceptors and processed by neurons in the central nervous system. However, there is a lack of direct information on how the natural vibrational environment is perceived by insects. As it has been suggested for spiders (Barth et al., 1988), field measurements of the constant background vibrational noise induced by low wind speeds (below 0.2 m/s) indicate that in hay meadows such abiotic noise at frequencies below 50 Hz is most likely not detected by insect vibroreceptors (Šturm et al., 2021). Although the measured amplitudes were in the range of 10^{-6} m/s, insect vibration receptors are relatively insensitive in this frequency range (reviewed in Čokl et al., 2006; Lakes-Harlan and Strauß, 2014). Field observations and laboratory behavioural experiments suggest that stronger wind has a negative effect on vibrational signalling and that insects time their signalling activity to periods with low wind levels (Akassou et al., 2021; Hamel and Cocroft, 2012; McNett et al., 2010; Tishechkin, 2007; Velilla et al., 2020a,b). Insects are also able to distinguish wind-induced vibrations from substrate vibrations emanating from other sources (Castellanos and Barbosa, 2006; Guedes et al., 2012; Hager and Krausa, 2019).

On the other hand, insect vibrational signals are in general adapted to specific transmission properties of their environment (e.g., Čokl, 2008; McNett and Cocroft, 2008; Virant-Doberlet and Čokl, 2004). Field recordings revealed that temporal distribution of vibrational signals was not random, because the signallers avoided masking by partitioning the communication space on a fine temporal scale (over seconds to minutes) (Šturm et al., 2021) and also avoiding the warmest time of the day (Akassou et al., 2021). In addition, vibroscape study in the hay meadow habitat showed that from the point of a receiver's individual experience, the natural vibrational environment was highly variable and depended, on the one hand, on the random appearance of vibrational signals of individual species within the vibroscape and, on the other hand, also on the position of the resident plant within the meadow (Šturm et al., 2021).

Laboratory behavioural experiments showed that spiders can detect bird songs as substrate vibrations and even discriminate between songs of predatory and non-predatory species (Gordon and Uetz, 2012; Lohrey et al., 2009). Measurements of the vibrational component of bird songs recorded on the plants in the field suggest that under studied conditions, where birds were sitting in a tree and were singing more than 10 m away from the recording position, the amplitude is most likely too low to be detected by insects (Šturm et al., 2021).

9. Behavioural manipulation for pest control

9.1 Semiochemicals and semiophysicals

The technique of pest control that operates through the deliberate manipulation of a pest's behaviour is called "behavioural manipulation" (BM) and it is based on the use of signals that either stimulate or inhibit the target behaviour, thus changing its expression (Foster and Harris, 1997). BM can directly or indirectly affect the behaviour of a pest. In the first case, BM can either modify the orientation of the target organism or prevent or elicit a specific behaviour (e.g., feeding, freezing). The indirect effect is achieved when BM disrupts the intra-specific communication of a pest, thus compromising its proliferation (e.g., reproduction) on the crop and consequently reducing the damage. Probably, the most popular BM technique, largely used worldwide to protect crops, is the mating disruption by pheromones, to control insect pests such as moths, flies and other noxious insects (Cardé, 1990). The basic concept of mating disruption is to release a huge quantity of sexual pheromones in the area of interest (e.g., orchards) to prevent mating partners, usually males locating females, to find each other, thus suppressing mating. An example is the European grapevine moth, *Lobesia botrana*, an economically important pest that lays eggs in grape clusters from which voracious larvae will eventually hatch (Ioriatti and Lucchi, 2016). The larvae cause injuries to the grapes by feeding, thus favouring the spread of pathogens that eventually ruin the cluster. In this case, BM targets adults, which do not cause direct damage, to reduce next-generation larval infestations. Semiochemicals (e.g., pheromones and allelochemicals), however, are not the only stimuli that can be employed to accomplish BM. Other non-chemical stimuli like lights, sounds and vibrations either are already used or have the potential to be applied with the same purpose. Non-chemical stimuli able to modify the behaviour of a pest have been recently grouped in a distinct category of physical signals, the semiophysicals (Nieri et al., 2021).

Regarding biotremology, vibrational signals perfectly fit the definition of semiophysicals in that they convey intra- and interspecific information that induce precise behavioural responses (Cocroft and Rodriguez, 2005). A first difference between them and semiochemicals is the medium through which the stimuli travel: whilst semiochemicals, similarly to sounds, travel through the air, and consequently can reach a considerably large active space (which depends on several environmental factors, especially wind), substrate-borne signals are restrained to considerably shorter distances (Mazzoni et al., 2019). They also are subjected to attenuation and distortion (Čokl and Virant-Doberlet, 2003; Mazzoni et al., 2014). In this view, semiochemicals are best suited for long-range and vibrational signals for short-range communication, a combination that indeed occurs in nature in many species' communication, including stink bugs (Hemiptera: Pentatomidae). Stink bugs are relatively good flyers and can move for long distances (Weber et al., 2017). Thanks to pheromones, they attract their own kind to the same plants, but once landed, the mating process is driven almost exclusively by vibrational signals (De Groot et al., 2011b; Laumann et al., 2017). Stink bugs provide a perfect example of multimodal communication where semiochemicals and semiophysicals are complementary by mediating two different parts of the mating process (Fig. 13B).

Another important peculiarity of substrate-borne vibrations compared to semiochemicals is that they are emitted in sequences of temporally defined bursts and the associated information disappears as soon as the emission is interrupted (Nieri et al., 2021). The detection of vibrations is necessary to locate the sender by homing in on the signal trace to the source. The interruption of the mating signal could be enough to compromise the ongoing search of one gender—more often males, whereas females remain stationary (Mazzoni et al., 2009a,b). The point is to know the behaviour of the target species and the exact role of associated vibrational signals before starting any plan of BM. Each species can emit many different signals and each of them potentially elicits in the receiver a distinct behavioural response, such as the acceptance to mate, attraction by orienting to the source, or interruption of communication. The first goal of the researcher

Fig. 13 Devices used for behavioural manipulation of insect pests with vibrational signals. (A) Shaker used for vibrational mating disruption in vineyards. (B) Vibrotrap prototype that combines semiochemicals and semiophysicals to attract *Halyomorpha halys* stink bugs. (C) Vibration exciter that generates vibration to induce behavioural responses in coleopterans on a tree. (D) Device for the detection of specific vibrations emitted by stored product insects. *Panel (C): Photo courtesy of Takuma Takanashi. Panel (D): Photo courtesy of Richard Mankin.*

should be to assess these roles and well characterize the associated signals to understand the spectral and/or temporal features that trigger the desired response. A preliminary selection within the signal repertoire of a species and a fine comprehension of the signal characteristics is the requisite to design an artificial "dummy" (e.g., Caorsi et al., 2021; Mazzoni et al., 2017). This requires identifying, by means of playback experiments, the parameters that best trigger the desired behavioural response within the natural range of the signal features. Playback trials should be performed under controlled conditions to reduce any possible influence of external factors; however, when the communication is multimodal, with the occurrence of other sensory modalities, isolating the role of only vibrational signals is not an easy task.

Things are much easier when one sensory modality is more relevant than others. In some insect groups, like in leafhoppers and planthoppers (Hemiptera: Cicadomorpha and Fulgoromorpha), intraspecific communication, especially that associated with mating, depends mainly on the correct emission and reception of species-specific vibrational signals (Čokl and Virant-Doberlet, 2003). For example, there are species that require that the mating signals strictly respect some temporal and/or spectral features without which the mating process would be stopped (Kuhelj et al., 2016; Polajnar et al., 2014). In the grapevine leafhopper *Scaphoideus titanus*, males send a signal composed of a series of single and double pulses at intervals of approximately 40 ms. If the females do not reply with their own signal within a time window of 20–30 ms from a male pulse, males will stop calling and the mating duet will not be established (Mazzoni et al., 2009a,b). In another case, it has been shown that males of the planthopper *Hyalesthes obsoletus* are overly sensitive to secondary spectral components of the female calling signal and they will mostly ignore individuals whose signals lack the highest frequency peaks (>600 Hz), even if the signal dominant frequency and the temporal pattern remain optimal (Mazzoni et al., 2015).

All these mechanisms of reciprocal recognition are important not only to ascertain the identity of the candidate partner but also to reveal its quality (i.e., age, health) and thus are basic to sexual selection. This implies that vibrational signals are important drivers in decision making and can represent a powerful tool of BM: not only do they provide relevant information to insects that rely on them for many fundamental behaviours, but they also are species-specific and, in many cases, they must be perceived in their integrity, from sender to receiver, to accomplish their biological role.

9.2 Behavioural manipulation strategies

Proposed BM techniques based on the use of substrate-borne vibrations can be divided into three main categories: (1) those interfering with the intraspecific communication of the pest (disruption); (2) those that elicit attraction or stimulate a reaction, either evoking the initiation or promoting the continuation of a behaviour (attraction/promotion); and (3) those that orient the pest away from the stimulus or inhibit a behaviour (repellence/inhibition) (Nieri et al., 2021). Indeed, the three categories can overlap, and more drivers can work at the same time in some cases. This also means that there is not a clear separation amongst categories and that the same signal could be disruptive for a species and repellent for another one.

To the first category, disruption, belong methods of mating disruption. To prevent or interrupt the mating process, one option, in fact, is to do masking by releasing vibrations with enough amplitude and the same frequency pattern of the target species signals into the crop (Eriksson et al., 2012; Mazzoni et al., 2019; Saxena and Kumar, 1984). In this way, individuals are not able to detect or properly recognize the partner signals. On the other hand, the same result can be achieved by releasing a huge number of conspecific signals, which impede the location of the real source of natural signals, into the host plant (Krugner and Gordon, 2018). If instead only one attractive signal is deployed, then the behavioural response that is elicited (i.e., search) could lead the target insect precisely towards an exciter, and if the exciter is placed inside a trapping device, the insect can be lured into the trap (Mankin et al., 2013; Mazzoni et al., 2017). However, the same signal, whilst inducing the activation of search, can also inhibit another behaviour (i.e., feeding) thus reducing direct damage to the crop (Avosani et al., 2020a). Substrate-borne vibrations can also be used to monitor insect pests, either stimulating the insects to emit their own signals to make them detectable by a recording device (i.e., active monitoring), or just detecting the natural occurrence of vibrational signals or cues in the infested substrate (i.e., passive monitoring) (Mankin et al., 2011) (Fig. 13D). Even if this strategy does not strictly manipulate the behaviour of the target species, the large number of studies across the insect class demonstrates its potentiality in pest monitoring (Table 2).

9.2.1 Disruption

The principal research study on disruption, and in particular on mating disruption, regards leafhoppers (Hemiptera: Cicadellidae) that, given their mainly unimodal communication, are the best candidates for the application of such techniques. The experimentation closer to an actual field application for commercial orchards is having as target the grapevine leafhopper, *S. titanus*. In Italy, three vineyards are currently managed with this method, the first one being established in 2017 in Trentino (Mazzoni et al., 2019) (Fig. 13A). The artificial masking signal employed to accomplish mating

Table 2 List of insect pest species in which behaviour manipulation (BM) strategies with vibrations have been tested either in laboratory or field conditions.

Order: Family	Species	Behaviour manipulation strategy	References
Blattodea: Rhinotermitidae	*Coptotermes acinaciformis*	Inhibition	Inta et al. (2009)
Hemiptera: Aleyrodidae	*Bemisia tabaci*	Disruption/ inhibition	Yanagisawa et al. (2021)
Hemiptera: Aleyrodidae	*Trialeurodes vaporariorum*	Disruption*/ repellence*	Fattoruso et al. (2021)
Hemiptera: Aphididae	*Acyrthosiphon pisum*	Repellence	Dill et al. (1990) and Losey and Denno (1998a,b)
Hemiptera: Aphididae	*Myzus persicae*	Inhibition	Lee et al. (2012)
Hemiptera: Aphrophoridae	*Philaenus spumarius*	Disruption*/ inhibition	Avosani et al. (2021a,b)
Hemiptera: Cicadellidae	*Amrasca devastans*	Disruption	Saxena and Kumar (1980)
Hemiptera: Cicadellidae	*Aphrodes bicincta*	Detection	Korinšek et al. (2016)
Hemiptera: Cicadellidae	*Empoasca vitis*	Disruption	Nieri and Mazzoni (2018)
Hemiptera: Cicadellidae	*Homalodisca vitripennis*	Disruption	Gordon et al. (2017), Krugner and Gordon (2018), and Mazzoni et al. (2017)
Hemiptera: Cicadellidae	*Psammotettix alienus*	Disruption	Derlink et al. (2018)
Hemiptera: Cicadellidae	*Scaphoideus titanus*	Disruption	Eriksson et al. (2012), Mazzoni et al. (2009a,b), Mazzoni et al. (2019), and Polajnar et al. (2016a,b)
Hemiptera: Delphacidae	*Nilaparvata lugens*	Disruption	Saxena and Kumar (1984)
Hemiptera: Lividae	*Diaphorina citri*	Disruption/ attraction/ detection	Hartman et al. (2017), Lujo et al. (2016), Mankin et al. (2013), and Rohde et al. (2013)

Table 2 List of insect pest species in which behaviour manipulation (BM) strategies with vibrations have been tested either in laboratory or field conditions.—cont'd

Order: Family	Species	Behaviour manipulation strategy	References
Hemiptera: Membracidae	*Echenopa binotata*	Disruption	McNett et al. (2010)
Hemiptera: Pentatomidae	*Euschistus heros*	Disruption/ attraction	Laumann et al. (2017, 2018) and Dias et al. (2021)
Hemiptera: Pentatomidae	*Euschistus servus*	Detection	Lampson et al. (2013)
Hemiptera: Pentatomidae	*Chinavia ubica*	Disruption	Dias et al. (2021)
Hemiptera: Pentatomidae	*Chinavia impicticornis*	Disruption	Dias et al. (2021)
Hemiptera: Pentatomidae	*Halyomorpha halys*	Attraction	Mazzoni et al. (2017)
Hemiptera: Pentatomidae	*Nezara viridula*	Detection/ disruption	Lampson et al. (2013) and Polajnar and Čokl (2008)
Hemiptera: Pentatomidae	*Plautia stali*	Disruption/ deterrence	Uechi and Takanashi (2021)
Hemiptera: Triozidae	*Bactericera cockerelli*	Disruption/ attraction	Avosani et al. (2020a,b)
Coleoptera: Cerambycidae	*Apriona japonica*	Disruption	Hosomi (1996)
Coleoptera: Cerambycidae	*Monochamus alternatus*	Inhibition	Takanashi et al. (2016) and Tsubaki et al. (2014)
Coleoptera: Chrysomelidae	*Leptinotarsa decemlineata*	Inhibition	Acheampong and Mitchell (1997)
Coleoptera: Curculionidae	*Dendroctonus frontalis*	Disruption/ inhibition	Hofstetter et al. (2014)
Coleoptera: Curculionidae	*Dendroctonus ponderosae*	Disruption	Hofstetter et al. (2016)
Coleoptera: Curculionidae	*Ips confusus*	Disruption/ deterrence	Lukic et al. (2021)
Coleoptera: Dryophthoridae	*Rhynchophorus ferrugineus*	Detection	Pinhas et al. (2008)

Continued

Table 2 List of insect pest species in which behaviour manipulation (BM) strategies with vibrations have been tested either in laboratory or field conditions.—cont'd

Order: Family	Species	Behaviour manipulation strategy	References
Coleoptera: Tenebrionidae	*Tribolium castaneum*	Inhibition	Mullen (1975), Hirashima et al. (1993), and Jinham et al. (2012)
Coleoptera: Tenebrionidae	*Tribolium confusum*	Inhibition	Mullen (1975)
Coleoptera: Tenebrionidae	*Tribolium freeman*	Inhibition	Hirashima et al. (1995)
Lepidoptera: Geometridae	*Macaria (Semiothisa) aemulataria*	Repellence	Castellanos and Barbosa (2006)
Orthoptera: Acrididae	*Schistocerca gregaria*	Repellence	Friedel (1999)

The asterisks indicate strategies that have been proposed for vibrational BM.

disruption is a species-specific signal, elaborated from the natural male rivalry signal that is used by males to compete for female access (Mazzoni et al., 2009a,b). This disturbance noise is continuously transmitted, 24 h a day, to grapevines by exciters (Tremos S/E, CBC (Europe) S.r.l.), which are screwed into iron trellis poles, with an efficiency distance of around 20–30 m from the source. This means that each device could cover a total distance of 40–60 m in the experimental vineyard. Besides the fine knowledge of the mating behaviour, it has been important to study the basal behavioural threshold of amplitude required to interrupt the mating process. This information was crucial to understanding the entity of the active space of the devices used to transmit the disruptive noise into the grapevines and thus calculate the needed device density. In particular, two different thresholds were measured, each one in association to the relative positions of the potential partners (Polajnar et al., 2016a,b). Lab and semi-field trials revealed that if a male and a female occur on the same leaf, it takes a substrate velocity of approximately 15 μm/s to impede the duet establishment. Honouring this threshold gives the highest chance of protection from the insect. However, the higher the desired amplitude, the higher the required devices' density, and the higher the system costs. A tradeoff is possible: because of signal attenuation, which was measured on the order of 10 dB in grapevine plants, whenever the insect calling signal passed from leaf to stem or to other leaves, a secondary threshold of 1.5 μm/s has been proposed (Mazzoni et al., 2019). The adoption of the latter threshold should suffice in inhibiting mating in the case of low-density populations, reducing the device density, but raising the risks.

Other leafhoppers that are objects of study are both grapevine pests: the green grapevine leafhopper, *Empoasca vitis*, and the glassy-winged sharpshooter, *Homalodisca vitripennis*. Since *E. vitis* and *S. titanus* can simultaneously occupy the same vineyards, the signal used to disrupt the *S. titanus* mating behaviour has been implemented to mask the main frequencies of *E. vitis* intraspecific communication (Nieri and Mazzoni, 2018). This signal has been used since 2018 in the vibrational vineyards established in Italy and is currently under evaluation for efficacy. A slightly different strategy has been used for the glassy-winged sharpshooter. A stimulating signal, the female call, has been used in semi-field trials to interfere with the male's location of the female (Krugner and Gordon, 2018). In this case, males are likely to be confused by the detection of the attracting signal coming from multiple sources and they may spend more time trying to locate the source of vibrations. Evidence of this mechanism of action has been reported also in BM experiments of the tomato/potato psyllid, *Bactericera cockerelli*, in which the mating behaviour was suppressed because males were attracted to the playback source (Avosani et al., 2020b). Other experiments, instead, used playbacks of female rival signals to reduce and thus disrupt mating communication in stink bugs (Dias et al., 2021).

9.2.2 Attraction/promotion

Attraction to substrate-borne vibrations is an example of how the three categories of BM may overlap. In fact, in several species, orientation is mediated by substrate-borne vibrations because vibrational signals are emitted by the mating partner and the emission of such signals mediate pair formation. This is the case of many plant-dwelling insects, such as leafhoppers, planthoppers, treehoppers, and stink bugs (Čokl and Virant-Doberlet, 2003). Usually the female is the stationary gender, who emits the attractive signal, and the male searches for the source of the vibrational stimulus. If the searching behaviour of the male is triggered by simple repetition of the female signal, it is possible to develop a trap that continuously emits the attractive signal and lures the male to it (Mazzoni et al., 2017) (Fig. 13B). However, often the temporal pattern of the male-female alternation of signals (i.e., the duet) is crucial for species identification and its failure interrupts the searching behaviour of the male (Kuhelj et al., 2016). If this is the case, the continuous repetition of the attractive signal may be insufficient, and an automated system that can play back the right species temporal features at the right time should be developed accordingly (Korinšek et al., 2019; Mankin et al., 2013). In the case of species that rely on both chemical and vibrational communication, the trap can be implemented by adding the semiochemicals that mediate attraction at long range. This happens in stink bugs (Laumann et al., 2017; Mazzoni et al., 2017) as well as in psyllids (Guédot et al., 2010).

9.2.3 Repellence/inhibition

Vibrational cues are often used in predator-prey interactions; they can both inform predators of the presence of a prey and, vice-versa, inform prey of approaching predators (Virant-Doberlet et al., 2011, 2014). In this regard, vibrational cues have been used in BM strategies as repellents to induce pests to abandon host plants. The aphid, *Acyrthosiphon pisum*, and the caterpillar, *Macaria (Semiothisa) aemulataria*, have been studied in particular for their dropping behaviour (Castellanos and Barbosa, 2006; Dill et al., 1990; Losey and Denno, 1998a, 1998b). Immobility and freezing are two other responses induced by vibrational stimuli that occur in several beetles, in fruit flies and in honeybees (Takanashi and Kojima, 2021). The possibility of inducing these behaviours by playing back specific frequencies in the substrate is of interest, in particular in pest control of fruit and ornamental trees (Takanashi et al., 2019) (Fig. 13C).

In general, most of the applications in this category are represented by the use of mechanical stimuli as a stressor that inhibits activities, but also affects the development of the insect. An example of the first case are wood pests, such as termites, whose feeding behaviour directly compromises wood quality and can cause severe problems to human facilities. Experimental trials showed that the foraging activity of *Coptotermes acinaciformis* is reduced when the wood block on which the insects are feeding is vibrated with a species-specific alarm signal (Inta et al., 2009). Another promising application is to target the development of insects. Mechanical stress, such as vibrations, seems to alter the production of hormones linked to metamorphosis and larval development (Hirashima, 2009; Hirashima et al., 1993). In beetles of the genus *Tribolium*, the survival rate was significantly reduced by exposing immatures to substrate-borne vibrations at low frequencies (Hirashima et al., 1995; Jinham et al., 2012; Mullen, 1975).

10. Methods for studying signalling behaviour in biotremology

The recording technology, in terms of acquisition devices, storage and processing of the recorded files, used to study biotremology is shared with bioacoustics methodologies. In particular, one of the main challenges of data handling in bioacoustics is linked to the need for data storage and processing of very large files (Gibb et al., 2019). On the contrary, transducers used to detect and reproduce substrate-borne vibrations represent the main distinction in methodology between the two disciplines. Because of the nature of the waves typical of vibrations, their detection and reproduction are more complex than sounds (Yack, 2016), nevertheless there are several transducers, both input and output transducers, that are able to detect or playback substrate-borne vibrations. The price range can vary greatly depending on the sensitivity of the sensor or the output amplitude of the actuator. It is important to point out that the advantages and disadvantages of choosing one over another are not dependent on the cost, but on the aim of the experiment. The way a transducer is used should be considered when choosing the right equipment (Nieri et al., in press). In this section, to give the reader a basic overview of the available instruments, the main sensors and actuators used in the study of vibrational communication of insects are reported with examples of applications, advantages and disadvantages. The list is not exhaustive and any alternative approach is encouraged to be used if scientifically valid.

10.1 Detection and recording of vibrations

Transducers able to detect surface vibrations can be divided into contact and non-contact sensors. Contact sensors are coupled to the substrate, whereas non-contact sensors use optic properties to detect the movement of the substrate. To date, the standards for vibration detection are the piezoelectric accelerometers and the laser Doppler vibrometres (LDVs), for each category, respectively.

Accelerometers consist of a piezoelectric crystal coupled with a seismic mass (Regtien and Dertien, 2018). They detect the displacement of the substrate to which they are attached and measure the acceleration; their response depends on the mass loaded and their specific resonant frequency (Aszkler, 2005). As for all contact sensors, the size and weight of the instrument must be considered carefully when choosing the equipment to use for a study, because large or heavy contact sensors placed on a small or lightweight structure can skew the acquired data and affect the accuracy of the results (Aszkler, 2005). Specifically, when mounting a sensor on a substrate, the latter is loaded mechanically, resulting in a change of acceleration and resonance frequency (Regtien and Dertien, 2018). This effect, known as "mass loading", is particularly relevant when studying plant-dwelling insects (Cocroft and Rodriguez, 2005). In fact, in this case, the leaves or the stem of a plant are the substrate through which the insects transmit and detect vibrations and the weight range of accelerometers, from a few 10^{-1} g to above 10^{2} g (Aszkler, 2005), can easily overlap or exceed that of the plant section. Nevertheless, lightweight accelerometers have been successfully used to record insects' vibrational signals and they are particularly advantageous in natural environments. Yet, as a rule of thumb, the mass of the transducer should not exceed 5% of the mass of the structure to which it is attached (Brennan, 2007; Cocroft and Rodriguez, 2005; Guedes et al., 2012; Korinšek et al., 2016). Accelerometers can be uniaxial or multiaxial, depending on how many directions of displacement can be detected simultaneously. Usually the smallest and cheapest options are uniaxial accelerometers, but multiaxial sensors can provide additional data about the substrate displacement (Elias and Mason, 2014). The accelerometer mounting (i.e., with a stud, an adhesive or a magnet) introduces variability in the acquisition process (Aszkler, 2005). In general, the highest degree of accuracy is provided by stud mounting the sensor to the substrate. However, this is often not possible in biotremology experiments, because the substrate is a living or thin structure, such as a leaf. In this case, wax offers a very convenient and easily removable approach, but it might be insufficient at high temperatures (Aszkler, 2005; Mankin et al., 2004). A magnet adds mass load to the accelerometer, increasing the risk for the "mass loading" effect, and for this reason, this solution is usually not used in biotremology setups (Aszkler, 2005).

Probably the ideal sensors to detect vibrations on lightweight substrates without affecting their frequency response are optical sensors such as laser Doppler vibrometres (LDV) (Cocroft and Rodriguez, 2005; Šturm et al., 2019). LDV detect the Doppler frequency shift that occurs when light is scattered by a moving surface (Rothberg et al., 2017). It was first developed to measure fluid velocity (Yeh and Cummins, 1964). Since then the applications of LDV to detect the vibrations of solid substrates have increased exponentially, from structural monitoring to micro-electronics, from medical sciences to acoustics, and last but not least, in biotremology (Cocroft and Rodriguez, 2005; Rothberg et al., 2017). LDVs have been successfully used also in field recordings; however, they encounter practical problems due to their weight, the need for a power supply, a tripod and a PC. They also are very sensitive to high temperatures and in summer, it is recommended to keep them under a beach umbrella or a tent (Akassou et al., 2021). For all these reasons, they have rarely been used in natural environments (but see Šturm et al., 2019).

Even if accelerometers and LDVs detect the same physical vibration, their output will differ in both waveform and amplitude spectra, because LDVs respond to the velocity of the vibration and accelerometers to the acceleration (Šturm et al., 2019). This different response affects the dynamic frequency range of the sensor, which will be biased towards high frequencies in the case of accelerometers. Accelerometers and LDVs are reliable equipment. They are usually calibrated and thus they can be used to measure the real intensity/amplitude of a stimulus. Another problem is the high cost of these devices; however, there are many custom-made transducers that can be made with cheaper sensors and that can be used when a precise quantification of the stimulus is not needed. Amongst these, there are piezoelectric phonograph cartridges, piezoelectric disks, loudspeaker membranes, the Linear Resonant Actuator (LRA), and guitar pickups (Blassioli-Moraes et al., 2005; Cocroft and Rodriguez, 2005; Henry et al., 1999; Mankin, 2019; Nieri et al., in press).

All the sensors listed so far detect vibrations from a single point source at a time. A scanning LDV can describe how a wide surface vibrates without the need to use multiple sensors, simultaneously; this approach could be useful to observe vibrational signals as they are emitted by the sender and as they are detected by the receiver, at the same time. However, their price is much higher than a normal LDV, making scanning LDV difficult to use in biotremology studies. At any rate, new technologies continue to be developed, such as the use of audio or video recordings to extrapolate substrate displacement (Davis et al., 2014; Javh et al., 2018), whilst at the same time there is a progressive reduction of the equipment costs.

10.2 Transmission of vibrations to the substrate

Because of the heterogeneous nature of the substrates through which substrate-borne vibrations are transmitted, delivering a vibrational stimulus at a specific location is more complex than playing back a sound

(Cocroft et al., 2014b). Even if there are a large number of devices that can be used to transmit vibrations to a substrate, their reliability depends mainly on how the playback file is filtered to compensate for three conditions: (i) the device properties, (ii) the way it is coupled to the substrate, and (iii) the filtering properties of the substrate itself (Cocroft and Rodriguez, 2005; Cocroft et al., 2014b; Nieri et al., in press). The filtering process can be done manually or by means of ready-to-use software (Michael et al., 2019).

Electromagnetic shakers made for industrial use are commercially available and they are considered the most reliable actuators. Industrial shakers, such as the mini-shaker Type 4810 by Brüel and Kjær Sound & Vibration A/S (Nærum, Denmark), come ready to be used and they can be driven with any audio application, with or without a pre-amplifier (Derlink et al., 2014; Mazzoni et al., 2009a,b; Nieri and Mazzoni, 2018). In general, the output is stable in time and reliable, but even with such a standard device, it should be considered that electromagnetic shakers are designed to have a flat frequency response with respect to acceleration. Therefore, if a vibrational stimulus has been recorded with a laser, when it is played back with a shaker, the output will be skewed towards low frequencies with a consistent loss of amplitude in the highest frequencies of the signal. Thus, a proper filtering of the original stimulus is crucial to ensure that the output is as desired.

Once acknowledged that any actuators need proper filtering of the playback file, many more options, cheaper than industrial shakers, are available and have been used to reproduce insect vibrational signals on plants, such as piezoelectric buzzers (Lujo et al., 2016; Mankin et al., 2013), piezoelectric stacks or actuators (Brownell, 1977; Cocroft et al., 2000; Elias et al., 2004), magnets (Conrad and Ayasse, 2015; McNett et al., 2010; Michelsen et al., 1982) and loudspeakers. Amongst these, magnets are probably the most difficult to control and, since they need to be loaded onto the substrate, they add mass to the substrate, thus affecting its frequency response. Loudspeakers can be easily found of variable power and can ensure playbacks of high amplitude. They can be used either unmodified or modified to reduce the airborne component (Michael et al., 2019). If not modified, the speaker is positioned next to the substrate. Once the emitted sound hits the solid substrate, it propagates into it as substrate-borne vibrations (Polajnar et al., 2014). However, in this case it is difficult to control the input of the vibrations into the substrate.

11. Conclusions and perspectives

Biotremology well embodies the perfect synthesis between basic and applied research in a scientific discipline. Thanks to a broad multidisciplinary approach, insect biotremology is inspired by many branches of classic entomology such as morphology, physiology, ecology, ethology and many others, but also by engineering disciplines (i.e., mechanics, informatics, electronics), only seemingly far distant from biology. Focusing on an ancient and widespread sensory channel, biotremology offers an opportunity to address general and modality-specific questions in animal communication (Cocroft et al. 2014b), whilst it also offers possibilities in the context of a world in search of smart solutions for agriculture and forestry. In terms of decision-making based on AI approaches and reduced environmental impact, biotremology is a very promising road towards sustainability. Nevertheless, many issues still hamper its progress and there is the need to invest in terms of both resources and interest, which passes through a full recognition of its role by other scientists and stakeholders. In particular, we are confident that biotremology will soon play an important role in ecology and plant protection. Currently, both basic studies and the applications are quickly developing whilst trying to overcome the lack of previous experience to acquire crucial information.

Ecological studies based on biotremology still suffer from the lack of knowledge of the natural vibrational environment. In fact, in order to understand the selection pressures imposed by the environment, it is necessary to determine the sources and their relevance in the evolution of signal and behavioural strategies of signallers and receivers under realistic ecological conditions. In biotremology, we have reached the stage where we can start building a comprehensive and realistic view of the vibratory world around us and recognizing not only the challenges that animals relying on vibrational signalling are facing in their environment but also adaptations that allow them to cope with these challenges. Efforts should be made to study vibroscape in different habitats and on different plant substrates and relate the characteristics of the natural vibrational environment where insects live and behave with their sensory mechanisms. So far, the only existing study (Šturm et al., 2021) has merely provided a preview into the natural vibrational environment at a single field location. Future research should aim at extending to a greater number of insect species in different taxonomic groups living in different ecological conditions. In a more general perspective, vibroscape studies should address themes of ecoacoustics where biodiversity is investigated, also including agroecosystems and considerations on human impacts.

Despite the fact that we already know a great deal about how vibration stimuli are detected at the level of insect receptor organs, many aspects of their function and evolution are still weakly understood. For example, even for a specialized vibration sensor such as SGO, it is unclear whether it has evolved once or more times independently in insects (Lakes-Harlan and Strauß, 2014), and what is the adaptation value of the diversity of functional morphologies and complexities of this organ (complex) seen across insects. In this respect, increasing the range of taxa with known functional morphology of vibratory receptor organs would help to elucidate their evolution on a finer phylogenetic scale. Furthermore, even in the model species extensively studied for the physiology of vibration detection, a direct link between morphology/peripheral origin, central projection patterns, and the physiology of individual vibratory sensilla has not been unequivocally established, which limits our understanding of the important aspects of the system's function. Another important but understudied topic in insect vibration detection is the role of body and organ mechanics, including postural effects (Stritih-Peljhan and Strauß, 2018).

Compared to other biotremology research topics, progress in understanding neuronal processing and perception of vibration stimuli in insects has been especially slow. Whilst a number of neuron elements have been described morphologically and physiologically in the two model taxa to date, and quite some insight has been gained into the input integration and filtering at the first synaptic levels of processing, we have no knowledge on how insect categorize and interpret vibration stimuli at final sensory processing stages in the brain. Together with that, our understanding of the link between neurophysiology and behaviour is still weak, as well. In addition, we still need to investigate how insects filter relevant signals in the presence of noise, and how they differentiate between signals of different origin in the complex transmission environments, such as plants. Besides that, we would highly benefit from extending the neuronal studies from Orthoptera and stink bugs to further insect lineages, both closely and distantly related, for comparative reasons. This may, amongst others, help to elucidate whether vibratory processing networks are homologous in insects or have evolved independently more times, to what degree may have their homologous elements diverged functionally during evolution, and also how may have variability in the signalling substrates influenced signal processing. The interest and expertise in neurophysiology thus need to be extended also outside the laboratories currently focused to these scarcely addressed, but highly interesting topics, in insect biotremology.

In the field of plant protection, biotremology offers innovation and the unique chance to explore brand new methods and strategies. Considering the large number of insect species that rely on substrate-borne vibrations to communicate, and that many agricultural insect pests fall in this category (e.g., Hemiptera), the breadth of potential for applied biotremology is extremely wide. Many relevant species in agroecosystems already have been reported to use vibrational signals in crucial behaviours, but applicative studies are missing. This is the case with some leafhoppers, spittlebugs, and aleyrodidae (Avosani et al., 2020a; Derlink et al., 2018; Fattoruso et al., 2021). There are many species whose behaviour needs a deep and accurate study with regards to vibrational communication. These include, for instance, aphids, Thysanoptera, and many Heteroptera that are economically relevant crop pests. Species like the spotted lanternfly, *Lycorma delicatula*, and the palm planthopper, *Haplaxius crudus*, are two examples of insect pests that are threatening viticulture (Lee et al., 2019) and palm crops (Dzido et al., 2020), respectively, and that could be treated with approaches from applied biotremology. The lists of potential invasive pest species, or pests that have already been enlisted as quarantine pests, is huge (for example, in Europe they are enlisted by EPPO, 2020a,b) as well as the potential research that can be conducted on them using approaches of biotremology. As for the practical part, one can argue that the transmission of vibrations into the field requires lots of energy and the deployment of electronic devices in the field could be too expensive, over acceptability from the market. However, experience teaches us that technology is always in progress whilst new solutions at progressively lower costs become rapidly available. For these reasons, we are optimistic and not afraid to foresee a bright future for biotremologists.

Acknowledgements

We thank Peggy Hill for her continuous support, which is one of the main reasons that the field of biotremology has become a successful discipline. We also thank her for helping to improve the present manuscript. We thank Dušan Devetak for the discussion of physiological data. We also thank Takuma Takanashi, Richard Mankin and Rok Šturm for providing images and/or audio recordings. Finally, we would like to thank Vittorio Veronelli (CBC-Europe) for the precious support over the years in the development of applied biotremology.

Funding information

JS—Funded by the Deutsche Forschungsgemeinschaft (DFG, German Research Foundation)—STR 1329/2-1; NSP and MVD received financial support from the Slovenian Research Agency (Javna agencija za raziskovalno dejavnost RS) (research core funding no. P1-0255 and research project J1-8142).

References

Acheampong, S., Mitchell, B.K., 1997. Quiescence in the Colorado potato beetle, *Leptinotarsa decemlineata*. Entomol. Exp. Appl. 82, 83–89.

Agrawal, S., Dickinson, E.S., Sustar, A., Gurung, P., Shepherd, D., Truman, J.W., Tuthill, J.C., 2020. Central processing of leg proprioception in *Drosophila*. eLife 9, e60299.

Ahi, J., Kalmring, K., Ebendt, R., Hellweg, J., 1993. Physiology and central projection of auditory receptor cells in the prothoracic ganglion of three closely related species of bush-crickets (Tettigoniidae, Insecta). J. Exp. Zool. 265, 684–692.

Akassou, I., Avosani, S., Caorsi, V., Verrastro, V., Ciolli, M., Mazzoni, V., 2021. Intrasexual vibrational behavior of *Philaenus spumarius* in semi-field conditions. Insects 12 (7), 584.

Albert, J.T., Göpfert, M.C., 2015. Hearing in *Drosophila*. Curr. Opin. Neurobiol. 34, 79–85.

Alt, J.A., Lakes-Harlan, R., 2018. Sensing of substrate vibrations in the adult cicada *Okanagana rimosa* (Hemiptera: Cicadidae). J. Insect Sci. 18, 16.

Aszkler, C., 2005. Acceleration, shock and vibration sensors. In: Wilson, J. (Ed.), Sensor Technology Handbook. Elsevier, Burlington, pp. 137–159.

Atkins, G., Pollack, G.S., 1987. Response properties of prothoracic, interganglionic, sound-activated interneurons in the cricket *Teleogryllus oceanicus*. J. Comp. Physiol. A. 161, 681–693.

Attenborough, K., Van Renterghem, T., 2021. Predicting Outdoor Sound, second ed. CRC Press.

Autrum, H., 1941. Über Gehör und Erschütterungssinn bei Locustiden. Z. vergl. Physiol. 28, 580–637.

Autrum, H., Schneider, W., 1948. Vergleichende Untersuchungen über den Erschütterungssinn der Insekten. Z. vergl. Physiol. 31, 77–88.

Avosani, S., Daher, E., Franceschi, P., Ciolli, M., Verrastro, V., Mazzoni, V., 2020a. Vibrational communication and mating behavior of the meadow spittlebug *Philaenus spumarius*. Entomol. Gen. 40, 307–321.

Avosani, S., Berardo, A., Pugno, N.M., Verrastro, V., Mazzoni, V., Cornara, D., 2021a. Vibrational disruption of feeding behaviors of a vector of plant pathogen. Entomol. Gen. https://doi.org/10.1127/entomologia/2021/1327.

Avosani, S., Franceschi, P., Ciolli, M., Verrastro, V., Mazzoni, V., 2021b. Vibrational playbacks and microscopy to study the signalling behaviour and female physiology of Philaenus spumarius. J. Appl. Entomol. 145, 518–529.

Avosani, S., Sullivan, T.E., Ciolli, M., Mazzoni, V., Suckling, D., 2020b. Can vibrational playbacks disrupt mating or influence other relevant behaviours in *Bactericera cockerelli* (Triozidae: Hemiptera)? Insects 11, 299.

Bacher, S., Casas, J., Wäckers, F., Dorn, S., 1997. Substrate vibrations elicit defensive behavior in leafminer pupae. J. Insect Physiol. 43, 945–952.

Bailey, W.J., 1991. Acoustic Behaviour of Insects: An Evolutionary Perspective. Chapman and Hall, London.

Ball, E.E., Hill, K.G., 1978. Functional development of the auditory system of the cricket, *Teleogryllus commodus*. J. Comp. Physiol. 127, 131–138.

Baroni-Urbani, C., Buser, M.W., Schilliger, E., 1988. Substrate vibration during recruitment in ant social organization. Insect. Soc. 35, 241–250.

Barth, F.G., 1986. Vibrationssinn und vibratorische Umwelt von Spinnen. Naturwissenschaften 73, 519–530.

Barth, F.G., Bleckmann, H., Bohnenberger, J., Seyfarth, E.A., 1988. Spider of genus *Cupiennius* Simon 1891 (Araneae, Ctenidae). II. On the vibratory environment of a wandering spider. Oecologia 77, 194–201.

Bell, P.D., 1980a. Multimodal communication by the black-horned tree cricket, *Oecanthus nigricornis* (Walker) (Orthoptera: Gryllidae). Can. J. Zool. 58, 1861–1868.

Bell, P.D., 1980b. Transmission of vibrations along plant stems: implications for insect communication. J. New York Entomol. Soc. 88, 210–216.

Benediktov, A., Korsunovskaya, O., Polilov, A., Zhantiev, R., 2020. Unusual mechanism of vibratory signals in pygmy grasshoppers *Tetrix tenuicornis* (Sahlberg, 1891) (Orthoptera: Tetrigidae). Sci Nat 107, 11.

Bennet-Clark, H.C., 1998. Size and scale effects as constraints in insect sound communication. Philos. T. Roy. Soc. B 353, 407–419.

Bennet-Clark, H.C., 1999. Resonators in insect sound production: how insects produce loud pure-tone songs. J. Exp. Biol. 202, 3347–3357.

Bertram, S.M., Rook, V.L.M., Fitzsimmons, L.P., 2010. Strutting their stuff: victory displays in the spring field cricket, *Gryllus veletis*. Behaviour 147, 1249–1266.

Bickmeyer, U.G., 1991. Electrophysiologische und neuroanatomische Untersuchung vibrosensitiver Neuronen im Bauchmark der Wanderheuschrecke *Locusta migratoria*. Doctoral dissertation, Philipps-Universität Marburg, Marburg, Germany.

Bickmeyer, U., Kalmring, K., Halex, H., Mücke, A., 1992. The bimodal auditory–vibratory system of the thoracic ventral nerve chord in *Locusta migratoria* (Acrididae, Locustinae, Oedipodini). J. Exp. Zool. 264, 381–394.

Blassioli-Moraes, M.C., Laumann, R.A., Čokl, A., Borges, M., 2005. Vibratory signals of four neotropical stink bug species. Physiol. Entomol. 30, 175–188.

Boyan, G.S., 1993. Another look at insect audition; the tympanic receptors as an evolutionary specialisation of the chordotonal system. J. Insect Physiol. 39, 187–200.

Brennan, B.J., 2007. Abdominal wagging in the social paper wasp *Polistes dominulus*: behavior and substrate vibrations. Ethology 113, 692–702.

Broder, E.D., Wikle, A.W., Gallagher, J.H., Tinghitella, R.M., 2021. Substrate-borne vibration in Pacific field cricket courtship displays. J. Orthop. Res. 30, 43–50.

Brown, W.D., 2016. Mating behavior of the endemic Hawaiian cricket *Leptogryllus elongatus* (Orthoptera: Gryllidae: Oecanthinae). J. Insect Behav. 29, 449–458.

Brownell, P.H., 1977. Compressional and surface waves in sand: used by desert scorpions to locate prey. Science 197, 479–482.

Brownell, P.H., van Hemmen, J.L., 2001. Vibration sensitivity and a computational theory for prey localizing behavior in sand scorpion. Am. Zool. 41, 1229–1240.

Buh, B., 2011. Morphological and Functional Characterization of Vibratory Receptor Neurons in Cave Crickets of the Genus *Troglophilus* (Orthoptera, Rhaphidophoridae). University of Ljubljana, Ljubljana, Diploma thesis.

Bullock, T.H., 1984. Comparative neuroethology of startle, rapid escape, and giant fiber-mediated responses. In: Eaton, R.C. (Ed.), Neural Mechanisms of Startle Behavior. Springer, Boston, pp. 1–13.

Büschges, A., 1994. The physiology of sensory cells in the ventral scoloparium of the stick insect femoral chordotonal organ. J. Exp. Biol. 189, 285–292.

Caldwell, M.S., 2014. Interactions between airborne sound and substrate vibration in animal communication. In: Cocroft, R.B., Gogala, M., Hill, P.S.M., Wessel, A. (Eds.), Studying Vibrational Communication. Springer, Berlin, pp. 65–92.

Caorsi, V., Cornara, D., Wells, K.E., Moser, D., Berardo, A., Miselli, R., Torriani, M., Pugno, N.M., Tasin, M., Maistrello, L., Mazzoni, V., 2021. Design of ideal vibrational signals for stinkbug male attraction, through vibrotaxis experiments. Pest Manag. Sci. https://doi.org/10.1002/ps.6590.

Cardé, R., 1990. Principles of mating disruption. In: Ridgway, R., Silverstein, R., Inscoe, M. (Eds.), Behavior-Modifying Chemicals for Insect Management. Marcel Dekker Inc., New York, pp. 47–71.

Castellanos, I., Barbosa, P., 2006. Evaluation of predation risk by a caterpillar using substrate-borne vibrations. Anim. Behav. 72, 461–469.

Cocroft, R.B., 1999. Parent-offspring communication in response to predators in a subsocial treehopper (Hemiptera: Membracidae: *Umbonia*). Ethology 105, 553–568.

Cocroft, R.B., 2001. Vibrational communication and the ecology of group-living, herbivorous insects. Am. Zool. 41, 1215–1221.

Cocroft, R.B., 2005. Vibrational communication facilitates cooperative foraging in a phloem-feeding insect. Proc. R. Soc. B 272, 1023–1029.

Cocroft, R.B., De Luca, P., 2006. Size–frequency relationships in insect vibratory signals. In: Drosopoulos, S., Claridge, M.F. (Eds.), Insect Sounds and Communication. Taylor & Francis, Boca Raton, FL, pp. 99–110.

Cocroft, R.B., Hamel, J.A., 2010. Vibrational communication in the "other insect societies": a diversity of ecology, signals and signal functions. In: O'Connell-Rodwell, C.E. (Ed.), The Use of Vibrations in Communication: Properties, Mechanisms and Function Across Taxa. Transworld Research Network, Kerala, pp. 47–68.

Cocroft, R.B., McNett, G.D., 2006. Vibrational communication in treehoppers (Hemiptera: Membracidae). In: Drosopoulos, S., Claridge, M.F. (Eds.), Insect Sounds and Communication: Physiology, Behaviour, Ecology and Evolution. Taylor & Francis, Boca Raton, FL, pp. 305–317.

Cocroft, R.B., Rodriguez, R.L., 2005. The behavioral ecology of insect vibrational communication. Bioscience 55, 323–334.

Cocroft, R.B., Tieu, T.D., Hoy, R.R., Miles, R.N., 2000. Directionality in the mechanical response to substrate vibration in a treehopper (Hemiptera: Membracidae: *Umbonia crassicornis*). J. Comp. Physiol. A. 186, 695–705.

Cocroft, R.B., Rodríguez, R.L., Hunt, R.E., 2010. Host shifts and signal divergence: mating signals covary with host use in a complex of a specialized plant-feeding insects. Biol. J. Linn. Soc. 99, 60–72.

Cocroft, R.B., Gogala, M., Hill, P.S.M., Wessel, A., 2014a. Fostering research progress in a rapidly growing field. In: Cocroft, R.B., Gogala, M., Hill, P.S.M., Wessel, A. (Eds.), Studying Vibrational Communication. Springer-Verlag, Berlin, Heidelberg, pp. 3–12.

Cocroft, R.B., Hamel, J.A., Su, Q., Gibson, J., 2014b. Vibrational playback experiments: challenges and solutions. In: Cocroft, R.B., Gogala, M., Hill, P.S.M., Wessel, A. (Eds.), Studying Vibrational Communication. Springer-Verlag, Berlin, Heidelberg, pp. 249–274.

Čokl, A., 1983. Functional properties of vibroreceptors in the legs of *Nezara viridula* (L.) (Heteroptera, Pentatomidae). J. Comp. Physiol. A. 150, 261–269.

Čokl, A., 2008. Stink bug interaction with host plants during communication. J. Insect Physiol. 54, 1113–1124.

Čokl, A., Amon, T., 1980. Vibratory interneurons in the central nervous system of *Nezara viridula* L. (Pentatomidae, Heteroptera). J. Comp. Physiol. A. 139, 87–95.

Čokl, A., Virant-Doberlet, M., 2003. Communication with substrate-borne signals in small plant-dwelling insects. Annu. Rev. Entomol. 48, 29–50.

Čokl, A., Virant-Doberlet, M., 2009. Vibrational communication. In: Resh, V.H., Cardé, R.T. (Eds.), Encyclopedia of Insects, second ed. Academic Press, Amsterdam, pp. 1034–1038.

Čokl, A., Kalmring, K., Wittig, H., 1977. The responses of auditory ventral-cord neurons of *Locusta migratoria* to vibration stimuli. J. Comp. Physiol. A. 120, 161–172.
Čokl, A., Otto, C., Kalmring, K., 1985. The processing of directional vibratory signals in the ventral nerve cord of *Locusta migratoria*. J. Comp. Physiol. A. 156, 45–52.
Čokl, A., Kalmring, K., Rössler, W., 1995. Physiology of atympanate tibial organs in forelegs and midlegs of the cave-living Ensifera, *Troglophilus neglectus* (Rhaphidophoridae, Gryllacridoidea). J. Exp. Zool. 273, 376–388.
Čokl, A., Virant-Doberlet, M., Zorović, M., 2006. Sense organs involved in the vibratory communication of bugs. In: Drosopoulos, S., Claridge, M.F. (Eds.), Insect Sounds and Communication—Physiology, Behaviour, Ecology and Evolution. CRC Press, Boca Raton, FL, pp. 71–80.
Čokl, A., Zorović, M., Millar, J.G., 2007. Vibrational communication along plants by the stink bugs *Nezara viridula* and *Murgantia histrionica*. Behav. Process. 75, 40–54.
Čokl, A., Zorović, M., Žunič-Kosi, A., Stritih, N., Virant-Doberlet, M., 2014. Communication through plants in a narrow frequency window. In: Cocroft, R.B., Gogala, M., Hill, P.S.M., Wessel, A. (Eds.), Studying Vibrational Communication. Springer-Verlag, Berlin, Heidelberg, pp. 171–195.
Conchou, L., Lucas, P., Meslin, C., Proffit, M., Staudt, M., Renou, M., 2019. Insect odorscape: from plant volatiles to natural olfactory scenes. Front. Physiol. 10, 972.
Conrad, T., Ayasse, M., 2015. The role of vibrations in population divergence in the Red Mason bee, *Osmia bicornis*. Curr. Biol. 25, 2819–2822.
Dambach, M., 1972a. Der Vibrationssinn der Grillen. I. Schwellenmessungen an Beinen frei beweglicher Tiere. J. Comp. Physiol. 79, 281–304.
Dambach, M., 1972b. Der Vibrationssinn der Grillen. II. Antworten von Neuronen im Bauchmark. J. Comp. Physiol. 79, 305–324.
Dambach, M., 1989. Vibrational responses. In: Huber, F., Moore, T.E., Loher, W. (Eds.), Cricket Behavior and Neurobiology. Comstock Publishing Associates, Ithaca, pp. 178–197.
Davis, A., Rubinstein, M., Wadhwa, N., Mysore, G.J., Durand, F., Freeman, W.T., 2014. The visual microphone: passive recovery of sound from video. ACM Trans. Graph. 33, 4. Article 79, 10 pages.
De Groot, M., Čokl, A., Virant-Doberlet, M., 2011a. Species identity cues: possibilities for errors during vibrational communication on plant stems. Behav. Ecol. 22, 1209–1217.
De Groot, M., Čokl, A., Virant-Doberlet, M., 2011b. Search behaviour of two hemipteran species using vibrational communication. Cent. Eur. J. Biol. 6, 756–769.
Debaisieux, P., 1938. Organes scolopidiaux des pattes d'insectes II. Cellule 47, 77–202.
Dejan, K.A., Fresquez, J.M., Meyer, A.M., Edgerly, J.S., 2013. Maternal territoriality achieved through shaking and lunging: an investigation of patterns in associated behaviors and substrate vibrations in a colonial embiopteran, *Antipaluria urichi*. J. Insect Sci. 13, 82.
Derlink, M., Abt, I., Mabon, R., Julian, C., Virant-Doberlet, M., Jacquot, E., 2018. Mating behaviour of Psammotettix alienus (Hemiptera: Cicadellidae). Insect Sci. 25, 148–160.
Derlink, M., Pavlovčič, P., Stewart, A.J.A., Virant-Doberlet, M., 2014. Mate recognition in duetting species: the role of male and female vibrational signals. Anim. Behav. 90, 181–193.
Devetak, D., 1985. Detection of substrate vibrations in the antlion larva, *Myrmeleon formicarius* (Neuroptera: Myrmeleonidae). Biol. Vestn. 33, 11–22.
Devetak, D., 1998. Detection of substrate vibration in Neuropteroidea: a review. Acta Zool. Fenn. 209, 87–94.
Devetak, D., 2014. Sand-borne vibrations in prey detection and orientation of antlions. In: Cocroft, R.B., Gogala, M., Hill, P.S.M., Wessel, A. (Eds.), Studying Vibrational Communication. Springer-Verlag, Berlin, Heidelberg, pp. 319–330.

Devetak, D., Amon, T., 1997. Substrate vibration sensitivity of the leg scolopidial organs in the green lacewing, *Chrysoperla carnea*. J. Insect Physiol. 43, 433–437.

Devetak, D., Pabst, M.A., 1994. Structure of the subgenual organ in the green lacewing, *Chrysoperla carnea*. Tissue Cell 26, 249–257.

Devetak, D., Gogala, M., Čokl, A., 1978. A contribution to the physiology of vibration receptors in the bugs of the family Cydnidae (Heteroptera). Biol. Vestn. 26, 131–139.

Devetak, D., Pabst, M.A., Lipovšek Delakorda, S., 2004. Leg chordotonal organs and campaniform sensilla in *Chrysoperla* Steinmann 1964 (Neuroptera): structure and function. Denisia 13, 163–171.

Devetak, D., Špernjak, A., Janžekovič, F., 2005. Substrate particle size affects pit building decision and pit size in the antlion larvae *Euroleon nostras* (Neuroptera: Myrmeleontidae). Physiol. Entomol. 30, 158–163.

Devetak, D., Mencinger-Vračko, B., Devetak, M., Marhl, M., Špernjak, A., 2007. Sand as a medium for transmission of vibratory signals of prey in antlions *Euroleon nostras* (Neuroptera: Myrmeleontidae). Physiol. Entomol. 32, 268–274.

DeVries, P.J., 1990. Acoustic enhancement of symbioses between butterfly caterpillars and ants. Science 248, 1104–1106.

Dias, A.M., Borges, M., Blassioli-Moraes, M.C., Lorran Figueira Coelho, M., Čokl, A., Laumann, R.A., 2021. Inhibitory copulation effect of vibrational rival female signals of three stink bug species as a tool for mating disruption. Insects 12, 177.

DiCaprio, R.A., Wolf, H., Büschges, A., 2002. Activity-dependent sensitivity of proprioceptive sensory neurons in the stick insect femoral chordotonal organ. J. Neurophysiol. 88, 2387–2398.

Dill, L.M., Fraser, A.H.G., Roitberg, B.D., 1990. The economics of escape behaviour in the pea aphid, *Acyrthosiphon pisum*. Oecologia 83, 473–478.

Djemai, I., Casas, J., Magal, C., 2001. Matching host reactions to parasitoid wasp vibrations. Proc. R. Soc. B 268, 2403–2408.

Dzido, J.L, Sánchez, R., Dollet, M., Julia, J.F., Narvaez, M., Fabre, S., Oropeza, C., 2020. Pest categorisation of *Haplaxius crudus*. Neotrop. Entomol. 49, 795–805.

Eberhard, M.J.B., Lang, D., Metscher, B., Pass, G., Picker, M.D., Wolf, H., 2010. Structure and sensory physiology of the leg scolopidial organs in Mantophasmatodea and their role in vibrational communication. Arthropod Struct. Dev. 39, 230–241.

Eggers, F., 1928. Die stiftführenden Sinnesorgane. Morphologie und Physiologie der chordotonalen und der tympanalen Sinnesapparate der Insekten. Borntraeger, Berlin.

Eibl, E., 1978. Morphology of the sense organs in the proximal parts of the tibiae of *Gryllus campestris* L. and *Gryllus bimaculatus* deGeer Insecta, Ensifera. Zoomorphologie 89, 185–205.

Elias, D.O., Mason, A.C., 2014. The role of wave and substrate heterogeneity in vibratory communication: practical issues in studying the effect of vibratory environments in communication. In: Cocroft, R.B., Gogala, M., Hill, P.S.M., Wessel, A. (Eds.), Studying Vibrational Communication. Springer-Verlag, Berlin, Heidelberg, pp. 215–247.

Elias, D.O., Mason, A.C., Hoy, R.R., 2004. The effect of substrate on the efficacy of seismic courtship signal transmission in the jumping spider *Habronattus dossenus* (Araneae: Salticidae). J. Exp. Biol. 207, 4105–4110.

Endler, J.A., 1993. Some general comments on the evolution and design of animal communication systems. Philos. Trans. R. Soc. Lond. B 340, 215–225.

Endler, J.A., 2019. Biotremology and sensory ecology. In: Hill, P.S.M., Lakes-Harlan, R., Mazzoni, V., Narins, P., Virant-Doberlet, M., Wessel, A. (Eds.), Biotremology: Studying Vibrational Behavior. Springer, Cham, pp. 27–41.

Endo, J., Takanashi, T., Mukai, H., Numata, H., 2019. Egg cracking vibration as a cue for stink bug siblings to synchronize hatching. Curr. Biol. 29, 143–148.

EPPO, 2020a. A2 List of Pests Recommended for Regulation as Quarantine Pests. https://www.eppo.int/ACTIVITIES/plant_quarantine/A2_list#insects.

EPPO, 2020b. A1 List of Pests Recommended for Regulation as Quarantine Pests. https://www.eppo.int/ACTIVITIES/plant_quarantine/A1_list.2020a.

Eriksson, A., Anfora, G., Lucchi, A., Virant-Doberlet, M., Mazzoni, V., 2011. Inter-plant vibrational communication in a leafhopper insect. PLoS One 6, e19692.

Eriksson, A., Anfora, G., Lucchi, A., Lanzo, F., Virant-Doberlet, M., Mazzoni, V., 2012. Exploitation of insect vibrational signals reveals a new method of pest management. PLoS One 7, e32954.

Evans, A.R., 1988. Mating systems and reproductive strategies in three Australian gryllid crickets: *Bobilla victoriae* Otte, *Balamara gidya* Otte, *Teleogryllus commodus* (Walker) (Orthoptera: Gryllidae: Nemobiinae; Trigonidiinae; Gryllinae). Ethology 78, 21–52.

Evans, T.A., Inta, R., Lai, J.C.S., Prueger, S., Foo, N.W., Fu, E.W., Lenz, M., 2009. Termites eavesdrop to avoid competitors. Proc. R. Soc. B 276, 4035–4041.

McKelvey, E.G.Z., Gyles, J.P., Michie, K., Pancorbo, V.B., Sober, L., Kruszewski, L.E., Chan, A., Fabre, C.C.G, 2021. Drosophila females receive male substrate-borne signals through specific leg neurons during courtship. Curr. Biol. 31, 1–11.

McNett, G.D., Cocroft, R.B., 2008. Host shifts favour vibrational signal divergence in *Enchenopa binotata* treehoppers. Behav. Ecol. 19, 650–656.

Fattoruso, V., Anfora, G., Mazzoni, V., 2021. Vibrational communication and mating behavior of the greenhouse whitefly *Trialeurodes vaporariorum* (Westwood) (Hemiptera: Aleyrodidae). Sci. Rep-UK 11, 6543.

Faure, P.A., Mason, A.C., Yack, J.E., 2009. Invertebrate Ears and Hearing. In: Binder, M.D., Hirokawa, N., Windhorst, U. (Eds.), Encyclopedia of Neuroscience. Springer, Berlin, pp. 2035–2042.

Fianco, M., Souza-Dias, P.G.B., Farias-Martins, F., Magro, S., Prasniewski, V.M., Ricci, J., Zefa, E., Szinwelski, N., 2018. Ethology of the cricket *Endecous* (*Endecous*) *chape* Souza-Dias de Mello, 2017 (Orthoptera: Grylloidea: Phalangopsidae) I: Agonistic and reproductive behavior. J. Orthop. Res. 27, 193–201.

Fiedler, K., Hölldobler, B., Seufert, P., 1996. Butterflies and ants: the communicative domain. Experientia 52, 14–24.

Field, L.H., Matheson, T., 1998. Chordotonal organs of insects. Adv. Insect Physiol. 27, 1–228.

Field, L.H., Pflüger, H.-J., 1989. The femoral chordotonal organ: a bifunctional orthopteran (*Locusta migratoria*) sense organ? Comp. Biochem. Physiol. A 93, 729–743.

Foster, S.P., Harris, M.O., 1997. Behavioral manipulation methods for insect pest-management. Annu. Rev. Entomol. 42, 123–146.

Friedel, T., 1999. The vibrational startle response of the desert locust *Schistocerca gregaria*. J. Exp. Biol. 202, 2151–2159.

Fuchs, S., 1976. The response to vibrations of the substrate and reactions to the specific drumming in colonies of carpenter ants (*Camponotus*, Formicidae, Hymenoptera). Behav. Ecol. Sociobiol. 1, 155–184.

Füller, H., Ernst, A., 1973. Die Ultrastruktur der femoralen Chordotonalorgane von *Carausius morosus* Br. Zool. Jb. Anat. 91, 574–601.

Gibb, R., Browning, E., Glover-Kapfer, P., Jones, K.E., 2019. Emerging opportunities and challenges for passive acoustics in ecological assessment and monitoring. Methods Ecol. Evol. 10, 169–185.

Godden, D.H., 1972. The motor innervation of the leg musculature and motor output during thanatosis in the stick insect *Carausius morosus*. Br. J. Comp. Physiol. 80, 211–225.

Gogala, M., 2006. Vibratory signals produced by Heteroptera — Pentatomorpha and Cimicomorpha. In: Drosopoulos, S., Claridge, M.F. (Eds.), Insect Sounds and Communication: Physiology, Behaviour. Ecology and Evolution. Taylor & Francis, Boca Raton FL, pp. 275–295.

Goodwyn, P.P., Katsumata-Wada, A., Okada, K., 2009. Morphology and neurophysiology of tarsal vibration receptors in the water strider *Aquarius paludum* (Heteroptera: Gerridae). J. Insect Physiol. 55, 855–861.
Gordon, S.D., Sandoval, N., Mazzoni, V., Krugner, R., 2017. Mating interference of glassy-winged sharpshooters, *Homalodisca vitripennis*. Entomol. Exp. Appl. 164, 27–34.
Gordon, S.D., Uetz, G.W., 2012. Environmental interference: impact of acoustic noise on seismic communication and mating success. Behav. Ecol. 23, 707–714.
Griffin, M.J., 1990. Handbook of Human Vibration. Elsevier Academic Press, London, UK.
Guedes, R.N.C., Matheson, S.M., Frei, B., Smith, M.L., Yack, J.E., 2012. Vibration detection and discrimination in the masked birch caterpillar (*Drepana arcuata*). J. Comp. Physiol. A. 198, 325–335.
Guédot, C., Horton, D.R., Landolt, P.J., 2010. Sex attraction in *Bactericera cockerelli* (Hemiptera: Triozidae). Environ. Entomol. 39, 1302–1308.
Hager, F.A., Kirchner, W.H., 2013. Vibrational long-distance communication in the termites *Macrotermes natalensis* and *Odontotermes* sp. J. Exp. Biol. 216, 3249–3256.
Hager, F.A., Krausa, K., 2019. Acacia ants respond to plant-borne vibrations caused by mammalian browsers. Curr. Biol. 29, 717–725.
Hager, F.A., Krausa, K., Kirchner, W.H., 2019. Vibrational behavior in termites (Isoptera). In: Hill, P.S.M., Lakes-Harlan, R., Mazzoni, V., Narins, P., Virant-Doberlet, M., Wessel, A. (Eds.), Biotremology: Studying Vibrational Behavior. Springer, Cham, pp. 309–327.
Hamel, J.A., Cocroft, R.B., 2012. Negative feedback from maternal signals reduces false alarms by collectively signaling offspring. Proc. R. Soc. B 279, 3820–3826.
Hanrahan, S.A., Kirchner, W.H., 1994. Acoustic orientation and communication in desert tenebrionid beetles in sand dunes. Ethology 97, 26–32.
Harris, C.M., Dinges, G.F., Haberkorn, A., Gebehart, C., Büschges, A., Zill, S.N., 2020. Gradients in mechanotransduction of force and body weight in insects. Arthropod Struct. Dev. 58, 100970.
Hartman, E., Rohde, B., Lujo, S., Dixon, M., McNeill, S., Mankin, R.W., 2017. Behavioral responses of male *Diaphorina citri* (Hemiptera: Liviidae) to mating communication signals from vibration traps in citrus (Sapindales: Rutaceae) trees. Fla. Entomol. 100, 767–771.
Hedwig, B., Stumpner, A., 2016. Central neural processing of sound signals in insects. In: Pollack, G.S., Mason, A.C., Popper, A.N., Fay, R.R. (Eds.), Insect Hearing. Springer, Cham, pp. 177–214.
Henry, C.S., Wells, M.M.L., Simon, C.M., 1999. Convergent evolution of courtship songs among cryptic species of the *carnea* group of green lacewings (Neuroptera: Chrysopidae: Chrysoperla). Evolution 53, 1165–1179.
Hehlert, P., Zhang, W., Göpfert, M.C., 2021. Drosophila mechanosensory transduction. Trends Neurosci. 44, 323–335.
Henry, C.S., Brooks, S.J., Duelli, P., Johnson, J.B., Well, M.M., Mochizuki, A., 2013. Obligatory duetting behaviour in the *Chrysoperla carnea*-group of cryptic species (Neuroptera: Chrysopidae): its role in shaping evolutionary history. Biol. Rev. 88, 787–808.
Hergenröder, R., Barth, F.G., 1983. Vibratory signals and spider behavior: how do the sensory inputs from the eight legs interact in orientation? J. Comp. Physiol. A. 152, 361–371.
Hill, P.S.M., 2008. Vibrational Communication in Animals. Harvard University Press, Cambridge, MA.
Hill, P.S.M., 2009. How do animals use substrate-borne vibrations as an information source? Naturwissenschaften 96, 1355–1371.
Hill, P.S.M., 2012. Do insect drummers actually drum? Studying vibrational communication across taxa. Mitt. Dtsch. Ges. Allg. Angew. Entomol. 18, 603–611.

Hill, P.S.M., 2014. Stretching the paradigm or building a new? Development of a cohesive language for vibrational communication. In: Cocroft, R.B., Gogala, M., Hill, P.S.M., Wessel, A. (Eds.), Studying Vibrational Communication. Springer-Verlag, Berlin, Heidelberg, pp. 13–30.

Hill, P.S.M., Shadley, J.R., 1997. Substrate vibration as a component of a calling song. Naturwissenschaften 84, 460–463.

Hill, P.S.M., Shadley, J.R., 2001. Talking back: sending soil vibration signals to lekking prairie mole cricket males. Am. Zool. 41, 1200–1214.

Hill, P.S.M., Wessel, A., 2016. Biotremology. Curr. Biol. 26, R187–R191.

Hirashima, A., 2009. Tyramine and octopamine receptors as a source of biorational insecticides. In: Ishaaya, I., Horowitz, A.R. (Eds.), Biorational Control of Arthropod Pests. Springer, Dordrecht, the Netherlands, pp. 83–109.

Hirashima, A., Nagano, T., Eto, M., 1993. Stress-induced changes in the biogenic-amine levels and larval growth of *Tribolium castaneum* Herbst. Biosci. Biotechnol. Biochem. 57, 2085–2089.

Hirashima, A., Takeya, R., Taniguchi, E., Eto, M., 1995. Metamorphosis, activity of juvenile-hormone esterase and alteration of ecdysteroid titres: effects of larval density and various stress on the red flour beetle, *Tribolium freemani* Hinton (Coleoptera: Tenebrionidae). J. Insect Physiol. 41, 383–388.

Hoch, H., Deckert, J., Wessel, A., 2006. Vibrational signaling in a Gondwanan relict insect (Hemiptera: Coleorrhyncha: Peloridiidae). Biol. Lett. 2, 222–224.

Hofmann, T., Bässler, U., 1982. Anatomy and physiology of trochanteral campaniform sensilla in the stick insect, *Cuniculina impigra*. Physiol. Entomol. 7, 413–426.

Hofmann, H.A., Schildberger, K., 2001. Assessment of strength and willingness to fight during aggressive encounters in crickets. Anim. Behav. 62, 337–348.

Hofmann, T., Koch, U.T., Bässler, U., 1985. Physiology of the femoral chordotonal organ in the stick insect, *Cuniculina impigra*. J. Exp. Biol. 114, 207–223.

Hofstetter, R.W., Dunn, D.D., McGuire, R., Potter, K.A., 2014. Using acoustic technology to reduce bark beetle reproduction. Pest Manag. Sci. 70, 24–27.

Hofstetter, R.W., McGuire, R., Dunn, D., 2016. Use of Acoustics to Disrupt and Deter Wood-Infesting Insects and other Invertebrates from and within Trees and Wood Products. U.S. Patent and Trademark Office, Washington, DC.

Hölldober, B., 1999. Multimodal signals in ant communication. J. Comp. Physiol. A. 184, 129–141.

Hosomi, A., 1996. Effect of vibration to the infestation of Apriona japonica (Thomson) (Coleoptera: Cerambycidae) adults on the fig. In: Proceedings of Japan Informal Group Meeting on Human Response to Vibration held at the Hokkaido Safety and Health Service, pp. 25–34.

Howse, P.E., 1962. The perception of vibration by the subgenual organ in *Zootermopsis angusticollis* Emerson and *Periplaneta americana* L. Experientia 18, 457–458.

Howse, P.E., 1964. An investigation into the mode of action of the subgenual organ in the termite, *Zootermopsis angusticollis* Emerson, and in the cockroach, *Periplaneta americana* L. J. Insect Physiol. 10, 409–424.

Howse, P.E., 1968. The fine structure and functional organization of chordotonal organs. Symp. Zool. Soc. Lond. 23, 167–198.

Hoy, R.R., Yack, J.E., 2009. Hearing. In: Resh, V.H., Cardé, R.T. (Eds.), Encyclopedia of Insects, second ed. Academic Press, Amsterdam, pp. 440–446.

Hrncir, M., Barth, F.G., 2014. Vibratory communication in stingless bees (Meliponini): the challenge of interpreting the signals. In: Cocroft, R.B., Gogala, M., Hill, P.S.M., Wessel, A. (Eds.), Studying Vibrational Communication. Springer-Verlag, Berlin, Heidelberg, pp. 349–374.

Hummel, J., Schöneich, S., Kössl, M., Scherberich, J., Hedwig, B., Prinz, S., Nowotny, M., 2016. Gating of acoustic transducer channels is shaped by biomechanical filter processes. J. Neurosci. 36, 2377–2382.

Hunt, J.H., Richard, F.J., 2013. Intracolony vibroacoustic communication in social insects. Insect. Soc. 60, 403–417.

Hustert, R., Pflüger, H.J., Bräunig, P., 1981. Distribution and specific central projections of mechanoreceptors in the thorax and proximal leg joints of locusts. III. The external mechanoreceptors: the campaniform sensilla. Cell Tissue Res. 216, 97–111.

Imaizumi, K., Pollack, G.S., 2001. Neural representation of sound amplitude by functionally different auditory receptors in crickets. J. Acoust. Soc. Am. 109, 1247–1260.

Inta, R., Evans, T.A., Lai, J.C.S., 2009. Effect of vibratory soldier alarm signals on the foraging behavior of subterranean termites (Isoptera: Rhinotermitidae). J. Econ. Entomol. 102, 121–126.

Ioriatti, C., Lucchi, A., 2016. Semiochemical strategies for tortricid moth control in apple orchards and vineyards in Italy. J. Chem. Ecol. 42, 571–583.

Jandt, J.M., Suryanarayanan, S., Hermanson, J.C., Jeanne, R.L., Toth, A.L., 2017. Maternal and nourishment factors interact to influence offspring developmental trajectories in social wasps. Proc. R. Soc. B 284, 20170651.

Javh, J., Slavič, J., Boltežar, M., 2018. Measuring full-field displacement spectral components using photographs taken with a DSLR camera via an analogue Fourier integral. Mech. Syst. Signal Pr. 100, 17–27.

Jeram, S., Čokl, A., 1996. Mechanoreceptors in insects: Johnston's organ in *Nezara viridula* (L.) (Pentatomidae, Heteroptera). Pflugers Arch. - Eur. J. Physiol. 431, R281–R282.

Jeram, S., Pabst, M.A., 1996. Johnston's organ and central organ in *Nezara viridula* (L.) (Heteroptera, Pentatomidae). Tissue Cell 28, 227–235.

Jinham, A.P., Solomon, K., Kumaran, J.T.T., Das, S.S.M., 2012. Efficacy of audible sound waves in inflicting tissue damage and mortality in *Tribolium castaneum* (Coleoptera: Tenebrionidae) larvae. Agric. Trop. Subtrop. 45, 32–36.

Kalmring, K., 1985. Vibrational communication in insects (reception and integration of vibratory information). In: Kalmring, K., Elsner, N. (Eds.), Acoustic and Vibrational Communication in Insects. Parey, Berlin, pp. 127–134.

Kalmring, K., Kühne, R., 1980. The coding of airborne-sound and vibration signals in bimodal ventral-cord neurons of the grasshopper *Tettigonia cantans*. J. Comp. Physiol. 139, 267–275.

Kalmring, K., Lewis, B., Eichendorf, A., 1978. The physiological characteristics of the primary sensory neurons of the complex tibial organ of *Decticus verrucivorus* L. (Orthoptera, Tettigonioidae). J. Comp. Physiol. 127, 109–121.

Kalmring, K., Kühne, R., Lewis, B., 1983. The acoustic behaviour of the bushcricket *Tettigonia cantans* III. Coprocessing of auditory and vibratory information in the nervous system. Behav. Process. 8, 213–228.

Kalmring, K., Kaiser, W., Otto, C., Kühne, R., 1985. Coprocessing of vibratory and auditory information in the CNS of different tettigoniids and locusts. In: Kalmring, K., Elsner, N. (Eds.), Acoustic and Vibrational Communication in Insects. Parey, Berlin, Hamburg, pp. 193–202.

Kalmring, K., Rössler, W., Unrast, C., 1994. Complex tibial organs in the forelegs, midlegs, and hindlegs of the bushcricket *Gampsocleis gratiosa* (Tettigoniidae): comparison of the physiology of the organs. J. Exp. Zool. 270, 155–161.

Kalmring, K., Hoffmann, E., Jatho, M., Sickmann, T., Grossbach, M., 1996. The auditory-vibratory sensory system of the bushcricket *Polysarcus denticauda* (Phaneropterinae, Tettigoniidae) II. Physiology of receptor cells. J. Exp. Zool. 276, 315–329.

Kalmring, K., Jatho, M., Rössler, W., Sickmann, T., 1997a. Acousto-vibratory communication in bushcrickets (Orthoptera: Tettigoniidae). Entomol. Gener. 21, 265–291.

Kalmring, K., Sickmann, T., Jatho, M., Zhantiev, R., Grossbach, M., 1997b. The auditory-vibratory sensory system of the bushcricket *Polysarcus denticauda*. (Phaneropterinae, Tettigoniidae) III. Physiology of the ventral cord neurons ascending to the head ganglia. J. Exp. Zool. 279, 9–28.

Kamikouchi, A., Ishikawa, Y., 2016. Hearing in *Drosophila*. In: Pollack, G.S., Mason, A.C., Popper, A.N., Fay, R.R. (Eds.), Insect Hearing. Springer, Cham, pp. 239–262.

Kavčič, A., Čokl, A., Laumann, R.A., Blassioli-Moraes, M.C., Borges, M., 2013. Tremulatory and abdomen vibration signals enable communication through air in the stink bug *Euschistus heros*. PLoS One 8, e56503.

Keil, T.A., 1997. Functional morphology of insect mechanoreceptors. Microsc. Res. Tech. 39, 506–531.

Keil, T.A., 2012. Sensory cilia in arthropods. Arthropod Struct. Dev. 41, 515–534.

Keuper, A., Kühne, R., 1983. The acoustic behavior of the bushcricket *Tettigonia cantans* II. Transmission of airborne-sound and vibration signals in the biotope. Behav. Process. 8, 125–145.

Keuper, A., Otto, C., Latimer, W., Schatral, A., 1985. Airborne sound and vibration signals of bushcrickets and locusts: their importance for the behavior in the biotope. In: Kalmring, K., Elsner, N. (Eds.), Acoustical and Vibrational Communication in Insects. Parey, Hamburg, pp. 135–142.

Kilpinen, O., Storm, J., 1997. Biophysics of the subgenual organ of the honeybee, *Apis mellifera*. J. Comp. Physiol. A 181, 309–318.

Kirchner, W.H., Broecker, I., Tautz, J., 1994. Vibrational alarm communication in the damp-wood termite *Zootermopsis nevadensis*. Physiol. Entomol. 19, 187–190.

Kittmann, R., Schmitz, J., 1992. Functional specialisation of the scoloparia of the femoral chordotonal organ in stick insects. J. Exp. Biol. 173, 91–108.

Kolmes, S.A., 1985. Surface vibrational cues in the precopulatory behavior of whirligig beetles. J. New York Entomol. Soc. 93, 1137–1140.

Korinšek, G., Derlink, M., Virant-Doberlet, M., Tuma, T., 2016. An autonomous system of detecting and attracting leafhopper males using species- and sex-specific substrate borne vibrational signals. Comput. Electron. Agric. 123, 29–39.

Korinšek, G., Tuma, T., Virant-Doberlet, M., 2019. Automated vibrational signal recognition and playback. In: Hill, P.S.M., Lakes-Harlan, R., Mazzoni, V., Narins, P., Virant-Doberlet, M., Wessel, A. (Eds.), Biotremology: Studying Vibrational Behavior. Springer, Cham, pp. 149–173.

Krahe, R., Budinger, E., Ronacher, B., 2002. Coding of a sexually dimorphic song feature by auditory interneurons of grasshoppers: the role of leading inhibition. J. Comp. Physiol. A 187, 977–985.

Krugner, R., Gordon, S.D., 2018. Mating disruption of *Homalodisca vitripennis* (Germar) (Hemiptera: Cicadellidae) by playback of vibrational signals in vineyard trellis. Pest Manag. Sci. 79, 2013–2019.

Kuhelj, A., de Groot, M., Blejec, A., Virant-Doberlet, M., 2016. Sender-receiver dynamics in leafhopper vibrational duetting. Anim. Behav. 114, 139–146.

Kühne, R., 1982a. Neurophysiology of the vibration sense in locusts and bushcrickets: response characteristics of single receptor units. J. Insect Physiol. 28, 155–163.

Kühne, R., 1982b. Neurophysiology of the vibration sense in locusts and bushcrickets: the responses of ventral-chord neurones. J. Insect Physiol. 28, 615–623.

Kühne, R., Lewis, B., Kalmring, K., 1980. The responses of the ventral cord neurons of *Decticus verrucivorus* (L) to sound and vibration stimuli. Behav. Process. 5, 55–74.

Kühne, R., Silver, S., Lewis, B., 1984. Processing of vibrators and acoustic signals by ventral cord neurones in the cricket *Gryllus campestris*. J. Insect Physiol. 30, 575–585.

Lakes, R., Schikorski, T., 1990. Neuroanatomy of tettigoniids. In: Bailey, W.J., Rentz, D.C.F. (Eds.), The Tettigoniidae: Biology, Systematics and Evolution. Crawford House Press, Bathurst, pp. 166–190.

Lakes-Harlan, R., Lefèvre, C., 2012. The femoral chordotonal organ of adult *Drosophila melanogaster* Meigen 1830. Mitt. Dtsch. Ges. Allg. Angew. Entomol. 18, 71–74.

Lakes-Harlan, R., Lehmann, G.U.C., 2015. Parasitoid flies exploiting acoustic communication of Insects—comparative aspects of independent functional adaptations. J. Comp. Physiol. A. 201, 123–132.

Lakes-Harlan, R., Strauß, J., 2014. Functional morphology and evolutionary diversity of vibration receptors in insects. In: Cocroft, R.B., Gogala, M., Hill, P.S.M., Wessel, A. (Eds.), Studying Vibrational Communication. Springer, Berlin, pp. 277–302.

Lakes-Harlan, R., Bailey, W.J., Schikorski, T., 1991. The auditory system of an atympanate bushcricket *Phasmodes ranatriformes* (Westwood) (Tettigoniidae: Orthoptera). J. Exp. Biol. 158, 307–324.

Lakes-Harlan, R., Stölting, H., Stumpner, A., 1999. Convergent evolution of insect hearing organs from a preadaptive structure. Proc. R. Soc. B 266, 1161–1167.

Lakes-Harlan, R., Jacobs, K., Allen, G.R., 2007. Comparison of auditory sense organs in parasitoid Tachinidae (Diptera) hosted by Tettigoniidae (Orthoptera) and homologous structures in a non-hearing Phoridae (Diptera). Zoomorphology 126, 229–243.

Lampson, B.D., Han, Y.J., Khalilian, A., Greene, J., Mankin, R.W., Foreman, E.G., 2013. Automatic detection and identification of brown stink bug, *Euschistus servus*, and southern green stink bug, *Nezara viridula*, (Heteroptera: Pentatomidae) using intraspecific substrate-borne vibrational signals. Comput. Electron. Agr. 91, 154–159.

Latimer, W., Schatral, A., 1983. The acoustic behavior of the bushcricket *Tettigonia cantans* I. Behavioural responses to sound and vibration. Behav. Process. 8, 113–124.

Laumann, R.A., Bottura Maccagnan, D.H., Čokl, A., 2017. Use of vibratory signals for stink bug monitoring and control. In: Čokl, A., Borges, M. (Eds.), Stink Bugs Biorational Control Based on Communication Processes. CRC Press, pp. 226–245.

Laumann, R.A., Maccagnan, D.H.B., Čokl, A., Blassioli-Moraes, M.C., Borges, M., 2018. Substrate-borne vibrations disrupt the mating behaviors of the neotropical brown stink bug, *Euschistus heros*: implications for pest management. J. Pest Sci. 91, 995–1004.

Lee, Y., Kim, H., Kang, T.J., Jang, Y., 2012. Stress response to acoustic stimuli in an aphid: a behavioral bioassay model. Entomol. Res. 42, 320–329.

Lee, D.H., Park, Y.L., Leskey, T.C., 2019. A review of biology and management of *Lycorma delicatula* (Hemiptera: Fulgoridae), an emerging global invasive species. J. Asia Pac. Entomol. 22, 589–596.

Lefebvre, P.C., Seifert, M., Stumpner, A., 2018. Auditory DUM neurons in a bush-cricket: a filter bank for carrier frequency. J. Comp. Neurol. 526, 1166–1182.

Leonhardt, S.D., Menzel, F., Nehring, V., Schmitt, T., 2016. Ecology and evolution of communication in social insects. Cell 164, 1277–1287.

Liang, X., Sun, L., Liu, Z., 2017. Mechanosensory Transduction in *Drosophila melanogaster*. Springer, Singapore.

Lin, Y., Rössler, W., Kalmring, K., 1994. Complex tibial organs in the fore-, mid- and hindlegs of the bushcrickets *Gampsocleis gratiosa* (Tettigoniidae): comparison of the morphology of the organs. J. Morphol. 221, 191–198.

Lin, Y., Rössler, W., Kalmring, K., 1995. Morphology of the tibial organs of Acrididae: comparison of subgenual and distal organs in fore-, mid-, and hindlegs of *Schistocerca gregaria* (Acrididae, Catantopidae) and *Locusta migratoria* (Acrididae, Oedipodinae). J. Morphol. 226, 351–360.

Lin, Y.-H., Liao, Y.-C., Yang, C.-C.Y., Billen, J., Yang, M.-M., Hsu, Y.-F., 2019. Vibrational communication between a myrmecophilous butterfly *Spindasis lohita* (Lepidoptera: Lycaenidae) and its host ant *Crematogaster rogenhoferi* (Hymenoptera: Formicidae). Sci. Rep. 9, 18548.

Lohrey, A.K., Clark, D.L., Gordon, S.D., Uetz, G.W., 2009. Antipredator responses of wolf spiders (Araneae: Lycosidae) to sensory cues representing an avian predator. Anim. Behav. 77, 813–821.

Losey, J.E., Denno, R.F., 1998a. Positive predator-predator interactions: enhanced predation rates and synergistic suppression of aphid populations. Ecology 79, 2143–2152.

Losey, J.E., Denno, R.F., 1998b. The escape response of pea aphids to foliar-foraging predators: factors affecting dropping behaviour. Ecol. Entomol. 23, 53–61.

Lujo, S., Hartman, E., Norton, K., Pregmon, E.A., Rohde, B.B., Mankin, R.W., 2016. Disrupting Mating Behavior of Diaphorina Citri (Liviidae). J. Econ. Entomol. 109, 1–7.

Lukic, I., Bedoya, C.L., Hofstetter, E.M., Hofstetter, R.W., 2021. Pinyon engraver beetle acoustics: stridulation apparatus, sound production and behavioral response to vibroacoustic treatments in logs. Insects 12, 496.

Ma, J.-Z., Lu, H.-Y., Li, X.-S., Tian, Y., 2020. Interfacial phenomena of water striders on water surfaces: a review from biology to biomechanics. Zool. Res. 41, 231–246.

Magal, C., Schöller, M., Tautz, J., Casas, J., 2000. The role of leaf structure in vibration propagation. J. Acoust. Soc. Am. 108, 2412–2418.

Mamiya, A., Gurung, P., Tuthill, J.C., 2018. Neural coding of leg proprioception in *Drosophila.* Neuron 100, 636–650.

Mankin, R.W., 2019. Vibrational trapping and interference with mating of *Diaphorina citri.* In: Hill, P.S.M., Lakes-Harlan, R., Mazzoni, V., Narins, P., Virant-Doberlet, M., Wessel, A. (Eds.), Biotremology: Studying Vibrational Behavior. Springer, Cham, pp. 399–413.

Mankin, R.W., Weaver, D.K., Grieshop, M., Larson, B., Morrill, W.L., 2004. Acoustic system for insect detection in plant stems: comparisons of *Cephus cinctus* in wheat and *Metamasius callizona* in Bromeliads. J. Agr. Urban Entomol. 21, 239–248.

Mankin, R.W., Hagstrum, D.W., Smith, M.T., Roda, A.L., Kairo, M.T.K., 2011. Perspective and promise: a century of insect acoustic detection and monitoring. Am. Entomol. 57, 30–44.

Mankin, R.W., Rohde, B.B., Mcneill, S.A., Paris, T.M., Zagvazdina, N.I., Greenfeder, S., 2013. *Diaphorina citri* (Hemiptera: Liviidae) responses to microcontroller-buzzer communication signals of potential use in vibration traps. Fla. Entomol. 96, 1546–1555.

Markl, H., 1965. Stridulation in leaf-cutting ants. Science 149, 1392–1393.

Markl, H., 1967. Die Verständigung durch Stridulationssignale bei Blattschneiderameisen. I. Die biologische Bedeutung der Stridulation. Z. Vergl. Physiol. 57, 299–330.

Markl, H., 1970. Die Verständigung durch Stridulationssignale bei Blattschneiderameisen. III. Die Empfindlichkeit für Substratvibrationen. Z. vergl. Physiol. 69, 6–37.

Markl, H., 1983. Vibrational communication. In: Huber, F., Markl, H. (Eds.), Neuroethology and Behavioral Physiology. Springer, Berlin, Heidelberg, pp. 332–353.

Markl, H., 1985. Manipulation, modulation, information, cognition: some of the riddles of communication. In: Hölldobler, B., Lindauer, M. (Eds.), Experimental Behavioral Ecology and Sociobiology. Gustav Fischer, Stuttgart, pp. 63–194.

Markl, H., Fuchs, S., 1972. Klopfsignale mit Alarmfunktion bei Roßameisen (Camponotus, Formicidae, Hymenoptera). Z. Vergl. Physiol. 76, 204–225.

Markl, H., Hölldobler, B., 1978. Recruitment and food-retrieving behavior in *Novomessor* (Formicidae, Hymenoptera). II. Vibration signals. Behav. Ecol. Sociobiol. 4, 183–216.

Markl, H., Wiese, K., 1969. Die Empfindlichkeit des Rückenschwimmers *Notonecta glauca* L. für Oberflächenwellen des Wassers. Z. vergl. Physiol. 62, 413–420.

Martinez, V., Nowbahari, E., Sillam-Dussès, D., Lorent, V., 2020. Antlions are sensitive to subnanometer amplitude vibrations carried by sand substrate. J. Comp. Physiol. A. 206, 783–791.

Mason, A.C., 1991. Hearing in a primitive ensiferan: the auditory system of *Cyphoderris monstrosa* (Orthoptera: Haglidae). J. Comp. Physiol. A. 168, 351–363.

Masoni, A., Frizzi, F., Nieri, R., Casacci, L.P., Mazzoni, V., Turillazzi, S., Santini, G., 2021. Ants modulate stridulatory signals depending on the behavioural context. Sci. Rep-UK 11, 1–12.

Matheson, T., Field, L.H., 1990. Innervation of the metathoracic femoral chordotonal organ of *Locusta migratoria*. Cell Tissue Res. 259, 551–560.

Matsuo, E., Kamikouchi, A., 2013. Neuronal encoding of sound, gravity, and wind in the fruit fly. J. Comp. Physiol. A. 199, 253–262.

Mazzoni, V., Lucchi, A., Čokl, A., Prešern, J., Virant-Doberlet, M., 2009a. Disruption of the reproductive behaviour of *Scaphoideus titanus* by playback of vibrational signals. Entomol. Exp. Appl. 133, 174–185.

Mazzoni, V., Prešern, J., Lucchi, A., Virant-Doberlet, M., 2009b. Reproductive strategy of the Nearctic leafhopper *Scaphoideus titanus* Ball (Hemiptera: Cicadellidae). Bull. Entomol. Res. 99, 401–413.

Mazzoni, V., Anfora, G., Virant-Doberlet, M., 2013. Substrate vibrations during courtship in three *Drosophila* species. PLoS One 8, e80708.

Mazzoni, V., Eriksson, A., Anfora, G., Lucchi, A., Virant-Doberlet, M., 2014. Active space and the role of amplitude in plant-borne vibrational communication. In: Cocroft, R.B., Gogala, M., Hill, P.S.M., Wessel, A. (Eds.), Studying Vibrational Communication. Springer-Verlag, Berlin, Heidelberg, pp. 125–145.

Mazzoni, V., Polajnar, J., Virant-Doberlet, M., 2015. Secondary spectral components of substrate-borne vibrational signals affect male preference. Behav. Process. 115, 53–60.

Mazzoni, V., Polajnar, J., Baldini, M., Rossi Stacconi, M.V., Anfora, G., Guidetti, R., Maistrello, L., 2017. Use of substrate-borne vibrational signals to attract the Brown Marmorated Stink Bug, *Halyomorpha halys*. J. Pest. Sci. 90, 1–11.

Mazzoni, V., Nieri, R., Eriksson, A., Virant-Doberlet, M., Polajnar, J., Anfora, G., Lucchi, A., 2019. Mating disruption by vibrational signals: state of the field and perspectives. In: Hill, P.S.M., Lakes-Harlan, R., Mazzoni, V., Narins, P., Virant-Doberlet, M., Wessel, A. (Eds.), Biotremology: Studying Vibrational Behavior. Springer, Cham, pp. 331–354.

Mc Cabe, S.I., Hrncir, M., Farina, W.M., 2015. Vibrating donor-partners during trophallaxis modulate associative learning ability of food receivers in the stingless bee *Melipona quadrifasciata*. Learn. Mem. 50, 11–21.

McNett, G.D., Luan, L.H., Cocroft, R.B., 2010. Wind-induced noise alters signaler and receiver behaviour in vibrational communication. Behav. Ecol. Sociobiol. 64, 2043–2051.

McVean, A., Field, L.H., 1996. Communication by substratum vibration in the New Zealand tree weta, *Hemideina femorata* (Stenopelmatidae: Orthoptera). J. Zool. (Lond.) 239, 101–122.

Meier, T., Reichert, H., 1990. Embryonic development and evolutionary origin of the orthopteran auditory organs. J. Neurobiol. 21, 592–610.

Mencinger-Vračko, B., Devetak, D., 2008. Orientation of the pit-building larva *Euroleon* (Neuroptera, Myrmeleontidae) to the direction of substrate vibrations caused by prey. Zoology 111, 2–8.

Menzel, J.G., Tautz, J., 1994. Functional morphology of the subgenual organ of the carpenter ant. Tissue Cell 26, 735–746.

Meyhöfer, R., Casas, J., 1999. Vibratory stimuli in host location by parasitic wasps. J. Insect Physiol. 45, 967–971.

Mhatre, N., Montealegre-Z, F., Balakrishnan, R., Robert, D., 2012. Changing resonator geometry to boost sound power decouples size and song frequency in a small insect. PNAS 109, E1444–E1452.

Michael, S.C.J., Appel, H.A., Cocroft, R.B., 2019. Methods for replicating leaf vibrations induced by insect herbivores. In: Gassmann, W. (Ed.), Plant Innate Immunity: Methods and Protocols. Methods in Molecular Biology. Springer Nature, pp. 141–157.

Michel, K., Amon, T., Čokl, A., 1983. The morphology of the leg scolopidial organs in *Nezara viridula* (L.) (Heteroptera, Pentatomidae). Rev. Can. Biol. Exp. 42, 139–150.
Michelsen, A., 2014. Mechanical signals in honeybee communication. In: Cocroft, R.B., Gogala, M., Hill, P.S.M., Wessel, A. (Eds.), Studying Vibrational Communication. Springer-Verlag, Berlin, Heidelberg, pp. 333–347.
Michelsen, A., Fink, F., Gogala, M., Traue, D., 1982. Plants as transmission channels for insect vibrational songs. Behav. Ecol. Sociobiol. 11, 269–281.
Michelsen, A., Kirchner, W.H., Lindauer, M., 1986a. Sound and vibrational signals in the dance language of the honeybee, *Apis mellifera*. Behav. Ecol. Sociobiol. 18, 207–212.
Michelsen, A., Kirchner, W.H., Andersen, B.B., Lindauer, M., 1986b. The tooting and quacking vibration signals of honeybee queens: a quantitative analysis. J. Comp. Physiol. 158, 605–611.
Mignini, M., Lorenzi, M.C., 2015. Vibratory signals predict rank and offspring caste ratio in a social insect. Behav. Ecol. Sociobiol. 69, 1739–1748.
Miles, R.M., Cocroft, R.B., Gibbons, C., Batt, D., 2001. A bending wave simulator for investigating directional vibration sensing in insects. J. Acoust. Soc. Am. 110, 579–587.
Montealegre-Z, F., Jonsson, T., Robson-Brown, K.A., Postles, M., Robert, D., 2012. Convergent evolution between insect and mammalian audition. Science 338, 968–971.
Morales, M.A., Barone, J.L., Hentry, C.S., 2008. Acoustic alarm signaling facilitates predator protection of treehoppers by mutualist ant bodyguards. Proc. Biol. Sci. 275, 1935–1941.
Morris, G.K., Mason, A.C., Wall, P., Belwood, J.J., 1994. High ultrasonic and tremulation signals in neotropical katydids (Orthoptera, Tettigoniidae). J. Zool. 233, 129–163.
Möss, D., 1971. Sinnesorgane im Bereich des Flügels der Feldgrille (*Gryllus campestris* L.) und ihre Bedeutung für die Kontrolle der Singbewegung und die Einstellung der Flügellage. Z. vergl. Physiol. 73, 53–83.
Moulins, M., 1976. Ultrastructure of chordotonal organs. In: Mill, P.J. (Ed.), Structure and Function of Proprioceptors in the Invertebrates. Chapman & Hall, London, pp. 387–426.
Mücke, A., 1991. Innervation pattern and sensory supply of the midleg of *Schistocerca gregaria*. Zoomorphology 110, 175–187.
Mücke, A., Lakes-Harlan, R., 1995. Central projections of sensory cells of the midleg of the locust, *Schistocerca gregaria*. Cell Tissue Res. 280, 391–400.
Mukai, H., Skals, N., Takanashi, T., 2020. Electrophysiological analysis of neural responses to substrate vibration in the Brown-Winged Green Bug, *Plautia stali* (Hemiptera: Pentatomidae). Jpn. J. Appl. Entomol. Zool. 64, 1–4.
Mullen, M.A., 1975. Infrasound retards development of *Tribolium castaneum* and *T. confusum*. J. Stored Prod. Res. 11, 111–113.
Murphey, R.K., 1971. Sensory aspects of the control of orientation to prey by the waterstrider, *Gerris remigis*. Z. Vergl. Physiol. 72, 168–185.
Murphey, R.K., 1973. Mutual inhibition and the organization of a non-visual orientation in *Notonecta*. J. Comp. Physiol. A. 84, 31–40.
Murphey, R.K., Mendenhall, B., 1973. Localization of receptors controlling orientation to prey by the back swimmer *Notonecta undulata*. J. Comp. Physiol. 84, 19–30.
Nebeling, B., 2000. Morphology and physiology of auditory and vibratory ascending interneurones in bushcrickets. J. Exp. Zool. 286, 219–230.
Nieh, J.C., 1993. The stop signal of honey bees: reconsidering its message. Behav. Ecol. Sociobiol. 33, 51–56.
Nieh, J.C., Tautz, J., 2000. Behaviour-locked signal analysis reveals weak 200-300 Hz comb vibrations during the honeybee waggle dance. J. Exp. Biol. 203, 1573–1579.
Nieri, R., Mazzoni, V., 2018. Vibrational mating disruption of *Empoasca vitis* by natural or artificial disturbance noises. Pest Manag. Sci. 75, 1065–1073.

Nieri, R., Anfora, G., Mazzoni, V., Rossi Stacconi, M.V., 2021. Semiochemicals, semiophysicals and their integration for the development of innovative multi-modal systems for agricultural pests' monitoring and control. Entomol. Gen. https://doi.org/10.1127/entomologia/2021/1236.

Nieri, R., Michael, S.C.J., Pinto, C.F., Orquizo, O.N., Appel, H.M., Cocroft, R.B., in press. Inexpensive methods for detecting and reproducing substrate-borne vibrations: advantages and limitations, in: Hill, P.S.M., Mazzoni, V., Stritih-Peljhan, N., Virant-Doberlet, M., Wessel, A. (Eds.), Biotremology: Physiology, ecology and evolution, Springer.

Nishino, H., 2000. Topographic mapping of the axons of the femoral chordotonal organ neurons in the cricket *Gryllus bimaculatus*. Cell Tissue Res. 299, 145–157.

Nishino, H., 2003. Somatotopic mapping of chordotonal organ neurons in a primitive ensiferan, the New Zealand tree weta *Hemideina femorata*: I Femoral chordotonal organ. J. Comp. Neurol. 464, 312–326.

Nishino, H., Field, L.H., 2003. Somatotopic mapping of chordotonal organ neurons in a primitive ensiferan, the New Zealand tree weta *Hemideina femorata*: II. Complex tibial organ. J. Comp. Neurol. 464, 327–342.

Nishino, H., Sakai, M., 1997. Three neural groups in the femoral chordotonal organ of the cricket *Gryllus bimaculatus*: central projections and soma arrangement and displacement during joint flexion. J. Exp. Biol. 200, 2583–2595.

Nishino, H., Mukai, H., Takanashi, T., 2016. Chordotonal organs in hemipteran insects: unique peripheral structures but conserved central organization revealed by comparative neuroanatomy. Cell Tissue Res. 366, 549–572.

Oberst, S., Bann, G., Lai, J.C.S., Evans, T.A., 2017. Cryptic termites avoid predatory ants by eavesdropping on vibrational cues from their footsteps. Ecol. Lett. 20, 212–221.

Oldfield, B.P., 1984. Physiology of auditory receptors in two species of Tettigoniidae (Orthoptera: Ensifera). J. Comp. Physiol. 155, 689–696.

Oldfield, B.P., Kleindienst, H.U., Huber, F., 1986. Physiology and tonotopic organization of auditory receptors in the cricket *Gryllus bimaculatus* DeGeer. J. Comp. Physiol. 159, 457–464.

Ota, D., Čokl, A., 1991. Mate location in the southern green stink bug, *Nezara viridula* (Heteroptera: Pentatomidae), mediated through substrate-borne signals on ivy. J. Insect Behav. 4, 441–447.

Otte, D., 1990. The relation between hearing and fling in insects. Ent. News 101, 29–34.

Otten, H., Wäckers, F.L., Isidoro, N., Romani, R., Dorn, S., 2002. The subgenual organ in *Pimpla turionellae* L. (Hymenoptera Ichneumonidae): ultrastructure and behavioral evidence for its involvement in vibrational sounding. Redia 85, 61–76.

Pearson, K.G., Hedwig, B., Wolf, H., 1989. Are the hind wing chordotonal organs of the locust elements in the flight pattern generator? J. Exp. Biol. 144, 235–255.

Pepiciello, I., Cini, A., Nieri, R., Mazzoni, V., Cervo, R., 2018. Adult–larval vibrational communication in paper wasps: the role of abdominal wagging in *Polistes dominula*. J. Exp. Biol. 221, jeb186247.

Pfannenstiel, R.S., Hunt, R.E., Yeargan, K.V., 1995. Orientation of a hemipteran predator to vibrations produced by feeding caterpillars. J. Insect Behav. 8, 1–9.

Pflüger, H.J., Bräunig, P., Hustert, R., 1988. The organization of mechanosensory neuropiles in locust thoracic ganglia. Philos. Trans. R. Soc. Lond. B 321, 1–26.

Pielström, S., Roces, F., 2012. Vibrational communication in the spatial organization of collective digging in the leaf-cutting ant *Atta vollenweideri*. Anim. Behav. 84, 743–752.

Pinhas, J., Soroker, V., Hetzroni, A., Mizrach, A., Teicher, M., Goldberger, J., 2008. Automatic acoustic detection of the red palm weevil. Comput. Electron. Agr. 63, 131–139.

Podlesnik, J., Klokočovnik, V., Lorent, V., Devetak, D., 2019. Prey detection in antlions: propagation of vibrational signals deep into the sand. Physiol. Entomol. 44, 215–221.

Polajnar, J., Čokl, A., 2008. The effect of vibratory disturbance on sexual behaviour of the southern green stink bug *Nezara viridula* (Heteroptera, Pentatomidae). Open Life Sci. 3, 189–197.

Polajnar, J., Svenšek, D., Čokl, A., 2012. Resonance in herbaceous plant stems as a factor in vibrational communication of pentatomid bugs (Heteroptera: Pentatomidae). J. R. Soc. Interface 9, 1898–1907.

Polajnar, J., Eriksson, A., Rossi Stacconi, M.V., Lucchi, A., Anfora, G., Virant-Doberlet, M., Mazzoni, V., 2014. The process of pair formation mediated by substrate-borne vibrations in a small. Insect. Behav. Process. 107, 68–78.

Polajnar, J., Eriksson, A., Virant-Doberlet, M., Lucchi, A., Mazzoni, V., 2016a. Developing a bioacoustic method for mating disruption of a leafhopper pest in grapevine. In: Horowitz, A., Ishaaya, I. (Eds.), Advances in Insect Control and Resistance Management. Springer, Cham, pp. 165–190.

Polajnar, J., Eriksson, A., Virant-Doberlet, M., Mazzoni, V., 2016b. Mating disruption of a grapevine pest using mechanical vibrations: from laboratory to the field. J. Pest. Sci. 89, 909–921.

Potamitis, I., Ganchev, T., Kontodimas, D., 2009. On automatic bioacoustics detection of pests: the cases of *Rhynchophorus ferrugineus* and *Sitophilus oryzae*. J. Econ. Entomol. 102, 1681–1690.

Prešern, J., 2007. Neurobiological fundamentals of orientation of the stink bug Nezara viridula (L.) (Heteroptera: Pentatomidae) towards the source of vibration signal. Doctoral dissertation, University of Ljubljana, Ljubljana, Slovenia.

Prešern, J., Triblehorn, J.D., Schul, J., 2015. Dynamic dendritic compartmentalization underlies stimulus-specific adaptation in an insect neuron. J. Neurophysiol. 113, 3787–3797.

Prešern, J., Polajnar, J., de Groot, M., Zorović, M., Virant-Doberlet, M., 2018. On the spot: utilization of directional cues in vibrational communication of a stink bug. Sci. Rep. 8, 5418.

Pringle, J.W.S., 1938. Proprioception in insects II. The action of the campaniform sensilla on the legs. J. Exp. Biol. 15, 114–131.

Rajaraman, K., Nair, A., Dey, A., Balakrishnan, R., 2018. Response mode choice in a multimodally duetting paleotropical pseudophylline bushcricket. Front. Ecol. Evol. 6, 172.

Regtien, P., Dertien, E., 2018. Sensors for Mechatronics. Elsevier, Amsterdam, pp. 245–265.

Rheinlaender, J., 1984. Das akustische Orientierungsverhalten von Heuschrecken. Eine vergleichende neuro- und verhaltensphysiologische Untersuchung. Habilitationsschrift, Universitat Bochum, Grillen und Fröschen.

Richard, F.J., Hunt, J.H., 2013. Intracolony chemical communication in social insects. Insect. Soc. 60, 275–291.

Rillich, J., Buhl, E., Schildberger, K., Stevenson, P.A., 2009. Female crickets are driven to fight by the male courting and calling songs. Anim. Behav. 77, 737–742.

Ritzmann, R.E., Pollack, A.J., Hudson, S.E., Hyvonen, A., 1991. Convergence of multimodal sensory signals at thoracic interneurons of the escape system of the cockroach, *Periplaneta americana*. Brain Res. 563, 175–183.

Roces, F., Hölldobler, B., 1995. Vibrational communication between hitchhikers and foragers in leaf-cutting ants (*Atta cephalotes*). Behav. Ecol. Sociobiol. 37, 297–302.

Roces, F., Hölldobler, B., 1996. Use of stridulation in foraging leaf-cutting ants: mechanical support during cutting or short-range recruitement signal? Behav. Ecol. Sociobiol. 39, 293–299.

Rohde, B., Paris, T.M., Heatherington, E.M., Hall, D.G., Mankin, R.W., 2013. Responses of *Diaphorina citri* (Hemiptera: Psyllidae) to conspecific vibrational signals and synthetic mimics. Ann. Entomol. Soc. Am. 106, 392–399.

Röhrig, A., Kirchner, W.H., Leuthold, R.H., 1999. Vibrational alarm communication in the African fungus-growing termite genus *Macrotermes* (Isoptera, Termitidae). Insect. Soc. 46, 71–77.

Rohrseitz, K., Kilpinen, O., 1997. Vibration transmission characteristics of the legs of freely standing honeybees. Zoology 100, 80–84.

Römer, H., 1985. Anatomical representation of frequency and intensity in the auditory system of Orthoptera. In: Kalmring, K., Elsner, N. (Eds.), Acoustic and Vibrational Communication in Insects. Paul Parey, Berlin, pp. 25–32.

Römer, H., Tautz, J., 1992. Invertebrate auditory receptors. In: Ito, F. (Ed.), Comparative Aspects of Mechanoreceptor Systems. Advances in Comparative and Environmental Physiology, vol. 10. Springer, Berlin, pp. 185–212.

Römer, H., Rheinlander, J., Dronse, R., 1981. Intracellular studies on auditory processing in the metathoracic ganglion of the locust. J. Comp. Physiol. A. 144, 305–312.

Römer, H., Lang, A., Hartbauer, M., 2010. The signaler's dilemma: a cost–benefit analysis of public and private communication. PLoS One 5 (10), e13325.

Römer, H., Smith, A.R., van Staaden, M., 2014. Hearing and sensory ecology of acoustic communication in bladder grasshoppers. In: Hedwig, B. (Ed.), Insect Hearing and Acoustic Communication. Springer, Berlin, Heidelberg, pp. 27–43.

Ronacher, B., Stumpner, A., 1988. Filtering of behaviourally relevant temporal parameters of a grasshopper's song by an auditory interneuron. J. Comp. Physiol. A. 163, 517–523.

Rössler, W., 1992. Functional morphology and development of tibial organs in the legs I, II and III of the bushcricket *Ephippiger* (Insecta, Ensifera). Zoomorphology 112, 181–188.

Rössler, W., Jatho, M., Kalmring, K., 2006. The auditory–vibratory sensory system in bushcrickets. In: Drosopoulos, S., Claridge, M.F. (Eds.), Insect Sounds and Communication. Taylor & Francis, Boca Raton, FL, pp. 35–69.

Rothberg, S.J., Allen, M.S., Castellini, P., Di Maio, D., Dirckx, J., Ewins, D.J., Halkon, B.J., Muyshondt, P., Paone, N., Ryan, T., Steger, H., Tomasini, E.P., Vanlanduit, S., Vignola, J.F., 2017. An international review of laser Doppler vibrometry: making light work of vibration measurement. Opt. Lasers Eng. 99, 11–22.

Rupprecht, R., 1968. Das Trommeln der Plecopteren. Z. vergl. Physiol 59, 38–71.

Sandeman, D., Tautz, J., Lindauer, M., 1996. Transmission of vibration across honeycombs and its detection by bee leg receptors. J. Exp. Biol. 199, 2585–2594.

Sarria, S.F.A., Chivers, D.B., Soulsbury, C.D., Montealegre, Z.F., 2017. Non-invasive biophysical measurements of travelling waves in the insect inner ear. R. Soc. Open Sci. 4, 170171.

Saxena, K.N., Kumar, H., 1980. Interruption of acoustic communication and mating in a leafhopper and a planthopper by aerial sound vibrations picked up by plants. Experientia 36, 933–936.

Saxena, K.N., Kumar, H., 1984. Acoustic communication in the sexual behaviour of the leafhopper, *Amrasca devastans*. Physiol. Entomol. 9, 77–86.

Scharf, I., Lubin, Y., Ovadia, O., 2011. Foraging decisions and behavioural flexibility in trap-building predators: a review. Biol. Rev. 86, 626–639.

Schmitz, J., Dean, J., Kittmann, R., 1991. Central projections of leg sense organs in *Carausius morosus* (Insecta, Phasmida). Zoomorphology 111, 19–33.

Schneider, W., 1950. Über den Erschütterungssinn von Käfern und Fliegen. Z. vergl. Physiol. 32, 287–302.

Schnorbus, H., 1971. Die subgenualen Sinnesorgane von *Periplaneta americana*: Histologie und Vibrationsschwellen. Z. vergl. Physiol. 71, 14–48.

Schön, A., 1911. Bau und Entwicklung des tibialen Chordotonalorgans bei der Honigbiene und bei Ameisen. Zool. Jb. Anat. Ontog. Tiere 31, 439–472.

Schöneich, S., Hedwig, B., 2010. Hyperacute directional hearing and phonotactic steering in the cricket (*Gryllus bimaculatus* deGeer). PLoS One 5, e15141.

Schul, J., Mayo, A.M., Triblehorn, J.D., 2012. Auditory change detection by a single neuron in an insect. J. Comp. Physiol. A. 198, 695–704.

Shaw, S.R., 1994a. Detection of airborne sound by a cockroach 'vibration detector': a possible missing link in insect auditory evolution. J. Exp. Biol. 193, 13–47.

Shaw, S.R., 1994b. Re-evaluation of the absolute threshold and response mode of the most sensitive known "vibration" detector, the cockroach's subgenual organ: a cochlea-like displacement threshold and a direct response to sound. J. Neurobiol. 25, 1167–1185.

Sickmann, T., 1997. Vergleichende funktionelle und anatomische Untersuchung zum Aufbau der Hör- und Vibrationsbahn im thorakalen Bauchmark von Laubheuschrecken. Doctoral dissertation, Philipps-Universität Marburg, Marburg, Germany.

Silver, S., Kalmring, K., Kühne, R., 1980. The responses of central acoustic and vibratory interneurones in bushcrickets and locusts to ultrasonic stimulation. Physiol. Entomol. 5, 427–435.

Smarandache-Wellmann, C.R., 2016. Arthropod neurons and nervous system. Curr. Biol. 26, R937–R980.

Song, H., Béthoux, O., Shin, S., Donath, A., Letsch, H., Liu, S., McKenna, D.D., Meng, G., Misof, B., Podsiadlowski, L., Zhou, X., Wipfler, B., Simon, S., 2020. Phylogenomic analysis sheds light on the evolutionary pathways towards acoustic communication in Orthoptera. Nat. Commun. 11, 4939.

Stein, W., Sauer, A., 1999. Physiology of vibration-sensitive afferents in the femoral chordotonal organ of the stick insect. J. Comp. Physiol. A. 184, 253–263.

Stewart, K.W., Zeigler, D.D., 1984. The use of larval morphology and drumming in Plecoptera systematics, and further studies of drumming behavior. Ann. Limnol. 20, 105–114.

Stiedl, O., Bickmeyer, U., 1991. Acoustic behaviour of *Ephippiger* Fiebig (Orthoptera, Tettigoniidae) within a habitat of Southern France. Behav. Process. 23, 125–135.

Stiedl, O., Kalmring, K., 1989. The importance of song and vibratory signals in the behavior of the bushcricket *Ephippiger* Fiebig (Orthoptera, Tettigoniidae): taxis by females. Oecologia 80, 142–144.

Stölting, H., Stumpner, A., 1998. Tonotopic organization of auditory receptor cells in the bushcricket *Pholidoptera griseoaptera* (Tettigoniidae, Decticinae). Cell Tissue Res. 294, 377–386.

Stölting, H., Moore, T.E., Lakes-Harlan, R., 2002. Substrate vibrations during acoustic signaling of *Okanagana rimosa* (Homoptera: Cicadidae: Tibicininae). J. Insect Sci. 2, 1–7.

Stölting, H., Stumpner, A., Lakes-Harlan, R., 2007. Morphology and physiology of the prosternal chordotonal organ of the sarcophagid fly *Sarcophaga bullata* (Parker). J. Insect Physiol. 53, 444–454.

Strauß, J., 2017a. The scolopidial accessory organ in the Jerusalem cricket (Orthoptera: Stenopelmatidae). Arthropod Struct. Dev. 46, 171–177.

Strauß, J., 2017b. The scolopidial accessory organs and Nebenorgans in orthopteroid insects: comparative neuroanatomy, mechanosensory function and evolutionary origin. Arthropod Struct. Dev. 46, 765–776.

Strauß, J., 2020. Neuronal innervation of the subgenual organ complex and tibial campaniform sensilla in the stick insect midleg. Insects 11, 40.

Strauss, J., Lakes-Harlan, R., 2008a. Neuroanatomy and physiology of the complex tibial organ of an atympanate ensiferan, *Ametrus tibialis* (Brunner von Wattenwyl, 1888) (Gryllacrididae, Orthoptera) and evolutionary implications. Brain Behav. Evol. 71, 167–180.

Strauss, J., Lakes-Harlan, R., 2008b. Neuroanatomy of the complex tibial organ of *Stenopelmatus* (Orthoptera: Ensifera: Stenopelmatidae). J. Comp. Neurol. 511, 81–91.
Strauß, J., Lakes-Harlan, R., 2009. The evolutionary origin of auditory receptors in Tettigonioidea: the complex tibial organ of Schizodactylidae. Naturwissenschaften 96, 143–146.
Strauß, J., Lakes-Harlan, R., 2010. Neuroanatomy of the complex tibial organ in the splay-footed cricket *Comicus calcaris* Irish 1986 (Orthoptera: Ensifera: Schizodactylidae). J. Comp. Neurol. 518, 4567–4580.
Strauß, J., Lakes-Harlan, R., 2013. Sensory neuroanatomy of stick insects highlights the evolutionary diversity of the orthopteroid subgenual organ complex. J. Comp. Neurol. 521, 3791–3803.
Strauß, J., Lakes-Harlan, R., 2014. Evolutionary and phylogenetic origins of tympanal hearing organs in insects. In: Hedwig, B. (Ed.), Insect Hearing and Acoustic Communication. Springer, Berlin, Heidelberg, pp. 5–26.
Strauß, J., Lakes-Harlan, R., 2017. Vibrational sensitivity of the subgenual organ complex in female *Sipyloidea sipylus* stick insects in different experimental paradigms of stimulus direction, leg attachment, and ablation of a connective tibial sense organ. Comp. Biochem. Physiol. A 203, 100–108.
Strauß, J., Stritih, N., 2016. The accessory organ, a scolopidial sensory organ, in the cave cricket *Troglophilus neglectus* (Orthoptera: Ensifera: Rhaphidophoridae). Acta Zool. 97, 187–195.
Strauß, J., Stumpner, A., 2015. Selective forces on origin, adaptation and reduction of tympanal ears in insects. J. Comp. Physiol. A. 201, 155–169.
Strauß, J., Stritih, N., Lakes-Harlan, R., 2014. The subgenual organ complex in the cave cricket *Troglophilus neglectus* (Orthoptera: Rhaphidophoridae): comparative innervation and sensory evolution. R. Soc. Open Sci. 1, 140240.
Strauß, J., Riesterer, A.S., Lakes-Harlan, R., 2016. How many sensory organs in the bushcricket leg? Neuroanatomy of the accessory organ in Tettigoniidae (Insecta: Orthoptera). Arthropod Struct. Dev. 45, 31–41.
Strauß, J., Lomas, K., Field, L.H., 2017. The complex tibial organ of the New Zealand ground weta: sensory adaptations for vibrational signal detection. Sci. Rep. 7, 2031.
Strauß, J., Stritih-Peljhan, N., Lakes-Harlan, R., 2019. Determining vibroreceptor sensitivity in insects: the influence of experimental parameters and recording techniques. In: Hill, P.S.M., Lakes-Harlan, R., Mazzoni, V., Narins, P.M., Virant-Doberlet, M., Wessel, A. (Eds.), Biotremology: Studying Vibrational Behavior. Springer, Cham, pp. 209–233.
Strauß, J., Moritz, L., Rühr, P.T., 2021. The subgenual organ complex in stick insects: functional morphology and mechanical coupling of a complex mechanosensory organ. Front. Ecol. Evol. 9, 632493.
Stritih, N., 2006. Response Properties, Morphology and Topographical Organisation of the Vibratory Neurones in the Prothoracic Ganglion of the Cave Cricket *Troglophilus neglectus* Krauss (Orthoptera, Rhaphidophoridae). Doctoral dissertation, University of Ljubljana, Ljubljana, Slovenia.
Stritih, N., 2009. Anatomy and physiology of a set of low-frequency vibratory interneurons in a nonhearing ensiferan (*Troglophilus neglectus*, Rhaphidophoridae). J. Comp. Neurol. 516, 519–532.
Stritih, N., Čokl, A., 2012. Mating behaviour and vibratory signaling in non-hering cave crickets reflect primitive communication of Ensifera. PLoS One 7, e47646.
Stritih, N., Čokl, A., 2014. The role of frequency in vibrational communication of Orthoptera. In: Cocroft, R.B., Gogala, M., Hill, P.S.M., Wessel, A. (Eds.), Studying Vibrational Communication. Springer, Berlin, Heidelberg, pp. 375–393.

Stritih, N., Stumpner, A., 2009. Vibratory interneurons of the non-hearing cave cricket indicate evolutionary origin of sound processing elements in Ensifera. Zoology 112, 48–68.

Stritih, N., Virant-Doberlet, M., Čokl, A., 2000. Green stink bug *Nezara viridula* detects differences in amplitude between courtship song vibrations at stem and petiolus. Pflugers Arch. - Eur. J. Physiol. 439, R190–R192.

Stritih-Peljhan, N., Rühr, P.T., Buh, B., Strauß, J., 2019. Low-frequency vibration transmission and mechanosensory detection in the legs of cave crickets. Comp. Biochem Physiol. A 233, 89–96.

Stritih-Peljhan, N., Strauß, J., 2018. The mechanical leg response to vibration stimuli in cave crickets and implications for vibrosensory organ functions. J. Comp. Physiol. A. 204, 687–702.

Stritih-Peljhan, N., Virant-Doberlet, M., n.d. Vibrational signaling, an underappreciated mode in cricket communication. Sci. Nat. In press.

Stumpner, A., 1996. Tonotopic organization of the hearing organ in a bushcricket. Naturwissenschaften 83, 81–84.

Stumpner, A., 1998. Picrotoxin eliminates frequency selectivity of an auditory interneuron in a bushcricket. J. Neurophysiol. 79, 2408–2415.

Stumpner, A., 2002. A species-specific frequency filter through specific inhibition, not specific excitation. J. Comp. Physiol. A. 188, 239–248.

Stumpner, A., Nowotny, M., 2014. Neural processing in the bush-cricket auditory pathway. In: Hedwig, B. (Ed.), Insect Hearing and Acoustic Communication. Springer, Berlin, Heidelberg, pp. 143–166.

Stumpner, A., Ronacher, B., 1991. Auditory interneurones in the metathoracic ganglion of the grasshopper *Chorthippus biguttulus*: I. Morphological and physiological characterization. J. Exp. Biol. 158, 391–410.

Stumpner, A., von Helversen, D., 2001. Evolution and function of auditory systems in insects. Naturwissenschaften 88, 159–170.

Stumpner, A., Lefebvre, P.C., Seifert, M., Ostrowski, T.D., 2019. Temporal processing properties of auditory DUM neurons in a bush-cricket. J. Comp. Physiol. A. 205, 717–733.

Stumpner, A., Gubert, S., Knorr, D.Y., Göpfert, M.C., 2020. Auditory DUM neurons in a bush-cricket: inhibited inhibitors. J. Comp. Physiol. A. 206, 793–807.

Šturm, R., Polajnar, R., Virant-Doberlet, M., 2019. Practical issues in studying natural vibroscape and biotic noise. In: Hill, P.S.M., Lakes-Harlan, R., Mazzoni, V., Narins, P., Virant-Doberlet, M., Wessel, A. (Eds.), Biotremology: Studying Vibrational Behavior. Springer, Cham, pp. 125–148.

Šturm, R., Rexhepi, B., López Díez, J.J., Blejec, A., Polajnar, J., Sueur, J., Virant-Doberlet, M., 2021. Hay-meadow vibroscape and interactions within insect vibrational community. iScience. https://doi.org/10.1016/j.isci.2021.103070.

Suryanarayanan, S., Hermanson, J.C., Jeanne, R.L., 2011. A mechanical signal biases caste development in a social wasp. Curr. Biol. 21, 231–235.

Takanashi, T., Kojima, W., 2021. Vibration-induced immobility in coleopteran insects. In: Sakai, M. (Ed.), Death-Feigning in Insects. Springer, Singapore, pp. 73–92.

Takanashi, T., Fukaya, M., Nakamuta, K., Skals, N., Nishino, H., 2016. Substrate vibrations mediate behavioural responses via femoral chordotonal organs in a cerambycid beetle. Zool. Lett. 2, 18.

Takanashi, T., Uechi, N., Tatsuta, H., 2019. Vibrations in hemipteran and coleopteran insects: behaviors and application in pest management. Appl. Entomol. Zool. 54, 21–29.

Tautz, J., 1977. Reception of medium vibration by thoracal hairs of caterpillars of *Barathra brassicae* L. (Lepidoptera, Noctuidae). I. Mechanical properties of the receptor hairs. J. Comp. Physiol. 118, 13–31.

ter Hofstede, H., Voigt-Heucke, S., Lang, A., Römer, H., Page, R., Faure, P., Dechmann, D., 2017. Revisiting adaptations of neotropical katydids (Orthoptera: Tettigoniidae) to gleaning bat predation. Neotrop. Biodivers. 3, 41–49.

Theophilidis, G., 1986. The femoral chordotonal organs of *Decticus albifrons* (Orthoptera: Tettigoniidae)—I. Structure. Comp. Biochem. Physiol. 84A, 529–536.

Tishechkin, D.Y., 2007. Background noises in vibratory communication channels of Homoptera (Cicadina and Psyllinea). Russ. Entomol. J. 16, 39–46.

Traue, D., 1978. Zur Biophysik der Schallabstrahlung bei Kleinzikaden Jassidae am Beispiel von *Euscelis incisus* Kb. (Homoptera-Cicadina: Jassidae). Zool. Beitr. N. F. 24, 155–164.

Travassos, M.A., Pierce, N.E., 2000. Aoustics, context and functions of vibrational signaling in a lycaenid butterfly-ant mutualism. Anim. Behav. 60, 13–26.

Tsubaki, R., Hosoda, N., Kitajima, H., Takanashi, T., 2014. Substrate-borne vibrations induce behavioral responses in the leaf-dwelling cerambycid, *Paraglenea fortunei*. Zool. Sci. 31, 789–794.

Tyrer, N.M., Gregory, G.E., 1982. A guide to the neuroanatomy of locust subesophageal and thoracic ganglia. Philos. Trans. R. Soc. Lond. B 297, 91–123.

Uechi, N., Takanashi, T., 2021. Behavioral control of *Plautia stali* (Hemiptera: Pentatomidae) using vibrations, and its application in pest management. Jpn. J. Appl. Entomol. Z. 65, 13–20.

Uetz, G.W., Roberts, J.A., 2002. Multisensory cues and multimodal communication in spiders: insights from video/audio playback studies. Brain Behav. Evol. 59, 222–230.

Uexküll, J., Kriszat, G., 1934. Streifzüge durch die Umwelten von Tieren und Menschen Ein Bilderbuch unsichtbarer Welten. Springer, Berlin, Heidelberg.

van Staaden, M.J., Römer, H., 1998. Evolutionary transition from stretch to hearing in ancient grasshoppers. Nature 394, 773–776.

Velilla, E., Muñoz, M., Quiroga, N., Symes, L., ter Hofstede, H.M., Page, R.A., Simon, R., Ellers, J., Halfwerk, W., 2020a. Gone with the wind: is timing in a neotropical katydid an adaptive response to variation in wind-induced vibratory noise. Behav. Ecol. Sociobiol. 74, 59.

Velilla, E., Polajnar, J., Virant-Doberlet, M., Commandeur, D., Simon, R., Cornelissen, J.H.C., Ellers, J., Halfwerk, W., 2020b. Variation in plant leaf traits affects transmission and detectability of herbivore vibrational cues. Ecol. Evol. 10, 12277–12289.

Virant-Doberlet, M., Čokl, A., 2004. Vibrational communication in insects. Neotrop. Entomol. 33, 121–134.

Virant-Doberlet, M., Čokl, A., Zorović, M., 2006. Use of substrate vibrations for orientation: from behaviour to physiology. In: Drosopoulos, S., Claridge, M.F. (Eds.), Insect Sounds and Communication. Taylor & Francis, Boca Raton, FL, pp. 81–97.

Virant-Doberlet, M., King, R.A., Polajnar, J., Symondson, W.O.C., 2011. Molecular diagnostics reveal spiders that exploit prey vibrational signals used in sexual communication. Mol. Ecol. 20, 2204–2216.

Virant-Doberlet, M., Mazzoni, V., de Groot, M., Polajnar, J., Lucchi, A., Symondson, W.O.C., Čokl, A., 2014. Vibrational communication networks and biotic noise. In: Cocroft, R.B., Gogala, M., Hill, P.S.M., Wessel, A. (Eds.), Studying Vibrational Communication. Springer, Berlin, Heidelberg, pp. 93–123.

Virant-Doberlet, M., Kuhelj, A., Polajnar, J., Šturm, R., 2019. Predator-prey interactions and eavesdropping in vibrational communication networks. Front. Ecol. Evol. 7, 203.

von Dohlen, P., Dambach, M., 1981. Vibration perception in the field cricket. Verh. Dtsch. Zool. Ges. 74, 216.

Weber, D.C., Morrison, W.R., Khriminian, A., Rice, K.B., Leskey, T.C., Rodriguez-Saona, C., Nielsen, A.L., Blaauw, B.R., 2017. Chemical ecology of *Halyomorpha halys*: discoveries and applications. J. Pest. Sci. 90, 989–1008.

Weidemann, S., Keuper, A., 1987. Influence of vibratory signals on the phonotaxis of the gryllid *Gryllus bimaculatus* Degeer (Ensifera: Gryllidae). Oecologia 74, 316–318.

Weissman, D.B., 2001. Communication and reproductive behaviour in North American Jerusalem crickets (*Stenopelmatus*) (Orthoptera: Stenopelmatidae). In: Field, L.H. (Ed.), The Biology of Wetas, King Crickets and their Allies. CABI Publishing, Wallingford, pp. 351–373.

Wessel, A., Mühlethaler, R., Hartung, V., Kuštor, V., Gogala, M., 2014. The Tymbal: evolution of a complex vibration-producing organ in the Tymbalia (Hemiptera excl. Sternorrhyncha). In: Cocroft, R.B., Gogala, M., Hill, P.S.M., Wessel, A. (Eds.), Studying Vibrational Communication. Springer, Berlin, Heidelberg, pp. 395–444.

Wiese, K., 1972. Das mechanorezeptorische Beuteortungssystem von Notonecta. I. Die Funktion des tarsalen Scolopidialorgans. J. Comp. Physiol. 78, 83–102.

Wiese, K., 1981. Influence of vibration on cricket hearing: interaction of low frequency vibration and acoustic stimuli in the omega neuron. J. Comp. Physiol. A 143, 135–142.

Wiese, K., Schmidt, K., 1974. Mechanorezeptoren im Insektentarsus. Z. Morph. Tiere 79, 47–63.

Yack, J.E., 2004. The structure and function of auditory chordotonal organs in insects. Microsc. Res. Tech. 63, 315–337.

Yack, J., 2016. Vibrational signaling. In: Pollack, G.S., Mason, A.C., Popper, A.N., Fay, R.R. (Eds.), Insect Hearing. Springer, Cham, pp. 99–123.

Yadav, C., Guedes, R.N.C., Matheson, S.M., Timber, T.A., Yack, J.E., 2017. Invitation by vibration: recruitment to feeding shelters in social caterpillars. Behav. Ecol. Sociobiol. 71, 51.

Yager, D.D., 1999. Structure, development, and evolution of insect auditory systems. Microsc. Res. Tech. 47, 380–399.

Yanagisawa, R., Suwa, R., Takanashi, T., Tatsuta, H., 2021. Substrate-borne vibrations reduced the density of tobacco whitefly *Bemisia tabaci* (Hemiptera: Aleyrodidae) infestations on tomato, *Solanum lycopersicum*: an experimental assessment. Appl. Entomol. Zool. 56 (2), 157–163. In this issue.

Yeh, Y., Cummins, H.Z., 1964. Localized fluid flow measurements with AN He-Ne laser spectrometer. Appl. Phys. Lett. 4, 176–178.

Young, D., 1970. The structure and function of a connective chordotonal organ in the cockroach leg. Phil. Trans. Royal Soc. London B 256, 401–426.

Zannini, D., Geurten, B., Spalthoff, C., Göpfert, M.C., 2014. Sound communication in Drosophila. In: Hedwig, B. (Ed.), Insect Hearing and Acoustic Communication. Springer, Berlin, Heidelberg, pp. 205–218.

Zill, S.N., Ridgel, A., DiCaprio, R., Frazier, S., 1999. Load signaling by cockroach trochanteralcampaniform sensilla. Brain Res. 822, 271–275.

Zill, S.N., Büschges, A., Schmitz, J., 2011. Encoding of force increases and decreases by tibial campaniform sensilla in the stick insect, *Carausius morosus*. J. Comp. Physiol. A. 197, 851–867.

Zorović, M., 2005. Morphological and Physiological Properties of Vibrational Neurones in Thoracic Ganglia of the Stinkbug *Nezara viridula* (Heteroptera: Pentatomidae). Doctoral dissertation, University of Ljubljana, Ljubljana, Slovenia.

Zorović, M., 2011. Temporal processing of vibratory communication signals at the level of ascending interneurons in *Nezara viridula* (Hemiptera: Pentatomidae). PLoS One 6, e26843.

Zorović, M., Ćokl, A., 2015. Laser vibrometry as a diagnostic tool for detecting wood-boring beetle larvae. J. Pest. Sci. 88, 107–112.

Zorović, M., Prešern, J., Čokl, A., 2008. Morphology and physiology of vibratory interneurons in the thoracic ganglia of the Southern Green Stinkbug *Nezara viridula* (L.). J. Comp. Neurol. 508, 365–381.

Zuk, M., Rotenberry, J.T., Tinghitella, R.M., 2006. Silent night: adaptive disappearance of a sexual signal in a parasitized population of field crickets. Biol. Lett. 2, 521–524.